MW00604482

THE LAST HUMAN JOB

THE LAST HUMAN JOB

THE WORK OF CONNECTING IN A DISCONNECTED WORLD

ALLISON J. PUGH

PRINCETON UNIVERSITY PRESS
PRINCETON & OXFORD

Published by Princeton University Press
41 William Street, Princeton, New Jersey 08540
99 Banbury Road, Oxford OX2 6JX

press.princeton.edu

Library of Congress Cataloging-in-Publication Data

Names: Pugh, Allison J., author.
Title: The last human job : the work of connecting in a disconnected world / Allison J. Pugh.
Description: Princeton : Princeton University Press, [2024] | Includes bibliographical references and index.
Identifiers: LCCN 2023040208 (print) | LCCN 2023040209 (ebook) | ISBN 9780691240817 (hardback) | ISBN 9780691240824 (ebook)
Subjects: LCSH: Labor—Social aspects. | Labor—Forecasting. | Industries—Social aspects. | Belonging (Social psychology) | Automation—Economic aspects. | Automation—Human factors. | Labor—Effect of technological innovations on. | BISAC: BUSINESS & ECONOMICS / Workplace Culture | PSYCHOLOGY / Interpersonal Relations
Classification: LCC HD4855 .P84 2024 (print) | LCC HD4855 (ebook) | DDC 306.3/6—dc23/eng/20230922
LC record available at https://lccn.loc.gov/2023040208
LC ebook record available at https://lccn.loc.gov/2023040209

British Library Cataloging-in-Publication Data is available

Editorial: Meagan Levinson and Erik Beranek
Production Editorial: Kathleen Cioffi
Text and Jacket Design: Karl Spurzem
Production: Erin Suydam
Publicity: Maria Whelan and Kathryn Stevens
Copyeditor: Natalie Jones

Jacket images: Ron Dale / Shutterstock

This book has been composed in Arno Pro with Bifocals

Printed on acid-free paper. ∞

Printed in the United States of America

10 9 8 7 6 5 4 3 2 1

CONTENTS

FIGURES

PREFACE

In my professional life I have interviewed more than a thousand people. First early on as a budding journalist, then later as a sociologist, I've sat beside dining tables and hospital beds, fishing ponds and lemonade stands, and talked with others about how they find meaning in woodworking, how much is too much for a child's birthday party, when to stay and when to leave a marriage, and what they think we owe our employers.

For me, interviews are unscripted encounters in which I try my best to hear not only what people say but also what they don't, reading their body language or attending to when they stutter or laugh, and reflecting back to them what I understand of their point of view. While we are talking, I closely mirror their perspective in order to elicit their truths, from tormented confessions of kicking out an errant teen to complex narratives about raising a Black child in a white suburb. "So you're saying something like this," I might start, and offer them a reflection of what I heard them say—or, often, what seemed to reverberate inside their words, even if they were not exactly saying it out loud. Most people I interview tell me that they find the experience one of being intensely seen for an hour or two, and that can be very rewarding. At its best the encounter can generate a kind of emotional resonance, a deep connectedness emanating from a sense of attunement with another human being.

After I wrote my last book, *The Tumbleweed Society* (2015), which relied on interviews with eighty people, many of whom had lost their jobs, I started to think about this close attention, and what it seemed to do for others and for me. Although I didn't have a word for it at the time, I also started to see it elsewhere—in my kids' schools and at the doctor's office. I noticed it when I was speaking to friends who were social workers or community organizers. I saw it when a colleague responded effectively to the unvoiced anxiety of a graduate student, when a youth minister heard and spoke to the underlying sadness beneath a teenager's story about a friend drama, when a disabled friend talked about how a hairdresser made her feel known and cherished. I began to see that it went far beyond care work or other helping professions, and instead was the secret ingredient of economic activity of all kinds, from journalism to management.

Everywhere, people seemed to rely on their capacity to connect to another person and to convey that they "see" the other to do their work. This capacity was like an additional layer of labor *beneath* the labor that everyone recognized, and few ever acknowledged this additional layer outright. It struck me, then, that we lacked a language to discuss what this work had in common across these many settings, and that its namelessness contributed to its invisibility among employers, programs, and institutions. We talk about the importance of teaching or coaching or leading, but we don't talk about this other work *underlying* these crucial tasks.

I began to focus on this foundational work in my research, reading up on how various fields approach it, if they do at all, and seeking out those who do this kind of work for a living. I ended up coining the phrase "connective labor" to refer to that work that involves forging an emotional understanding with another person to create the outcomes we think are important. Over the course of my research, I saw that the empathic attention at work here is more than how it's usually described—an individual skill or tendency, a person's capacity to understand the other, some sort of "intelligence." Instead the work is a collaboration with the other. Whatever people make in these encounters, they make together.

As soon as I noticed the existence and prevalence of connective labor, I could also see that it was increasingly being subjected to new systems of data analytics, apps, and artificial intelligence that tried to make it more predictable, measurable, efficient—and reproducible. Increasingly, we were treating this work to the same sort of industrial logic that one might see on an assembly line. And while some celebrate these trends and others critique them, most are still talking about the benefits and costs as if they accrue to individuals—the individual worker who might lose their job, the individual patient or client who might gain access they don't have already. There is little to no sense of how we all are implicated in these changes. Yet the social dimensions of these interactions—the collaborative magic they create that lies between people, securing us in relationship to others around us—are not entirely reducible to the individuals involved. Instead they accumulate, bit by bit, to build a kind of connective tissue for ourselves and our communities, and it is this connective tissue that is at stake.

I have already seen how giving connective labor a name can be powerful. Whenever I give a talk about my research, there are always people in the audience who do this work. They start nodding early on, and they often come up afterward to ask me when the book will come out because they need to give it to their boss or their brother. In other words, connective labor is recognizable to those who do it, yet it is still hard to talk about. Having a name for it helps practitioners identify their unique contribution for the administrators, employers, insurers, and others who are at one remove from the connective labor encounter, and it can focus a conversation that has remained vague and scattered. It enables us to talk about the work we assume women do, and the work we ignore that men do. It allows us as a society to consider its conduct and impact, encourage its incidence, and, if we choose, organize for its protection.

This book shines a spotlight on the ways humans help make each other human, and how these ways are both precious and unequally distributed. The experience of someone witnessing us, of being seen by another person, is undeniably powerful, as many examples in the next pages will attest. I won't argue that these relationships are an unvarnished good. Indeed, connective labor can also serve as a form of domination, as others choose what parts of us get seen; a tool of coercion; or a means for the most intimate surveillance, as when feelings or identities that we thought were private are named by others. But it is this profound power—as a source of meaning, change, and control—that makes connective labor warrant a name, and that makes it ripe for our attention.

I spent years investigating what this work is and how the rise of data analytics and automation affect it. What I found is the focus of this book.

THE LAST HUMAN JOB

1

INTRODUCTION

The Power of Seeing the Other

Erin Nash[1] was an apprentice chaplain in an East Coast hospital when she allowed me to shadow her as she moved through her day. A tall, middle-aged white woman with chunky jewelry and a direct but kindly manner, she was at the end of a long year, with weekly sleepless nights on call and intense sessions with peers and supervisors to process shifts replete with sorrow and death. That afternoon, I sat with her and the other chaplains as they gathered around a seminar table, talking with each other about what it meant to be simply "present" with patients. When it came time for Erin to share, she told the others of a moment when she had been called to help with a patient named Hiram, who had been intubated "even though he really didn't want to be intubated, and the doctors were saying he would die if he was extubated, and that he might die even if he was intubated." She continued, "And I was just sitting there with him trying to be with him, reading his body signals, and he was full of anger, and screaming 'Why, why, why' through his tube, and he couldn't really write because he was on too many pain meds."

At that moment, Erin said, she grabbed a box of Kleenex and handed it to him, telling him to throw it against the wall and that doing so would make him feel better. "And then I reached out my hand, and I thought he was going to hold my hand, and he ended up grabbing me by the arm and pulling me in and holding on to me for fifteen minutes."

The moment was powerful for both Hiram and herself. "The next time I saw him he was not intubated, and not dying," Erin told us, putting her hands together as if she were praying, "and he said: 'There is nothing like being in the worst moment of your life and being met with comfort by someone you don't even know, when you feel like someone understands you.'" Erin related that story, and then looked around for a moment, while a few of those listening nodded. "As chaplains, we can bring that kind of presence that allows people to see that they can bring those kinds of connections and we won't turn away," she told the others.

"That kind of presence" is what many people—not just chaplains—bring to their jobs. Erin managed to "see" Hiram and understand that frustration was boiling inside him, enough to think he might want to throw the Kleenex box. Hiram let her know that he felt "seen" by pulling her in like a life vest, and then later with his fervent avowal about "when . . . someone understands you." The exchange had considerable impact on them both, and while such a powerful human interaction reverberates wherever it takes place, it had particular meaning for Erin as she drew on the vignette as testimony to convince her colleagues of the comfort they could offer. What Erin managed to do with Hiram deserves its own name, so that we can think better about its conduct and consequences: I've taken to calling it *connective labor*.

The crux of this labor involves "seeing" the other and reflecting that understanding back, and many workers—from therapists to coaches to teachers to managers to personal assistants to sales staff—depend on this process. Yet it is work that is essentially invisible, only partially understood, and not usually recognized, reimbursed, or rewarded, despite its ubiquity and importance. It has also long been associated with femininity, presumed to be part of women's nature, and more frequently linked to jobs that women tend to hold, like teaching or nursing, while ignored or downplayed when found in jobs where men predominate, like police or the law.

For five years, I have been observing and interviewing all kinds of people who engage in connective labor at work, and I've come to see that it often serves as an underlying catalyst and conduit for the tasks for which people are explicitly hired, from healing to motivating to teaching to persuading. A corporate manager, for example, may be hired for her capacity to organize and lead others, but if she cannot see and reflect her subordinates well enough to shepherd them effectively, her team will not produce and her own performance will suffer. Ostensible tasks like organizing or leading are important, to be sure, but they are also shiny objects, distracting us from the connecting beneath that makes them possible.[2]

Instead, connective labor is central to millions of jobs, including people working not just in health care, counseling, or education, but also in the legal, advertising, and entertainment industries, in management, in real estate, in tourism, even in security. By one estimate, 12 percent of the US paid labor force is likely engaged in a form of "interactive care work," and this number is but a partial count of the contemporary army of connective laborers, because many of those who deploy it are not always devoted to other people's well-being. For example, consulting, lobbying, and high-end sales are cases in which we might consider connective labor to be in service to persuasion, while parole officers, prison guards, hostage negotiators, spies, and detectives deploy the capacity to see the other, using it in service to control. The work of connective labor may require knowing and reflecting what someone thinks and desires, but it does not always involve holding that knowledge tenderly.[3]

The spread of connective labor accompanies the rising importance of such socioemotional skills in many kinds of work in the United States and globally; by some accounts, the US is moving from a "thinking economy" to a "feeling economy." Labor economists debate whether jobs emphasizing such "soft skills" have increased because new jobs like wedding planning or social media marketing take up a greater share of the US economy, or whether the importance of such skills has simply expanded within old jobs like consulting or the law; both appear to be true. Researchers analyzing a sample of 7.8 million job ads from 1950 to 2000, for example, found that more recent ads were much more likely to emphasize interactive tasks; they estimated that most of the change toward "feeling" took place within a given occupation, such as managers. The researchers wrote: "Our finding is important because it implies that the transformation of the US labor market has been far more dramatic than previous research has found." In other words, the feeling economy is even bigger than we thought.[4]

In my research for this book, people often used the word "magic" to describe what connecting created—reflecting their sense of not just its wondrous mystery but its power. Studies in many different occupations attest to its impact. Reviewing a battery of randomized controlled trials, for example, medical researchers found that the patient-clinician relationship has a detectable effect on healthcare outcomes—an impact they described as stronger than that of taking aspirin to ward off heart attacks. Psychologists report that the therapist-client relationship, or the "therapeutic alliance," is what matters for successful treatment. A small mountain of education research documents that student learning depends not only upon their engagement or academic achievement but also teachers' caring support for, awareness of, and interest in students' emotional and academic needs. Relationship, alliance, rapport, caring, interest—these studies might not be using the same words, but the phenomena they are observing have strong similarities, and together they suggest that connecting with others can have significant effects.[5]

Moneyball Comes for the Chaplains

Despite the cultural resonance, increasing economic importance, and mighty impact of connective labor, however, it is clear that in many clinics and classrooms, we take this form of work for granted, assuming it will be available on demand no matter what manner of impediments we might place in its way. As I followed Erin while she made her rounds, for example, I was struck not only by the panoply of human drama she witnessed or the desperate needs she met constantly, but also by the continual nagging requirements of collecting, reporting, and analyzing data metrics in her daily tasks.

Erin kept a record of her patient visits in no fewer than three different tracking systems. One was the standard EPIC electronic health record,

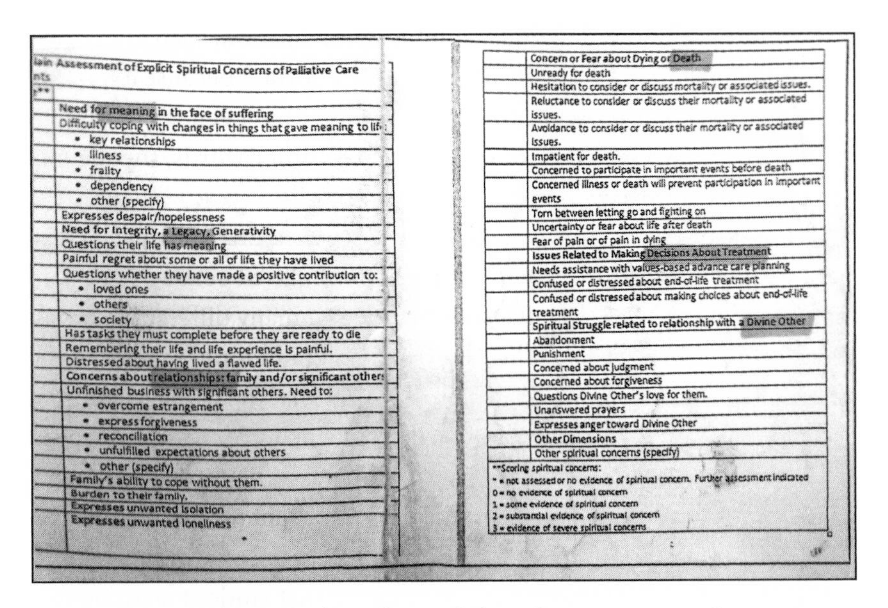

lain Assessment of Explicit Spiritual Concerns of Palliative Care
nts
**

Need for meaning in the face of suffering
Difficulty coping with changes in things that gave meaning to lif
- key relationships
- illness
- frailty
- dependency
- other (specify)

Expresses despair/hopelessness
Need for Integrity, a Legacy, Generativity
Questions their life has meaning
Painful regret about some or all of life they have lived
Questions whether they have made a positive contribution to:
- loved ones
- others
- society

Has tasks they must complete before they are ready to die
Remembering their life and life experience is painful.
Distressed about having lived a flawed life.
Concerns about relationships: family and/or significant other
Unfinished business with significant others. Need to:
- overcome estrangement
- express forgiveness
- reconciliation
- unfulfilled expectations about others
- other (specify)

Family's ability to cope without them.
Burden to their family.
Expresses unwanted isolation
Expresses unwanted loneliness

Concern or Fear about Dying or Death
Unready for death
Hesitation to consider or discuss mortality or associated issues.
Reluctance to consider or discuss their mortality or associated issues.
Avoidance to consider or discuss their mortality or associated issues.
Impatient for death.
Concerned to participate in important events before death
Concerned illness or death will prevent participation in important events
Torn between letting go and fighting on
Uncertainty or fear about life after death
Fear of pain or of pain in dying
Issues Related to Making Decisions About Treatment
Needs assistance with values-based advance care planning
Confused or distressed about end-of-life treatment
Confused or distressed about making choices about end-of-life treatment
Spiritual Struggle related to relationship with a Divine Other
Abandonment
Punishment
Concerned about judgment
Concerned about forgiveness
Questions Divine Other's love for them.
Unanswered prayers
Expresses anger toward Divine Other
Other Dimensions
Other spiritual concerns (specify)

**Scoring spiritual concerns:
* = not assessed or no evidence of spiritual concern. Further assessment indicated
0 = no evidence of spiritual concern
1 = some evidence of spiritual concern
2 = substantial evidence of spiritual concern
3 = evidence of severe spiritual concerns

FIGURE 1.1. A chaplain's cheat sheet to help code patient spiritual concerns.

which featured a stream of prompts so complex she carried her own cheat sheet with her to help her navigate it. (See figure 1.1.) After a visit with the family of a young woman who had died unexpectedly of a Tylenol overdose, Erin left the small hospital room where the woman's husband, aunt, and nephew had gathered in stunned silence, and went to find an available computer at a nearby nurses' station. She sat down at the hard-backed chair, clicking and typing to open up the EPIC chart as I looked on. Trying to translate what she had just seen into the chart's standardized parlance, Erin pecked away in response to the program's demands, explaining her notes to me: "Asking for a prayer is a resource, family together is a resource . . ." The computer kept freezing momentarily, and Erin spent fifteen minutes wrestling with it until it finally stopped responding altogether. "Yeah, it didn't save anything," she told me later, after powering up another computer to check. "It's OK. If this had been the end of the day, I would have been crying."

In addition to EPIC, the chaplains filled out a monthly statistics form required by their supervisor to keep track of their "units of service," which Erin greeted with a laughing resignation, saying she understood the rationale behind it. "EPIC doesn't capture everything we want to capture," she explained to me. Last was a spreadsheet about the in-hospital calls they answered over a six-week period, a list maintained separately by each individual chaplain. Over the course of the shift, she held hands, prayed, hugged, and even sang with patients and the people who cared about them; but she also consulted with nurses about buggy technology, conferred with colleagues about what to label a particular service in the spreadsheet, and made decisions about when to visit particular units in the hospital based on whether there were reliable computers nearby. Over the course of my research for this project, I listened to scores of doctors lamenting how the electronic health record (EHR) was optimized for billing rather than medical care, which made watching Erin's struggles with the computer even more striking. The hospital did not bill anybody for her "units of service," so why were the chaplains charting in triplicate?

Erin's capacity to see her patients was an emotional, spiritual, and intimate practice, but her job also offered a potent illustration of the measurement regimes now sweeping even these deeply personal occupations. In many industries, counting and assessing and applying all kinds of data is on the rise, crowding out the time that people like Erin have to pursue human connections. Even further, however, these campaigns have spread to counting connective labor itself, reflecting the cultural ascendance of data as authority. As I went from unit to unit in the hospital, watching as Erin pecked away at different computers, I could see that the data had its own insistent presence in her work, as it does in even the most low-tech connective labor jobs. (See figure 1.2.) *Moneyball* had come for the chaplains.[6]

They are not alone. Across the economy, connective labor is increasingly being subjected to new systems that try to make it more efficient, measurable, and reproducible. As connective labor increases in importance—and to be sure, in labor costs—firms in many industries have sought to get it under managerial control, introducing systems of data collection and analytics, imposing manuals and checklists, and

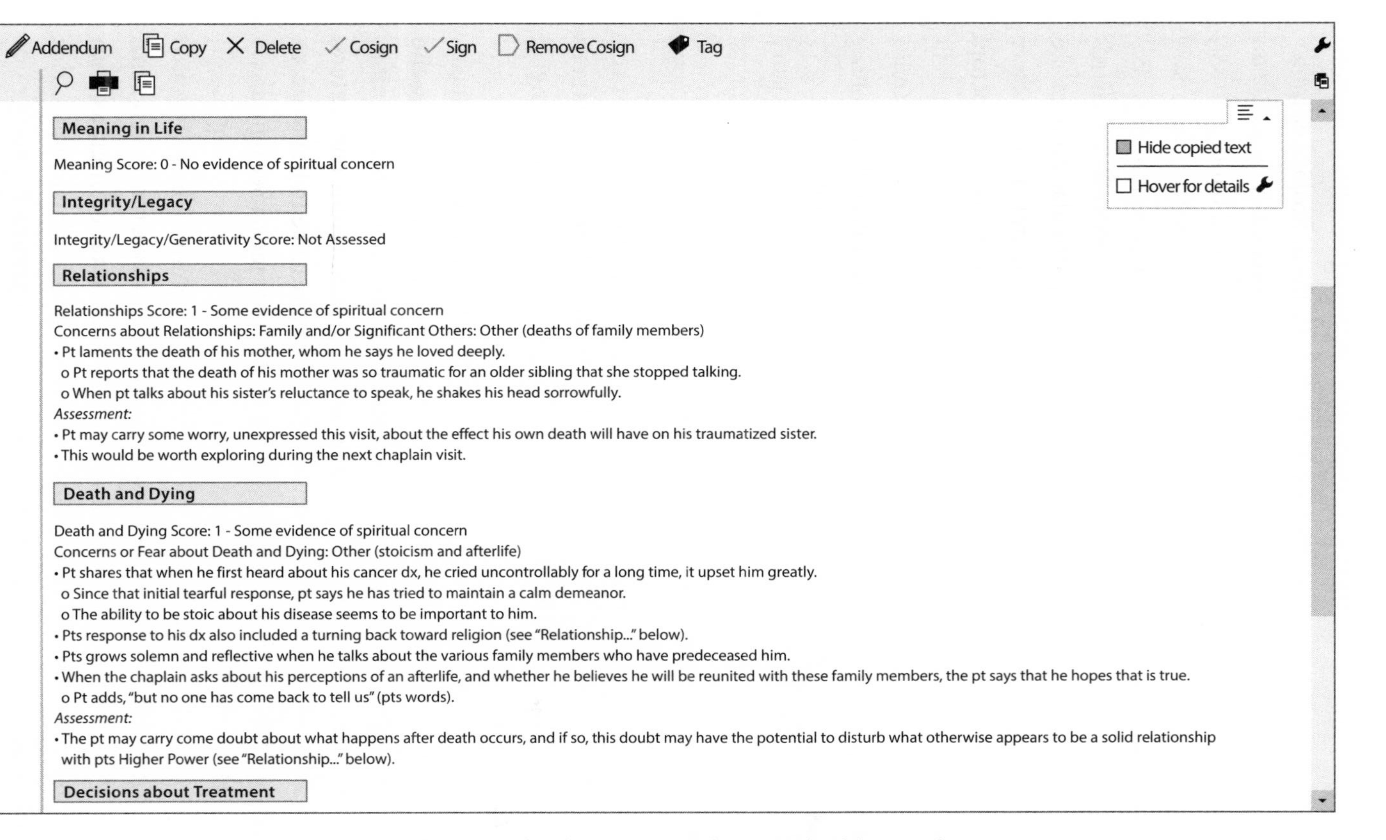

FIGURE 1.2. Chaplain notes on the EPIC health record.

implementing evaluation and assessment plans. At best, they do so assuming that such interventions will not impede their employees' capacity to forge the connections on which their work relies. At worst, firms ignore or dismiss those connections in the first place.[7] School districts, for example, are adopting "teacher-proof" curricula with step-by-step guidance for what children should read on a given day. Counseling centers are requiring therapists to administer surveys and offering clients graphs of particular clinicians' impact. Occupations are being transformed, as even these complex interpersonal jobs are reorganized to make them more predictable, through efforts to gather information and assessment data, and to introduce technology.[8]

We can certainly have sympathy for the goals underlying these changes, as they in part demonstrate a vision of a society where getting a good teacher or doctor would be less dependent on being lucky or affluent. Research finds that checklists and manuals can confer greater legitimacy upon many kinds of service work, and hedge our bets against incompetence and discrimination, while also protecting practitioners from demanding or chaotic situations and clients. Transparency and predictability can allow for greater coordination, mobility, and efficacy, studies have shown. As sociologists Stefan Timmermans and Steven Epstein point out, standardization does not inevitably lead to a "world of gray sameness."[9]

Yet these changes are coming about not just because program administrators or engineers want to improve access or performance, but because modern capitalism and modern bureaucracies converge on the priorities of data, accountability, and standardization in service to imperatives of efficiency and productivity—in the private sector with the goal of extracting profit, in the public sector with the goal of managing austerity. These twin domains—so often framed as opposites: the firm as efficient or flexible or rapacious, the bureaucracy as wasteful or immobile or dedicated to public goods—actually impose very similar pressures on interpersonal work, the human connections squeezed by an industrial logic in both settings. The squeeze takes place in the diminished time and space that firms set aside for it, in its framing as not a public good but a private luxury, in individualistic settings that emphasize outcome over process: in other words, in the configuration of resources and culture shaping human connection that we might call a "social architecture."[10]

And the squeeze matters. We know that the scripting of interactive service work threatens creativity and autonomy; transforms clients or patients into standardized "industrial objects"; and demoralizes workers, alienating them from their own feeling. Researchers have found that paperwork and repetitive tasks performed with little autonomy contribute to burnout and job dissatisfaction. Even before the coronavirus pandemic, more than 50 percent of physicians said they were burnt out and overwhelmed by data entry, with some of the highest rates for primary care doctors like pediatricians and family care providers. In a Gallup survey, about half of teachers and the same percentage of nurses reported experiencing high levels of job-related stress. Changes in the social architecture are transforming the work of people in connecting jobs, and along the way extracting enormous costs.[11]

The Last Human Job

Accompanying the spread of data analytics, checklists, and manuals has come the dawn of an AI spring, with a heralded rush of apps and automation. To many providers, connective labor is not very measurable, not very predictable, and not very automatable, yet engineers forge ahead in their creations, from AI couples counselors to virtual preschool to apps that advise diabetes patients. Of course, many of these forays remain in the lab, and critics caution against believing too much of the hype about AI's capacities to "disrupt" caregiving and other interpersonal jobs. Yet there are more than 350,000 health-related apps available, downloaded more than four billion times; the global market was valued at $38.2 billion in 2021. Plenty of these innovations are in use today, and the market appears likely to balloon even further.[12]

In late 2022, the company OpenAI released to the public ChatGPT-3, the first of a series of experimental bots featuring a new level of fluency and creativity, although still based on analyzing existing text from the internet. Michael Barbaro, host of the podcast *The Daily*, asked the bot why he tended to be critical of others, and read aloud the response.

BARBARO (*reading aloud*): "Being overly critical can also be a sign of low self-esteem or lack of self-confidence. It may be that you are using criticism of others as a way of feeling better about yourself—"

BARBARO (*interjects*): Ooh, I'm feeling seen—

BARBARO (*continues reading*): "Or try to control a situation that you feel anxious or uncertain about—"

BARBARO (*interjects*): Really seen!

When he finished reading, his guest, journalist Kevin Roose, asked him, "How does that land?" "It lands!" Barbaro responded. "Yeah, I mean it's conventional and a little rote, but it also feels like if it came out of the mouth of a relatively high-paid psychotherapist I would take it very seriously." Roose said he had had the same experience. "And it doesn't always do it perfectly, and it certainly doesn't know me in the way that a human therapist would after many sessions, but for something that is free and instantaneous and available on your phone at all hours a day, it actually is capable of some pretty remarkable kinds of advice and guidance." Automated connective labor had arrived.[13]

The public conversation about AI has so far been generally limited to three areas: algorithmic bias, surveillance/privacy, and job loss. We hear how AI turns historical correlations, often based on bias and stereotyping, into built-in assumptions, so that sentencing algorithms are more likely to predict recidivism for Black defendants than white ones, for example. We hear that apps track whether Amazon drivers look away from the road, that Baltimore's police deployed facial recognition cameras to monitor and arrest protestors, and that the Chinese government has deployed a "social credit" algorithm to assign citizens a risk score determining their ability to book a train ticket or take out a loan. We hear that AI will radically reduce many occupations, dermatologists and truck drivers alike. These are all worthy concerns.[14]

Missing in these discussions, however, is the impact these systems might have on moments in which we express and experience our humanity, on the emotional understandings we build of ourselves and others, and on the resonant meanings we create together that contribute to a social fabric. When we think of workers as individuals, we think about how innovations like ChatGPT or its inheritors might replace them; what is at risk, though, is more than an individual or his or her job, but instead the connections that are a mutual achievement between and among humans.

These are the latest stakes of crossing the automation frontier—the moving line that demarcates human work as more or less available for automation. That line has been contested ever since machines were invented; in 1589, Queen Elizabeth I once apparently rejected the application for a patent for a knitting machine, the first attempt to mechanize textile production. "Thou aimest high, Master Lee," she told the inventor. "Consider thou what the invention could do to my poor subjects. It would assuredly bring to them ruin by depriving them of employment, thus making them beggars."[15]

Notwithstanding the Queen's efforts, for a long while the automation frontier lay between physical labor and cognitive tasks; people thought technology replaced the jobs of manufacturing workers but that those of white-collar workers were safe. After it became clear that machines could do cognitive work as well, scholars drew a new line, pointing to the difference between tasks that were routine and those that presented more spontaneous challenges; a chambermaid's task of changing a bed, for example, is nonroutine and, though physical, challenging for technology to master, while contracts and even some laws that used to be drafted by attorneys can now be written by ChatGPT or its successors. As scholars draw and redraw the line, however, researchers begin to express some frustration with the seeming arbitrariness of the designation "routine." Most recently, the latest debate centers on social-emotional skills, with scholars arguing that such skills—such as those found in leadership, cooperation, and empathy—form the basis of a new frontier.[16]

While scholars and pundits may debate where the frontier is, AI researchers are not waiting for permission; socioemotional AI research has been burgeoning for the past decade. Of course, there is a big technological leap from a checklist or manual to an app that delivers therapy or a virtual nurse, and some people might not want to think about all these changes in the same viewfinder. Yet underlying the rise of data analytics and the dawn of the AI spring is the common assumption that these interpersonal jobs can be broken down into a series of tasks, in which abstract principles can be measured, taught, and, if need be, fed into an algorithm. On some level, the manual, the checklist, and the app all rely on an abstracting process that standardizes the worker's part of the encounter, stripping out the personal or the idiosyncratic—those unique qualities that connect people to each other and that shape what those connections look and feel like. In short, these trends are all examples of a growing *depersonalization.*

Depersonalization is a patchy fog across our social relations, its impact unevenly distributed. On one end of the US economy, low-income people receive connective labor that is harried, scripted, or, increasingly, automated, as engineers and policymakers embrace the notion that being seen by machines is "better than nothing." Even before the pandemic, for example, nearly half of Utah's four-year-olds were enrolled in "virtual preschool." On the other end, however, we see that one of the fastest-growing sets of occupations before the coronavirus hit was what economists call "wealth work," personal services that workers provide for rich people, from personal trainers to personal chefs to personal counselors, all of which depend on robust connective labor. As interpersonal work becomes ever more scripted and automated, being able to have a human attend to your needs has become a luxury good. Meanwhile, for the pressured middle class comes the proliferation of online platforms offering up connective labor on the fly, such as Care.com and UrbanSitter, with disenfranchised workers providing care work as a gig while technology, algorithms, and satisfaction scores mediate the relationship between employer and employed.[17]

The unevenness here reflects that we haven't acknowledged connective labor as worthwhile, either for its own sake or as the kind of activity that facilitates the ostensible work of teaching or primary care. Because of this omission, the distribution of connective labor is profoundly unequal across these populations, and likely to become even more so; in the future, the people on the top may get their connective labor from the people on the bottom, who in turn may get theirs from a bot.[18] While we each may occupy different spots in this landscape, we are all bearing witness to a collision in slow motion—the expansion and growth of connective labor in occupations across the economy, and the spread of systems to contain and control it—a colliding intensification in both the demand for feeling seen and the dictates to shape its supply. As a result, we are facing what looks like a depersonalization crisis, a social malady on several fronts.[19]

The Depersonalization Crisis

There is evidence of such a crisis in prevalent indicators of rising social alienation and isolation. Pundits and scholars refer to "the trust gap," "deaths of despair," or the "Great Pulling Apart," declaring social isolation "the problem that undergirds many of our other problems." In 2018, the British government appointed a Loneliness Minister, followed a few years later by the Japanese. Concerns mushroomed in the pandemic; in 2020, the United States Surgeon General Vivek Murthy published a book diagnosing "the current crisis of loneliness." These are diverse trends, and analysts have pointed to multiple causes, from segmented media to the unequal distribution of good jobs to the decline of traditional solidarity-building institutions like unions, churches, and bowling leagues. What each of these trends has in common, however, is a fragmentation of social connectedness.[20]

There's no question that connectedness matters. Studies show that both subjective and objective measures of it can have biological effects: feelings of loneliness and the objective measure of the size of one's social network each predict one's immune response to vaccination, for example. Loneliness has weighty negative effects on health and well-being—akin to smoking fifteen cigarettes a day, according to researchers—and a 2023 review of ninety studies with more than two million people found that being or feeling socially isolated are each linked to a higher risk of mortality. Belongingness is crucial to human thriving, psychologists say, "almost as compelling a need as food."[21]

But social scientists disagree about how much fragmentation the trends actually show, with one scholar in exasperation calling social isolation and loneliness "the headless horseman" of a story, forever riding on into the night. In the United States, for example, in contrast to the research lamenting a decline in Americans' social time (à la *Bowling Alone* [2000]), the sociability trend is apparently flat rather than downward. Sociologist Claude Fischer compared the percentage of Americans who "spent a social evening at least several times a month" from 1970 to 2020, and found essentially no change in those spending time with their relatives and friends, although he noted a 10 percent decline in sociability with neighbors. He concluded that "the total volume of personal contact has, in net, increased." Even the coronavirus pandemic—though it may have felt cataclysmic for our daily routines—appears to have mixed effects: people report being closer to their relatives and neighbors but more distant from some friends.[22]

At the risk of losing my sociologist's badge, however, I suggest that the persistent worry about social isolation may be more than a myth overturned by numbers that can't lie, and instead this headless horseman of a story keeps riding for a reason. The statistics are missing some important facts, and thus may not be capturing what it feels like to live in the United States and other modern industrialized societies. Most important, sociability data tell us primarily about close relationships like family or friends, but not about everyday encounters with weaker ties that—it turns out—play a significant role in well-being, a web of relations that we might call "social intimacy."

While sociologists have documented the "strength of weak ties" for getting a job, as it happens such ties—at the café, the classroom, or the salon—end up giving personal succor as well, with some people serving as confidants and counselors to others simply because they are there at the right time. Furthermore, contrary to the notion that money necessarily corrupts relations, this sense of closer-than-expected applies even to those service people with whom we exchange conversation throughout the day. One study in the United Kingdom found that people who talked to their barista derived well-being benefits more than those who breezed right by them; the authors titled their study "Is Efficiency Overrated?" Of course, these can often be perfunctory exchanges; indeed, retailers often make changes to control the spontaneity of human connection and increase efficiency, efficiency that busy people often say they prefer.

Nonetheless, the upshot here is that the gains of connection stem not just from the sacred family hearth, or even some convivial "tribe" of close friends, but rather from an extensive web of civic and commercial relations through which we make our way every day: from social intimacy. Research suggests that to understand sociability trends we need to reckon with a much broader terrain of human connection, the very terrain that has been radically transformed by metrics and technology. We need to reckon with depersonalization.[23]

Among those who do the work of human connection, the depersonalization crisis is experienced as a certain intensification of need. During my research for this book, a pediatrician told me that she was inundated by patient demands. "Healing happens through relation, and we're just not doing the kind of caring that people need to heal," she said. "But, you know, we've lost our village, all of us. We've lost the auntie who gave you the healing tip and the sister-in-law who taught you how to breastfeed, and the, you know, cousin who took care of your kids while you ran out to do errands or whatever. And that's what—I feel like that a lot of the health problems could be solved by somehow recreating that village." As a result, people were coming to her yearning for support that any good listener could provide, she thought. Other practitioners, such as teachers and librarians, reported encountering the same desperate thirst for being seen. In a crisis of depersonalization, the demand for human connection at the point of service can feel frantic.

Heeding the Depersonalization Crisis

How we address the depersonalization crisis depends on how we diagnose it. Some tech enthusiasts recognize that people feel unseen, invisible in a mass society where standardization has erased individual differences. The solution they offer, however, invites even more data and technology to step in. They urge a strategy that is widely called "personalization," involving a process of ever more precise tailoring, in which data is harnessed by technology to analyze someone's health history, how a person likes to drive, or even the content of one's sweat. Firms capitalize on that data to sell goods matching perceived need;

one such company collects information about babies' ages, for example, in order to funnel to their parents products aimed at stages of development as they pass from infants to toddlers, breathlessly touted in *Forbes* as "Enfagrow Personalizes Advice for Babies." Yet it is not just commercial outfits that deploy technology in this manner; "personalized medicine" and "personalized education"—sometimes called "precision medicine" or "precision education"—are parallel efforts to assess health or learning needs and produce recommendations tailored to the individual. While rigorous analyses tell us that these well-funded campaigns have made some modest gains in pharmaceutical treatments or student learning, the terms are a bit Orwellian because there is no "person" involved in the seeing they promise; surely this approach is not so much personalization as it is customization.[24]

Such customization makes sense if the problem is simply that people do not feel seen; as Michael Barbaro's experience with ChatGPT-3 suggested, machines can do something like "seeing," if they feed on the proper data to produce the right responses. The mechanized solution is a perfect example of what Marx called "commodity fetishism," where we fixate on the end result—feeling seen—without thinking about all the effort that goes into producing that feeling, or the people behind that effort. But what if the root of the depersonalization crisis is not only that people feel unseen, but that they do not feel seen *by another human being*? What if we have misunderstood what is lacking: the human connection that people make together, in interactive moments not reducible to just one person's skills or temperament? More data or technology is likely not the treatment for this kind of lack. What might be called for instead is a renewed infusion of human contact, a commitment to connection, even a *re*-personalization.[25]

Actual Personalization

There are certainly a wide range of human connections that matter, but by "connective labor" I mean something quite specific: the forging of an emotional understanding with another person to create valuable outcomes. While there might be many paths to that emotional understanding, they all seem to require some form of empathic listening, in which one cultivates a sense or a vision of the other person, and witnessing, in which that vision is reflected back to the other. Another crucial ingredient is the ability to regulate one's own emotions, to get out of the way of hearing and understanding the other. Most important, this process is deeply interactive, not least because for connective labor to land successfully, the other must assent—to some degree—with the vision that is being reflected their way. Connective labor is how we see the other, and how we convey to the other that they are seen.[26]

Not every job involves connective labor, of course, and some involve it more than others. Manufacturing jobs that don't ask workers to interact with other people, low-wage retail jobs or call centers that involve extensive scripts, highly technical jobs like surgeons or engineers with very little client-facing work, entertainers who put a lot out there about themselves without taking in much about the other—these are all jobs without much or any connective labor. The question I ask myself, when evaluating a given profession, is: How much do workers have to convey that they see, know, and understand the other person? In the answer lies the degree to which they perform connective labor in their job.[27]

There are some broader concepts in use that might seem related to connective labor, but these fall short for a number of reasons. Most popular, perhaps, particularly among business readers, is the notion of "emotional intelligence," a term that not only carries a whiff of innateness but also focuses solely on the individual worker's skills and talents. Yet connective labor is first and foremost an interaction between people that generates an emotional experience, and creates particular meanings, greater than the sum of the individual parts. Words like "skills" train our focus on what workers can or cannot do on their own, but being able to talk about the connections forged between and among people allows us to see what else is at risk when we start replacing people with machines.[28]

We might think of connective labor instead as a combination of recognition and emotional labor. Philosophers have long understood the power of recognition, although they have focused more on political and social recognition than the domain of sentiment. When we marry this concept to Arlie Hochschild's idea of "emotional labor," which captures how people use feeling to create and sustain relationships for a wage, we can better grasp how directing our emotional antennae toward the project of recognizing the other person can elicit a profound experience.[29]

Words like "see," "understand," or "recognize" are immodest ones, however, blithely promising far too much exactitude, for what is actually often a more imperfect match. Clinics and classrooms are littered with misrecognition and near-misses, as people come together across sometimes very great social distances, their vision of the other clouded by preconceived notions, by their own backgrounds and histories, by their inability to listen well enough, by the other's refusal to share. We know that gender and race shape the expectations people have of others' emotional performance at work, with women and Black men limited by these expectations and also punished for transgressing them; these expectations shape the kind of connective labor they are able to provide to and receive from others. Working conditions and climate, as well as social inequalities, can get in the way of people's capacity to see the other. Furthermore, under the best of circumstances, we are none of us perfectly legible to the other. Even when it is going well, then, connective labor often seems to involve a measure of grace, as people acquiesce to being seen at best partially, to a kind of "good-enough seeing." Given these limitations, both inherent and manufactured, connective labor's impact is remarkable indeed.[30]

Connection, Culture, and Ambivalence

There is good evidence to suggest that these effects are not biological universals, inherent in all humans, but instead culture- and time-specific. On the Micronesian island of Yap, for example, anthropologists tell us that people value those who control their emotions, who keep their composure, who are unreadable. The Yapese use the expression *feal awochean* (good face) to signify an impenetrable mien, and they view people who cannot maintain good face as immature or childish, akin to a fruit whose bright colors tell anybody who wants to know that it is ready to eat. When they want to point out that someone is too easily seen by others, the Yapese have a felicitous phrase *ke luul ni baabaay*, or "it ripened, the papaya."[31]

In the United States, too, seeing the other and being seen has apparently not always been valued. Before the early twentieth century, we did not even have a word for empathy. As empathy historian Susan Lanzoni tells us, the concept traveled from German artists and psychologists—who used the term *Einfuhlung* to describe "feeling-into objects"—to American psychologists who coined the word "empathy" in 1908. While early on empathy meant projecting yourself into a form or shape, by midcentury it meant the ability to understand the feelings and perspectives of others. With the help of popular psychology and journalists, the concept broke out of labs and into the general lexicon, finding its way into advice columns and everyday marketing. Empathy transformed into the meaning it has today—of an emotional understanding of the other—in part because we also developed a sense of the other (and ourselves) as even understandable.[32]

Furthermore, despite the demonstrated benefits of emotional recognition for many, some people in the contemporary US are still ambivalent about it. Connective labor, even that which is warm and competent, can burrow into the private emotional terrain that people may want to keep from others. Some people regard interpersonal recognition as an invasion of privacy or as a threat, particularly when they are tired of fending off disrespectful or harmful stereotypes, or when it is being deployed to control them by those with power to assess or punish. It can feel disturbing to be "seen."[33]

Despite this ambivalence, the contemporary United States—as well as other advanced industrialized nations—are rife with what the Yapese would probably consider "ripe papayas." "Seeing" and "being seen" is broadly invoked, and not just in the therapy clinic (i.e., see figure 1.3). US schools spend at least $21 billion annually teaching children empathy in socioemotional learning programs. Psychometric testing, often including measures of emotional regulation or relational traits, has come to dominate hiring; one report suggests eight out of ten of the top US employers use such tests. The turn to the "feeling economy" is more than a latter-day response to the threat of job disruption by AI or

FIGURE 1.3. Being seen in the market.

automation, but instead reflects much broader cultural trends that affect how we relate to each other on and off the market. As traditional social roles that used to give people's lives structure and meaning erode, people came to prize authenticity, or the "voice of an innate, primary nature that had been muffled." Therapeutic culture has shaped how we communicate about what we value at work and at home. Seeing each other is the currency of our time and place.[34]

Ultimately, connective labor is a social process, an interaction between people that is encouraged or impeded by their surroundings, by culture, politics and inequality, by the social architecture at work and the imperative of profit. It has considerable impact, not least of which upon people's experience of themselves in community, upon their social intimacy. The conditions under which people do this work matter in helping to shape its conduct and experience. And as indicated by all the typing that Erin the chaplain had to do at the beginning of this chapter, we have been attempting to measure, evaluate, and scale up this work without actually understanding what it is, what makes it valuable, and what's at stake in its transformation. This book aims to fill in those gaps.

The Research behind This Book

In 2015, I set out to understand this work, how people do it, what they get out of it, and how it is shaped and altered by the systems that try to measure, predict, and evaluate it. Rather than a traditional ethnography, in which an observer might embed themselves in a single community or two, I considered this project more like a conceptual ethnography, in which I went where the idea of connective labor took me.

I interviewed more than a hundred people for this book (assistants interviewed about ten of those), most of them people who actively practice connective labor, including therapists, physicians, teachers, chaplains, hairdressers, and community organizers. The bulk of these interviews took place with three groups of professionals (therapists, teachers, and primary care physicians) who varied in their time scarcity, as well as the centrality of relationship in their training and work; a fourth group—of those without college degrees—enabled me to explore how inequality shaped this labor. I also interviewed people who supervised, evaluated, or automated this work—from principals and program heads to the engineers working with robotics and AI—as well as a handful of those on the receiving end as patients, clients, and students.

I conducted more than three hundred hours of observations in doctor's offices and schoolrooms, therapy sessions and squad cars, in California, Virginia, Massachusetts, and—for a ten-day visit to a roboticist's lab—Japan. Among many examples, my observations included eight months participating in a weekly group devoted to humanistic medicine; six months watching physicians, nurses, and patients in an HIV clinic; a semester witnessing a class for aspiring school counselors; many hours observing videotaped therapy sessions with supervisors giving commentary to apprentice clinicians; three months sitting in on a hospital chaplain residency program; a weekend spent at a workshop that used horsemanship lessons to teach medical students about the doctor-patient relationship; and twelve hours of ride-along in a squad car in a distressed Western city with a community policing initiative. Most of this research took place before the onset of the coronavirus pandemic, but I touched base with a number of informants throughout the lockdown and its aftermath, to hear how they were thinking about their work anew.

I started this work fairly agnostic about the impact of systems like data analytics, AI, and robotics. While I knew about the troubles of burnout, alienation, and threatened job loss, I had also read about the gains promised by systems that could act as bulwarks against capricious unprofessionalism or demanding customers. While it was difficult to discern what was real about the AI spring from afar, it was also possible to see progress in innovations, particularly those that would enable people to gain new access to therapy or teaching. But after years of talking to and observing those who provide connective labor, I came to a new appreciation for what they manage to accomplish, for what was uniquely human about this work, and for how precious—and fragile—are the conditions that enable people to make powerful meaning together.

A Map of What Is to Come

The structure of the book reflects its argument: that connective labor is a valuable human practice under siege by systems that to some degree enable but often impede it. But what makes connective labor valuable exactly? Stories abound about its profound impact, but what does it in fact do? Chapter 2 explores what is valuable about connective labor, far beyond the ostensible tasks of teaching algebra or coaching soccer. We hear what neuroscientists and other researchers have uncovered about the mysteries of what happens when human beings mirror each other, and how they do not quite understand its powerful effects. By chapter's end, we see just what we are risking when we introduce new systems into this work.

Chapter 3 outlines its greatest threat, exploring how we have responded to the crisis of connection by doubling down on depersonalization, developing automation and AI in connective labor. This chapter outlines the profound impacts of the automation frontier for those who see and those who are seen.

Yet if it is under threat from automation, why is connective labor "the last human job"? Chapter 4 explores five uniquely human characteristics that make connective labor hard to systematize. We also hear about the shame, distrust, and vulnerability that bedevil these encounters, and how practitioners try to meet those challenges to somehow still produce the "magic."

The next three chapters outline the contours of the current crisis of depersonalization and what it portends for connective labor. Chapter 5 shows how an organization's social architecture can hamper or support the connective labor people are able to give, forcing practitioners to choose between work that is sustaining, sustainable, or subservient. Chapter 6 goes deep inside these organizations, revealing how scripting and counting degrade connective labor, a degradation that paradoxically makes its automation more appealing. Chapter 7 tells the story of what happens when connective labor goes awry. It investigates connective labor against the backdrop of stark social inequality, particularly of race and class, and how it derives power from the very disparities that make it dangerous. After hearing about all the ways in which organizations are doing it wrong, chapter 8 offers a closer look at those who are doing it right. We see how an organization's social architecture can lead to a connective labor that works.

Finally, in the book's conclusion, I extend the discussion of why we should care about connective labor by exploring its broader impacts on social intimacy. What does it mean for a community when it is not just material goods that are distributed unequally, but the capacity to see and be seen? I urge us to create a social movement for connection, arguing we should work to foster it not just for its capacity to act as a sort of grease for the ostensible tasks that we value, but because of its capacity to forge the social intimacy upon which we all rely.

—

In German, the word *Herzensbildung* means "training one's heart to see the humanity of another." Having words for things matters, not least because it enables us to identify when something valuable is under inadvertent threat. I call for a new awareness of connective labor and a social movement to protect it. The stakes are too high for us to shrink, strangle, or automate connective labor without knowing what it is and what it creates between and among people.[35]

2

THE VALUE OF CONNECTING

Karl Brandt was an affable white man in his midthirties and in his last year as a Fellow at an HIV clinic in California when I started shadowing him as he met with patients. His appointments sounded like any other primary care visit—complete with mundane conversations about flu shots and colonoscopies—except when he talked about T-cell counts and dosages of Biktarvy, the drug regimen that effectively contained the deadly virus to untransmissible levels. One day, I sat against the wall in the tiny examination room while Karl talked to Mr. Shiflett, who was coming in for the first time since his wife had died a few months earlier of complications from cancer surgery.

As soon as Karl sat down on his stool, he told Mr. Shiflett he was sorry about his wife. They talked about her surgery, with Mr. Shiflett saying, without much heat, that she had likely died from some sort of medical malpractice. Only after this conversation did Karl ask about his smoking, drug use, and acid reflux, telling him he was not worried about some recent weight gain because Mr. Shiflett might be responding to the grief.

Mr. Shiflett was in his midsixties, but he looked a lot older, his long white hair crowned by a bandana, like a rangy, muscled Willie Nelson. A contractor, he worked six days a week from eight in the morning until dark, he said, and when he came home, he worked on his own house until eleven o'clock at night, just trying to stay busy. On his lap he held a very large and bedraggled sheaf of papers, hoping someone there would help him sort out the insurance correspondence. His wife had been the one to manage all that, he said; she was the one who knew how to go online and how to use phone apps, and he just did not have time to learn it all. The sadness hovered around him as he said this, his papers a crackling pile of testimony to his absent wife.

It turned out he also had hepatitis C, which was Karl's main concern. Mr. Shiflett asked Karl how he could have gotten it—"Well, through blood contact," Karl said, "so from IV needles, from toothbrushes, from sexual contact"—and the conversation went down the path of which of these it might have been, with Mr. Shiflett insisting that he had "never been in the back of nobody's meal or drinking, and never was in back of someone's toothbrush," and although he used to use IV drugs, his brothers had taught him "do it the best or don't do it at all," so he "bought them from Walmart new," by which he seemed to mean that his needles were clean. The conversation was quite unusual for the clinic—most of the time, practitioners there stayed away from "how did you get it?" talk and focused more on disease management—but Mr. Shiflett was the one driving the exchange, clearly concerned. After a short time, Karl broke in. "It doesn't matter how you got it," he said mildly. "It's more important to think about how to control it and manage it in the present and the future. And you're doing great—taking the [Biktarvy] pill every day, quitting smoking." Later on, Karl told me he thought Mr. Shiflett felt "very guilty" because the man assumed he had given these diseases to his wife before she died.

He and his wife had been together since ninth grade, Mr. Shiflett said to Karl, and the tough, grizzled man teared up. "I sleep on my wife's side of the bed," he said, and then added, as an aside, "Well, it was my side of the bed, but she insisted it was her side, so I moved over, but now that she's gone, I sleep on her side of the bed." His marriage was even now clearly a living thing to him still, as he continued arguing, with gentle humor, about whose side of the bed was whose.

But some nights, Mr. Shiflett told Karl, his wife came to him in visions. "The urn is right there next to the bed," he said, "and I kiss the urn, and talk to it, and sometimes a big orb comes in from the living room, and it is glowing, and it comes into the room, near me, and it gets bigger, and I can see her inside it, as clear as I can see you, and I talk to her and she talks to me."

Karl was watching him while he was recounting this story; he had started the appointment turned slightly away and working at the computer as he clicked through the electronic health record, but now he faced Mr. Shiflett, his hand on the man's arm for comfort.

"I can tell you don't believe me," Mr. Shiflett said.

"It's not that I don't believe you, but I want to know: Does that help you?" Karl asked.

"Yes," Mr. Shiflett said. "It does make me feel better."

Karl was calm and kind throughout, very courteous in his demeanor, listening to Mr. Shiflett's claims about the medical malpractice, and reframing the issue of the widower's ghostly visions as a matter of comfort rather than credulity. It was not that he reflected back Mr. Shiflett's experience exactly—later he discussed how to handle the visions with the clinic psychologist—but he nonetheless conveyed respect. Ultimately, his body did most of the connective labor, reflecting the almost palpable emotional currents more than did his statements about blood contact or taking the pill. Karl took care in checking in about Mr. Shiflett's grief before anything else, he turned to the man and touched him when he was relating a deeply emotional story, he declined to contradict a fantastic narrative about the spectral visit by Mr. Shiflett's deceased wife—in all of these ways, the physician reflected back to the widower a witnessing presence.[1]

Maintaining that presence also involved managing the vestiges of inequality that threatened to puncture the encounter. Twice during the visit Mr. Shiflett mentioned, "You're smarter than me." "I have to listen to you about this [the hepatitis C] because I don't know anything about this," he said, although he added parenthetically after the second time "but if there's something that needs fixing up in a million-dollar house, I can do that well." (He also mentioned that he answered his phone without fail "because it is either you guys or someone calling me to fix their million-dollar house for a thousand dollars—it could be a thousand-dollar phone call.")

Yet despite his self-deprecation, Mr. Shiflett also found the mettle to ask Karl whether the drug he was taking had side effects. His best friend's wife had looked it up and noticed that it could cause liver problems, he said. Karl nodded. "Yes, that it can," he said, "But only in a tiny percentage of patients, so that's why you come in every three months and we check your liver, we check your blood counts." Mr. Shiflett replied that he was grateful to Karl for spending time with him and explaining things.

Mr. Shiflett was clearly a bit uncomfortable in the visit, unsure of his credibility (saying twice, "You're not going to believe me") and of his stature (as he repeatedly deferred to Karl's education and "smarts," and being "busy"). Yet at the same time, Mr. Shiflett managed to assert that he was a high-end contractor and mount a small challenge regarding the drug's side effects. In the throes of grief, he was able to tell a story of seeing his dead wife while Karl listened carefully. In these small moments, his humanity permeated the encounter.

The kind of witnessing presence that Karl offered is a vital part of many jobs. Therapists, teachers, coaches, primary care physicians, sex workers, even business managers and high-end sales staff—many depend on their ability to connect to others to make their contribution: clients healing, students learning, employees motivated and engaged, or customers satisfied. And there is plenty of research to suggest that their connection matters. At its best, connective labor is profoundly affecting, for both the person doing the seeing and the one who is seen.

But while many people might intuitively know that connecting is important for these jobs, they might not quite understand why. Employers who hire people for the "ostensible tasks"—those that are their central mission, such as teaching algebra or providing diabetes care—might recruit for "interpersonal sensitivity" or "emotional intelligence," on the premise that such characteristics help get the algebra learned or the treatment followed. In doing so, they are relying on a model of connecting-as-engine-grease, in which it provides some sort of psychological lubricant to enable the *real* work of teaching math or lowering blood sugar. The efforts to automate math instruction or diabetes care stem from the same perspective: that what really matters is the algebra or the treatment, not the process of getting there. Yet if we reduce connective labor to mere "soft skills," the fluff surrounding tasks that really matter, we ignore a wide range of potentially profound effects in the social or emotional realm. While the instrumental gains of math learned or insulin levels are often more measurable, that does not mean they are the only—or even the most—consequential outcomes. What *else* does connective labor do?

Whatever it does, it does in two directions. Karl was able to witness Mr. Shiflett's grief and guilt, and when he stopped typing to roll his chair over in front of him, putting a hand on his arm while the man recounted sleeping on his wife's side of the bed, Karl was letting him know that he saw him and his feelings. We also see that Mr. Shiflett felt some safety there in being able to share his story, and that while he may not be sure Karl believed him, he understood that Karl cared enough about him to let him have that comfort. Mr. Shiflett perceived Karl as caring, and mentioned how much he appreciated the physician "spending time with me, explaining things to me." Emerging from their encounter was mutual purpose, dignity, understanding—doctor and patient made something together that day, above and beyond the patient's medical treatment.

What they made is the subject of this chapter.

The Mystery of Finding Each Other

There are certain practices—like handshakes—in which people find their way to each other. Choral singing is a good example, where singers adjust their voices in minute ways as they make their way through the bars of a song to be able to blend with others. Another might be amateur swing dancing, in which people feel each other out, reading tiny signals to ascertain and coordinate their next moves together. In my spare time, I have been rowing in crews of four or eight people for more than thirty years, and I love the sport because of just this kind of magic, in which people connect to each other's rhythms based on mysterious clues like the sway of bodies nearby and the gurgle of the boat cutting through the water.[2]

Neurologists have a term for just these sorts of practices: perceptual crossing. Mutual touch and meeting someone's eye are examples of having to coordinate with others, and the study of these phenomena has been called "second-person neuroscience," to call attention to the fact that these are interactive encounters. A team of researchers in 2011 argued that too often, neurological research into recognition focuses on just one person's perception—say, using MRIs to measure their "mirror neurons." But the experience of acting in concert with another person and feeling their impact requires a different approach, they urged, one that takes seriously the unpredictability of the other, that recognizes a "second person."[3]

When it is successful, connective labor—seeing the other, and being seen—is another occasion in which people find their way, however mysteriously, to each other. Let's say a teacher offers up their witnessing of a student, reflecting her understanding of the student's perspective. The student receives that vision, perhaps modifies it, and sends it back to the teacher, along with a reflection of how the student perceives the teacher and her effort. Karl's experience with Mr. Shiflett illustrates how the work of connective labor is a two-way operation, involving recognition of the other by the worker, as well as recognition of the worker, or what he is doing, by the other. For it to "work" as a conduit of connective labor, such recognition is a mutual achievement, like an emotional handshake. We cannot be our own mirrors.[4]

As in a handshake, clients have to be active participants, in that they must offer up a bit of themselves to be seen. There may be prior expectations built upon experiences and beliefs, expectations that feed the interaction on both sides. Most important, people have to "receive," buy into, or confirm the emotional reading that workers offer, or the interaction falls flat or worse. Because of this receiving, the mirror analogy is not a perfect one, as providers do not simply reflect whatever they see, especially if what they see does not match the self-understanding of the person they are seeing. The reflected image has to sound at least partially true to the other person for it to land successfully; otherwise, it can feel more like being misunderstood.

In this vein, practitioners told me about not just reflecting the "ugly truth" of what they saw, but instead actively altering their reflections, to offer instead a gentler, more hopeful reframing. I observed a semester of training sessions for psychology graduate students with a master therapist, Russell Gray, for example, and one day he talked to them about conducting therapy with family groups. The therapists' primary task, Russell said, was the reframing work that he called "turning the radio down," shifting the negative stories families told themselves about themselves. "[That radio] can be so controlling," Russell said. The therapist's job is in some way to "undercut that bad mood—the 'our family sucks, we don't care about each other' message." The goal was to reframe the message positively but also plausibly, so that people could take the exact same experience and see it somewhat differently.

"What you're trying to do," he explained to the students, "is take behaviors that exist and find the positive intent, find the positive aspects of them, and highlight that. The trick in doing it well, is doing it meaningfully and plausibly."

> You can't just be Pollyanna. You have to acknowledge the negative affect first. [I could say]: "You two are really fighting, because you care about each other"—[meaning] that "people who don't care about each other just check out, the intensity is because she cares about you." [But] if I say that *first*, then they disagree. I have to say first—"you're really angry, furious, you feel like she's a jerk . . . But you know why you are fighting? It's because you're really caring about each other." Put the two together. What you want is other people in the family to see the same behaviors and instead of seeing it with blame, to see it with empathy. That's your ultimate goal.

But the plausibility was crucial, Russell said. "We're not trying to be car salesmen—'Hey, everything's great.' It's just 'Here's the positive spin on what that person's doing,'" he said. "Looking on the bright side is adaptive. You push it as far as you can do it, stopping short of where you lose your credibility." Witnessing—in this case of a positive family narrative that might not align at first with how its troubled members might see it—required someone to receive it for it to work properly.

The "Aha" of Reflective Resonance

There's a word for this kind of reframing: resonance, a blend of old and new that gives rise to the "aha" moment of epiphany. Particular ideas resonate, or reverberate in a deep way, when they pluck at the strings of belief and knowledge within us. For an idea to reverberate, recognizer and recognized must have some strings in common, a shared vocabulary of meaning. And the meaning they make is emotional: when resonant ideas make new sense out of old ideas, people can feel a sense of relief, peace, or joy, paradoxically even when the sense they make is daunting or tragic. When this particular kind of epiphany stems from seeing the other and being seen, many people call it "magic"; we might dub it "reflective resonance."[5]

Sometimes it is possible to watch the creation of reflective resonance, when reframings leave the listener visibly transformed. For example, I observed a group of the apprentice chaplains in an evaluation session, one taking place a few weeks before their graduation. Three apprentices and two supervisors—Ezekiel and Anne—were sitting on couches in Anne's living room, sharing tea while Anne's two small dogs barked periodically at passersby and resettled themselves on the cushions.

First up was Paula, a white woman in her thirties whose calm, even features belied an inner intensity. Paula had decided to leave the chaplaincy after her graduation, but she was also grieving the decision, partly because she was unsure about where to go next. She started off the session talking about her fear that she was some sort of dilettante, unable to stick with things, somehow "self-indulgent." "I just don't know what to do with my sense of calling," she said through tears. The last Tuesday, she had visited a patient who had managed her own anxiety by embracing uncertainty, through "an acceptance of mystery," Paula said, "[a stance of] 'I don't know why this is, I'm just going to trust.'" She was trying to adopt this posture as she faced her uncertain future, but "I know enough that I know it's going to be painful."

When it came time for the other trainees to offer end-of-program feedback to Paula, however, they told her about the growth they had seen in her, about her courage in facing her self-doubts. "This doesn't feel self-indulgent, I'm actually surprised to hear that word," said Brooke, another trainee. "You have been a model of emotional availability and honesty to me." "You've been a real gift to me as a student," added supervisor Anne, a white woman normally rather contained and not given to effusive praise. "You've taught me through your journey. I've really loved watching you in your own relentlessness to find truth. It's been a joy." Ezekiel, the other supervisor, a warm African American man, told her: "I'm glad you're moving from messiness to mystery."

Paula was overcome by the feedback from these trusted others. "[This year] has been a sacred space that nobody else has got to share," Paula told the group. "I can't even share this with my husband." Weeping, she said, "What I'm hearing is that through my own earnest seeking, I have given something. And that's what I'm hearing."

With their comments, Anne, Ezekiel, and the other trainees offered a different understanding of Paula's journey through the chaplaincy program, not one of indecision and self-indulgence but instead one of bravery and honesty. Their observations helped reframe Paula's own sense of her experience, because they felt true to Paula and at the same time offered a different lens from the one she was holding herself. The power of their witnessing lay in its capacity to help Paula interpret her own journey differently. In her reply to them, Paula referred to a phrase by the poet Mary Oliver, who wrote: "Who knows anyway what it is, that wild, silky part of ourselves without which no poem can live?"[6] "Thank you for seeing me," Paula told the group, visibly moved. "I mean, me, to use Mary Oliver's [phrase], that 'wild, silky part,' it's easier for me to see it reflected back. It's a gift you have given me, so thank you."

We can observe in this example some core characteristics of witnessing. First, it is interactive, both in process and effect. The capacity of the supervisors and other trainees to "see" Paula in this moment depended on not just their sensitivity and warmth but also upon her revealing herself to them—both as an emotionally courageous risk-taker, and as someone who dreaded that she was actually just indulgent or unreliable. That revelation—opening up to another person oneself, as well as one's fears about that self—can be radically vulnerable, and thus deeply distressing, even among a close-knit group, as we see with Paula's tears as she faced her own perceived shortcomings. On the other hand, the effect of this witnessing also shapes the people doing the seeing. Paula said she learned something about herself, but her seers also got something out of the experience: at the very least, in addition to her gratitude, a sense of their own efficacy and of their identities as people who see others effectively. Their connective labor "landed," and together they created a reflective resonance.

From Recognition to Social Intimacy

Emotional recognition of the kind between Karl and Mr. Shiflett, or Paula and her evaluators, occurs every day, between near-strangers and friendly acquaintances, in paid and unpaid interactions, in encounters of seeing and being seen. Psychologists view this kind of recognition as crucial in infancy, but have a bit less to say about its role away from the familial hearth. What the midcentury psychoanalyst Winnicott called "good-enough mothering"—meant not as a slight but instead as a generous phrase, making room for all kinds of imperfection—described mothers being able to reflect at least some version of what the infant was feeling. According to this thinking, babies who feel mostly reflected develop a sense of themselves as connected to others but also as individuals, understanding where mother ends and baby begins. Most psychoanalysts focus on these early childhood dynamics, although some note how healing it is when therapists offer this recognition to their adult clients. "Through witnessing the patient's . . . inner world, . . . the analyst promotes the patient's burgeoning self-definition," wrote one such account.[7]

Meanwhile, philosophers and sociologists also talk about recognition, but they are most interested in the kind that confers legitimacy and legibility upon particular social groups. Recognition is how people obtain civil rights due to their legal status, such as LGBTQ marriage rights, or how they are awarded particular esteem for traits or achievements; it is the site of increased jockeying by diverse groups bidding for rights to dignity and respect (e.g., women, people of color) while others fulminate about their increased invisibility (working-class white men). Scholars debate how important this kind of cultural recognition is, compared to, say, enduring poverty or job discrimination.[8]

Yet these arguments focus on either the micro level (the familial hearth) or the macro level (the political coliseum), and they ignore the vital middle space of civic life and commerce, of mundane interactions taking place every day between people: a middle space that I am calling social intimacy. Connective labor happens in this middle space, a context whose expansion reflects historical trends of shrinking families and growing markets. Furthermore, recognition is more than simply a message of esteem for a person or a social group; it is instead an emotional nod to the other. It is expressly this emotional register that helps it deliver such a powerful message of belonging. Against a backdrop of stark inequality, seeing the other—particularly when the other is markedly different—makes the potential impact of each small encounter reverberate.[9]

When I shadowed Erin, for example, the hospital chaplain whose story opened this book, I witnessed how seeing and being seen can cut across boundary lines and social categories like race or gender. I watched as she visited with Jordan, an African American patient, and his wife, Mariah, in their shared hospital room. Jordan had brain cancer, a terminal diagnosis, but he was chatty when Erin sat down with him. He mostly lamented his medical care—he didn't want radiation ("There's no cure, I should be able to choose whether or not to do anything, just keep on moving to the next person"), they brought him the wrong specialist ("Why you sending me a heart doctor, he don't know nothing about the brain"), and they kept getting his medication wrong, he said. I observed Mariah as she sat quietly to the side, her eyes askance, her face closed; she shook her head without meeting Erin's eyes when the chaplain asked her whether there was anything she could do to help her during this difficult time. I got the sense that she was not that interested in or hopeful about whatever "help" this white stranger thought she was going to offer her.

But when they stood up to pray, clasping hands in a circle in what was clearly a familiar ritual to all, Erin showed how closely she had been listening. "Help these people get the best medicine he needs," she said, "and help them get more focused care from their care team, and help him ease his frustration, and help them both as they go through this journey." Jordan said a quiet "amen" to Erin when she finished, and as I watched, Mariah's face just completely opened up at this moment, brightening dramatically as she smiled widely. We may not know for sure why she responded in this way, but Erin's spoken prayer told Mariah and Jordan that she had heard not only the specifics of Jordan's issues but also their emotional meaning, and she reflected this understanding to both of them directly, without showing any impatience or skepticism about his litany. For Mariah in that moment, Erin's words seemed to resonate like a church bell.

Recognition can nonetheless be dangerous terrain. Chapter 7 explores the risk of substantial harm that resides in misrecognition, when people read the other wrongly; shaming, when negative judgment demeans; or even when well-intentioned providers feel a duty to rescue. But when it is successful, seeing and being seen involves people coming together in at least partial agreement, creating something that employers have discovered has value.[10]

While such empathic reflection takes place in families and among friends, it is the large-scale movement of connective labor into the marketplace that has brought it within the purview of contemporary capitalism. If, as Nancy Fraser suggests, there are three different stages of how capitalist societies treat the labor of "sustaining connections," in the early stages it was generally ignored as women's work, paid "in the coin of love." It is in the latter stage of today's financialized capitalism that we find the massive commodification of care work in modern societies, largely in response to women's mass entry into the paid labor force. Yet the marketization of connective labor also reflects the rise of what Eva Illouz calls an "emotional capitalism," in which, starting in the early twentieth century, economic relations became deeply emotional, anchored by a growing therapeutic culture. As a result, connective labor came to underlie many ostensible tasks at work, understood as the "soft skills" that help workers achieve strategic goals. These twin developments—the movement of care into the market and the new emotional style at the workplace—explain not only the expansion of connective labor but also why it is subject to increasing pressures of efficiency and profit, as employers attempt to extract value from the bonds people create, further explored in chapters 5 and 6. For now, I will simply point out that connective labor generates considerable benefit that, despite all the attention, we are only just coming to understand.[11]

What Do We Know about the Effects of Seeing and Being Seen?

The power of a good listener is written onto the body. When someone hears us and lets us know, it has actual physiological effects that scientists have been able to document. In 2015, for example, a group of Finnish researchers had pairs of people tell stories to each other and measured the emotional arousal of both teller and listener as testified by electrical changes in their skin. The researchers found that when the listener conveyed that they heard and understood the other—through nods, facial expressions, encouraging noises, and even cheers or cries—the storytellers benefited noticeably: when they felt heard by their listeners, they felt calmer such that their emotional arousal decreased, and the more their listeners conveyed that, the stronger the impact. Meanwhile, the more the listeners were allied with the storytellers, the more they themselves experienced increased arousal; when one walks in someone else's shoes, it is apparently like taking on some of their energy. In other words, listening well "spreads the emotional load."[12]

Listening well is an act of emotional attunement, when we not only hear what the other is saying but also grasp at an undercurrent of feeling. People use many different words to convey this kind of activity, including common ones such as "empathy," "compassion," or "sympathy." Psychologists invoke concepts like rapport, empathic reflection, or the therapeutic alliance, while they sometimes call the result of these efforts "felt understanding." "Emotional intelligence," "high-quality connections," or "interpersonal sensitivity" are phrases bandied about in business contexts. In my view, despite small differences, the vast array of these concepts largely describe the same broad phenomenon. Regardless of what we call it, research tells us that extensive benefits accrue to those who are seen, as well as to those who do the seeing.

Those who demonstrate skill in reading others' emotions are more effective at work, for example, according to studies conducted with principals in school settings, corporate executives, and counselors. Children who demonstrate an ability to decode nonverbal meanings do better in academic settings. Managers whose subordinates report feeling understood by them contribute to improved workplace climates. Meanwhile, the experience of feeling understood is equally powerful. In medicine, patients are more likely to comply with treatments prescribed by practitioners they believe understand them; clients who feel like their therapists "get them" feel more satisfied with therapy. Research shows that if one feels understood, one feels closer to others and is more likely to listen to their advice or directions. Ultimately, these two-way effects explain why so much research backs up the importance of relationship in a variety of fields.[13]

Instrumental Gains, Ostensible Tasks

While these studies confirm the interactivity at work here—how connective labor has impact on both giver and recipient—many of them are in service to the connection-as-engine-grease model of connective labor, where it serves to ease the accomplishment of what really matters: the learning or healing, for example. And to be sure, many practitioners echoed this model when telling me stories about how connective labor helped them do their jobs well—in other words, its instrumental value.

"The relationship, I think, is crucial," said Charles Jiang, a primary care doctor. "The relationship makes it easier to talk about things. I find that patients, as we get to know each other, they disclose more and that's a product of the relationship. It helps with my diagnosis. It helps with the amount of time that I have to ask questions."

"I don't think you can do any type of therapy if the relationship component isn't there," said Gerard Juliano, a therapist with the VA. "You're listening and you're following up with questions and you're reflecting. 'That sounds really difficult, I can imagine how hurtful that must be for you.' Demonstrating empathy and focused on listening in a way that helps the client to know that 'this person really cares what I have to say and is really invested in my problem.'"

Teachers also talked about connective labor as a core conduit for learning. "I think each kid needs to be seen, like really seen," said Bert Juster, a teacher and principal who founded an independent school in Oakland. "I don't think a kid really gets it on a deep level. They might get it for the moment or they might get it for the test, but I don't think they are really bitten by the information or the content until they feel seen by the person they're learning from." He came to this notion as a teacher, but also as a student. "I'm thinking back to those teachers who taught me and who I learned most from, and I think I learned most from the people who really saw me." Wilson Parker, a teacher trainer in an education school, agreed. "You know, if you're not able to connect with students, then you will never be able to get to the content," he said.

Connecting also worked to get wayward students back on track so they could apply themselves in school. Mutulu Acoli, a special education teacher at an Oakland middle school, told me that the school knew to call on him when a particular student known for behavior problems started acting out. "Usually, like if I see him like that, I know I can go up to him and stop him, even if he's, like, torn through a couple other adults," Mutulu told me. "I'm like, 'Nope. Look at me. Who am I? You know I got you, so listen to me. What's going on, and let's talk so we can get you back in class.'" For Mutulu, "I got you" meant that he knew the boy well enough to be aware of the backstory, to realize the boy had his rationale for acting out, even as Mutulu worked to get him back on track. "Basically we have the trust that I'm not trying to just say he did something bad without reasoning, and I'm also going to be behind him in how he's feeling," Mutulu said, a connection that enabled him to calm the boy down.

These stories demonstrate some of the purely instrumental gains to be had—the ostensible tasks enabled—by deploying interpersonal sensitivity or listening well. There are indicators, however, that this view actually underestimates what is going on.

Beyond the Connection-as-Grease Model

Existing research points to connective labor's broader effects, but it tells that story primarily for just one set of participants: for those on the receiving end. Feeling understood helps people in their self-concept, their intimate relationships, and their broader sense of belonging. The more people feel understood, the more they feel satisfied with life. It enhances their relationships with others, deepening closeness and intimacy with friends and partners as well as practitioners like therapists and physicians. Researchers have also found that sharing truths generates belongingness; people feel more socially connected when they have had deeper conversations and divulge more during their interactions. "Our findings reinforce the idea that felt understanding is a core feature of human existence," wrote one group of researchers. "Without it, people cannot survive."[14]

Being connected in this way has a powerful impact, clearly, yet we do not know much about how it affects practitioners, aside from those more instrumental benefits—the more effective school principals and therapists, the managers with improved workplace climates, for example. Two veins of research have hinted at some broader effects, however: one focused on emotional labor, the other on social status.[15]

Hochschild originally coined the term "emotional labor" to mean managing one's feelings via "deep" or "surface" acting, while more recently it has become more of an umbrella concept to refer to all kinds of ways people use emotions as part of a job, including—we might add—connective labor. People manage their feelings by following social rules that reflect race, gender, and other inequalities: Who is expected to show deference? Who is expected not to cry? Who is proscribed from expressing anger?

On the one hand, rules about emotions undergird common courtesy, as when we feign feelings so as not to hurt others. Yet when workers such as flight attendants smile at the behest of their employers, they risk becoming alienated from their own selves, Hochschild warned. This kind of emotional estrangement is more likely under certain working conditions, such as when workers labor under time pressure, when they find themselves required to perform undue deference to others, or when they are oppressed by demands of data collection or technology. Still, service workers find relationships rewarding, particularly if conditions permit. Indeed, low-income, "low-skilled" service workers gain a certain dignity from emotional connections, even when other aspects of their job seem designed to withhold it.[16]

Another line of research argues that workers derive status or a sanctioned moral identity from interpersonal service work, particularly in the helping professions. These accounts report how practitioners use such work to accumulate authority or rank, to tell themselves they are "good," "useful," or heroic, or to "give back" to others less fortunate; scholars debate whether this labor mitigates or creates social boundaries with the people they are helping. While this research is important, the focus on status concerns feels thin to me, as if that was all that mattered, and the tone is sometimes that of a moral police, exposing providers with the *real* reasons for their supposed altruism. This research also sometimes conflates motivation for interpersonal work—which surveys boil down to feeling good by doing good, being with others, or enacting values—with its impact.[17]

But if we look further, there are some indicators that connective labor can produce more profound meaning for workers. By bearing witness to people with AIDS during the 1980s, for example, gay caregivers were drawn out of themselves, discovering a sacredness in community that reconciled them to their own identities, according to Seton Hall sociologist Philip Kayal. Rebecca Anne Allahyari's study of charity volunteers goes beyond whether or not they got to *claim* for themselves a moral identity and instead reports their desire to *change* their moral selves; she explores the complexities of cultivating a more virtuous personhood in a process she calls "moral selving."

Sometimes these deeper conclusions come out when researchers refuse to stop at simple accounts of motivation. A study of nursing assistants, for example, found that they liked "feeling useful," to be sure, but then went on to explore their moral imagination; workers reported sophisticated visions of what healing meant, that "healing a resident does not always mean to cure her, that healing, when it does occur, is usually reciprocal, and that healing comes through touch, often with those most broken." These sorts of accounts are rare, but indicate that there are depths to be plumbed when we ask what connective labor does, not just for the seen but for the seer.[18]

Connective labor is thus both emotional and social, a form of recognition and a kind of work. It is shaped by and linked to inequality, and yet extends up and down the class ladder. It is neither restricted to the familial hearth, nor to the cacophony of the political sphere, but takes place in the spaces of everyday commerce and civic life, between soccer coaches, hairdressers, managers, bartenders, and the people they serve. While research reports important effects for clients and patients, it does not give us very satisfying answers as to why those effects happen, nor does it tell us much about connective labor's potential deeper impacts upon workers. If connective labor is acknowledged at all, it is often understood as a conduit for the "ostensible tasks" for which such workers are hired. Yet what else happens when people see or are seen? What else does emotional recognition do?

The Threads of Connection

When we talk to the practitioners themselves, they offer some clues. It turns out that connective labor is not just a spoonful of sugar, helping to make the "real" medicine of teaching or counseling go down. Instead, above and beyond the other tasks it helps accomplish, it offers its own transformative effects. When circumstances permit, an encounter of seeing and being seen spins threads that define and bind the people involved. These threads generate dignity, purpose, and understanding, moral identities that reverberate, that create a reflective resonance. In such small moments is our social intimacy woven.

Here is where we must acknowledge that these interactions can also cause harm: when we feel people should see us and they do not, it can feel like violence, the memory of a particularly bad encounter lingering like a wound that won't heal. Our focus in the sections that follow, however, is on what happens with "good-enough" connective labor, when someone bears at least partial witness to another who receives it.

Making Dignity

Birdie Mueller was a white nurse practitioner in California with a bustling manner and a high-beam smile full of cheery warm pragmatism. Sitting at a picnic table outside her community clinic, she told me that she had assumed she was going to be a doctor like her father until she failed organic chemistry at her elite private college, and how even as a nurse practitioner she struggled with what she called "ego issues." But as a nurse, she could focus on the "human element" like being present, saying, "I think we all just want acknowledgment of our sufferings even if you can't cure it or do anything about it." She recalled a homeless man who came into a clinic she used to work for, who had been on the streets for years, "never really in a shelter, probably cross-country homeless back and forth." She remembered the moment. "He had some wounds on his feet, just gnarly, calloused, and [I] just sat and did wound care for his feet."

> He was so hunched over probably from years of osteoporosis and just walking and being hunched over. There's very few people who are going to make eye contact with him, because he physiologically can't really look up, so to be there and help wash and clean his wounds. It wasn't going to do much. He was still going to be on his feet all the time. That was not going to be the intervention, he was still going to be—He was so resistant to going into any shelter, it was just a band-aid over a really big problem.

But nonetheless, she washed his feet, in an instant that captured what nursing was about for her: the humility, the service, the witnessing. "Just to give him that moment of, 'I'm seeing you, I'm acknowledging you, this is me caring for you,'" she said. "It was powerful for both of us. Powerful to just realize, I mean that was the transition for me also . . . [it was] good for my ego. Just being like, 'Oh yeah, I'm not above [this].'"

This is the first connecting thread that this work offers: binding workers and their charges together on the basis of their shared membership in the vast collective of humankind, one that warrants a right to being seen. When Birdie washed the man's feet, she recognized his core humanity, a fact that she imagined was ignored by so many who passed him on the street.

Melia Santos, an optometrist who worked in a community clinic in Virginia, was talking about that humanity when she explained why she felt so displeased by her experience shadowing a particularly mercenary provider, one who saw patients only as income streams. In one breath, she explained what was wrong there: "Because we're all human beings and we all deserve to feel cared for, and if you have a problem with your health, you deserve to have access to someone you feel that cares about you and is willing to go the extra mile to heal you."

At its most basic, connective labor helps convey to the other simply that they are worthy of being seen by another person. People regaled me with stories of their clients or students coming to understand their own capacity to make claims on others' time and attention. Russell Gray, the master therapist, told his students that their job was to build that sense in their clients. One day he was talking to them about connecting to shy and withdrawn adolescents. Those are sometimes the toughest kids to work with, he told his listeners, because they are used to not having a voice, and as they are really quiet they don't give you much to work with. "Everything they say is gold," he said, and advised the student therapists to find it all "really interesting." Pretending to be in therapy with one such teenager, he said with mock enthrallment, "Tell me more about *Grand Theft Auto*."

His students laughed, but Russell continued in seriousness. "You're building a relationship where they're going to learn 'Somebody wants to hear what I have to say. I'm an interesting person to this therapist.' The relationship is hugely valuable, it's undervalued in our society—the value of the human connection is under-recognized in our society." Through relationship, he said, "you're teaching the kid the skills to start talking. When you get to the point when they're talking, then they are done with therapy and they're ready for the world."

It did not bother me to hear Russell talking about faking an interest in *Grand Theft Auto* to convince his client that they were worth listening to; I was not worried that his connective labor was just a performance. I suppose I recognized the compromises that a parent might make to stay connected to a four-year-old who wanted to talk about nothing but dinosaurs, or a thirteen-year-old with a passion for fashion magazines. Connective labor can be a reciprocal encounter and at the same time involve some degree of performance. But I was discomfited for other reasons that seemed to hit me periodically during this project while I listened to practitioners talk about their work.

First, they might overestimate the degree to which they understand or see the other person, because patients or clients might conceal it successfully when they do not feel understood, perhaps as part of an effort to hide from the world at large, or so as not to make the doctor or therapist feel bad. Hearing about the trainee therapists pretending to be interested in *Grand Theft Auto* made me worry a bit about their future teenager clients who might pretend to feel seen. And to be sure, these encounters misfire with some frequency, as most of us can probably remember misrecognition by a doctor or teacher in our lives. Nonetheless, there is some research to suggest that while people often do not understand others as well as they think they do, they do not have to do very much for people to feel understood; benefits come from what the attachment theorist John Bowlby has dubbed "tolerably accurate reflections of reality."[19]

A second concern bubbling up from the *Grand Theft Auto* story is about its underlying moral: Was its point that disadvantaged or underconfident people need to look to others to know their worth? Such a reading implied a dehumanizing abasement, and that the path to redemption is through being seen by another, who would turn out to be some kind of savior.

Some of the workers I talked to were certainly the star of their own rescue dramas, and I explore the savior role in chapter 7. But ultimately the concern about saviors, while valid, does not capture the ambivalence common among the practitioners I spoke to, many of whom continually asked themselves how to be useful without imposing too much of themselves. Furthermore, many workers, especially those in trauma settings, never see the "end results" of their labor. Others never really get thanked by those whom they labor assiduously to "see": how many of us who feel grateful for our elementary school teachers have taken the opportunity to tell them so? The meaning of connecting work often comes less from the payoff of someone else's gratitude and more from the sense that one is contributing. Most important, for many, from teachers to therapists to hairdressers, their witnessing offers a thread of connection to the other, actually binding them both. If any "saving" is happening, it seems to go both ways.

Birdie's story of washing the man's feet exemplifies this point. The moment was about more than simply giving to another person; it also represented her coming to terms with her own place in the world, that she was "not above [this]." His dignity was hers to realize as well. When practitioners read a morality into their work, they described it as if they were finding in themselves what they were insisting upon for others. In the process of treating someone else like a person, of conveying their very humanity, workers discovered their own.

Making Purpose

The second connecting thread that spins out of this work is a sense of purpose, one that accrues to both seers and seen. If the first thread links people on the basis of their core human dignity, as part of a vast human collective, then the second one binds people to each other as meaningful individuals who matter.

This was the first lesson he learned in teaching, said Conrad Auerbach, now a middle school teacher in a large Virginia city. "If you connect with [students], they will put effort into their work in class much more," he said. "If you don't connect with them, they just do what they want to do." He first absorbed this axiom as a student teacher while he was still in college, he said. "I learned in that TA class, when I was a senior, that hey, if you just chit chat and talk to the student about what's going on, then they'll much easier allow you to help them on their math. So, yeah, it kind of started there." For students, clients, and others who were seen, "purpose" looked like motivation.

"A lot of therapy is motivation and persuasion," said Penelope Mason, a therapist. "Getting someone who is feeling very worthless to go to the social event anyway, convincing them that the social experience is uncomfortable but is going to be worth it anyway because they will learn to tolerate the negative thoughts, 'and if you show up enough, you'll have a different experience,'" she said. "Humans help humans feel motivated."

Relationships worked in an almost subliminal fashion to persuade others, workers believed. "That's the thing, that is motivating everything. Everything," said Rebecca Rooney, a primary care physician in the San

Francisco Bay Area. "So, your low back pain, and your willingness to go to a physical therapist, or willingness to take an NSAID, or your willingness to work on your sleep. After they just had a session with me, or a session with someone who is really skillful in this, people say 'I don't know what just happened there. But I feel better, and I'm motivated and I trust you and I'm not sure what it was, but I'm ready to work on my health.'" Rebecca smiled, adding: "I feel like I just told you my secrets."

Yet purpose was not simply injected into clients and students, the relationship acting like a hypodermic needle. Instead, like dignity, it traveled both ways. Many workers told me that they gained a sense of purpose themselves from having an impact, witnessing vulnerability, and connecting with individuals. The range of people I spoke to and observed for this project was considerable, yet when they talked about what made their work meaningful, it was this similarity that stood out.

Many practitioners found purpose in the sense that they were making an impact in someone else's life. A number of teachers sought to be "the teacher that I wanted, and that I needed, and that I finally got," as Pamela Moore, an exuberant African American middle school teacher put it. Pamela had grown up in an unceasingly transient family, but at one point a teacher was persistent enough with her middle school self to see that her selective mutism was not a sign of special needs. "She was a wonderful teacher and [also] just [a] relationship. Just sat and just listened and that made a world of difference. I want to be that teacher for my kids in [middle school]." Practitioners also said they gained a sense of purpose simply from the privilege of observing people at their most defenseless, and they used words like "honored" or "humbled," and "getting to be there for it." They felt trusted, at moments that mattered.

Jenna, a West Coast physician, had a part-time osteopathic practice, which allowed her to offer lengthy appointments; she talked about the profundity of observing patients in "a very vulnerable position." "It's a trusting relationship," she said. "Because of that trust it imbues the relationship with—I don't know—almost like a power, a sanctity. There's just something about it. I feel really honored and really lucky that I get to do that. You know [it] gives me—it gives me just as much as I give to people. It's like a feedback loop."

Some practitioners found purpose simply in other people's resilience, the strength they witnessed in the face of suffering. "When I get really miserable, I go and spend time with a patient," said Ruthie Carlson, a primary care physician in the Southeast. Not long ago, for example, a patient was diagnosed with cancer, and Ruthie had been "devastated by the news." Ruthie fought all day to get the woman prior authorization from insurance for a procedure, along the way tangling with her practice's office manager about how she was filing the order. "I got very frustrated and realized I wasn't getting any support, and I just called the patient and said, 'How are you doing?'" Ruthie said. "I called her and she said, 'I decided to come into work. I'm not feeling that bad. I think I'm doing OK.'"

Ruthie was floored by her example. The woman knew she was likely to have some sort of cancer, but nonetheless was "doing OK," enough to get herself to work, Ruthie recalled. "She was amazing. What restored me on this day was that gal's going to work, and she's feeling pretty good, with metastatic stomach cancer. You know what? My stuff is not that bad. I am going to get up and go into my office and take care of patients. They have the strength to do what they're doing. It put things back in perspective. They give you the strength because they're going on."

The motivation sparked in others is the flip side of the purpose they find within themselves. Once again, the rewards of connective labor extend out in both directions, in this case spinning out a connecting thread that bring worker and client into relationship not just with a larger human collective but with each other.

Making Understanding

The third thread that connective labor spins is perhaps the most complex: a greater understanding of self and other. Ellis Gable was a white funeral home director in a small Southern town. He said that bereaved family members were sometimes unhappy about how a body looked, but that "sometimes there's not really even anything wrong. Sometimes it's the grief taking over." He told a story of a husband who had forced the funeral home to call the embalmer to revisit his wife's body again and again. Finally, the husband was confronted by the owner. Gable narrated the encounter:

> [The owner] said, "You know, I've had one of the best embalmers in the city here five or six times today. Every time you come up with something, he comes up and addresses it." And he said, "You're just mad because your wife died." He said, "You have nobody else to take it out on so you're taking it out on us." And the guy literally turned around, he could see him shaking. He was crying and he turned back, looked at [the owner], and said, "You're exactly right." He said, "I'm mad because my wife died." He said, "Everything's fine."

The story of the wronged funeral home director had the sound of one that had been repeated again and again among the home employees, confirming as it did a truth they held dear, that "sometimes it's the grief taking over." To be sure, this narrative likely enabled the employees to feel better when families were dissatisfied or upset. In the story they tell themselves about this event, the funeral home director performs a witnessing that expands someone else's self-understanding. Confronted by the owner's reflection, the husband recognized an emotional truth. Ellis's story illustrates the last kind of connection offered by this kind of emotional recognition: connective labor helps others make sense of their experience. By reflecting the other's perspective, by framing and reframing their account, workers give it back suffused with meaning, so that—they believed, and I sometimes observed—people come to a new understanding about themselves.

Practitioners from all kinds of fields talked about how they helped others see and understand themselves. Vivian Vaughan, an Asian American middle school teacher at a private school, told me she thought of this dimension as "where's your line?," asking students to figure out who they were and who they were not. She was worried that kids were increasingly so focused on content that they were becoming lost to themselves.

She had actually switched schools in search of a culture that would value student independent thinking, she said. At her old school, she used to teach speech class, and one of her final assignments was for the students to identify something in society that they thought was beautiful but that society in general did not think was beautiful. She recalled: "And so some kids would come up with speeches like on smoke and how beautiful smoke was or spider webs or the Pythagorean Theorem, but there were a fair amount of kids who just could not do it, and who would come to me and just say, 'Just tell me what I should do my speech on.' [I would say,] 'That's not the point. The point is I'm wanting you to think what is beautiful to you. That's what I want to know.' It just broke my heart."

At her new school, she was able to lead a seminar where she could put the questions to the students: "Where is your line? Where are your boundaries? And why is that important? Where do you say, 'I'm not going there'?" She had told the kids that when former students come back after college to visit her, the question "where's your line?" represents their number-one problem. "They don't know where their line is, and so when they tell me about [some] financial [difficulty] or a boyfriend or a parent, they don't know where to draw their own—it's basically the same. Where is your line? Where do you say, 'I'm not doing that'?" Being able to give this kind of lesson was her "dream come true," Vivian said.

Like the first two connecting threads, this one goes both ways, as workers come to their own understanding in the process of recognizing another. People find their moral identities in this work of reflecting another person, and part of that discovery is when those whom they see reflect back to them something of their own self.[20]

Wendy Peters was a white educator who had taught for decades in public schools that were majority African American students. "They didn't know who I was, they see this white woman, right? Coming in, it's like 'What are you doing here?,' right?" Wendy recalled.

> I'd tell them my background. "Look, yes, I'm white, I grew up with two brothers. Do you know what a fictive sibling is? I have two fictive brothers and they're African American and we were all raised together. And then, I married, my husband is African American. Then we had a biracial kid," you know, I go through—And there was a girl that I had in class, I had her and her sister, and there was one

point in the year and she goes, "You're not the white person I expected you to be."

For Wendy, it was clearly important that she not be seen as a typical white teacher, oblivious to race and racism, perhaps teaching canonical texts and inadvertently dealing in microaggressions. She told me the story of the student's observation—"You're not the white person I expected you to be"—as a statement of great honor, a moment when she clearly felt seen by a student, in an identity-making moment of which she was proud.

Even when the relationships are asymmetrical, then, as teacher and student or therapist and client, people do not just "see" their clients and students; they are also "seen" by them. This reciprocity is perhaps most difficult for practitioners to admit, as they are there to help others and not necessarily to be helped themselves. Yet the importance of such recognition becomes particularly clear when it is absent.

I spent the better part of a year observing Karl and his colleagues at the HIV clinic, for example, and one day a patient—a Mr. Fox—came in with very high levels of the virus, yet uninterested in taking any medication. Depressed and ashamed after telling his sexual partners that they had been exposed, Mr. Fox talked about how he would take his own life if there were a way to do it that would not cause pain or inconvenience. Karl tried to no avail to convince him to get on the antiretroviral (ARV) drugs that effectively control the virus.

After his initial visit with Mr. Fox, Karl came into the tiny office where two senior attending physicians—Dr. Martin, short, dark-haired, and circumspect, and Dr. Taylor, a large man with a more expansive air—sat and bantered as usual in front of their computers. Karl asked the group: "Which of you attendings over there are good at challenging interactions?" "Dr. Martin," joked Dr. Taylor, and they all chuckled at his buck-passing.

Karl described the problem of Mr. Fox's recalcitrance. "I talked to him a lot about it, telling him without harping on [the serious medical issues that were] coming," he told Dr. Martin. "I asked him, 'What if I just prescribed it, would you fill it and take it?' and he said no. I really tried hard to come at it in different directions, and he just has a long history of mental health issues, self-esteem problems, worthless feelings, and I think that's going to be a big impairment." The man had given a urine sample for chlamydia, and he had told the doctors he had had recent oral and anal sexual contact as well, but he had declined other treatments Karl suggested. "I don't want to break him, and just say 'do this' and 'do that,'" Karl said, in a worried voice. Dr. Martin did not seem as concerned, saying, "I think the trick is have him come regularly." As they walked off together, Karl continued: "I don't want him to stop engaging with me because I'm pressuring him too much, but . . ." Yet when they returned to the office a short while later after seeing Mr. Fox, Karl was voicing some outrage. "It's like he thinks it's selfish that we want him to get better!" he exclaimed. "Yeah, I'm like, 'Yeah, I do find it satisfying.'"

Later, Dr. Martin told the team psychiatrist, Dr. Hansen, what had happened. "We just told him, 'Our goal is to get you on ARV drugs, we don't want to twist your arm, that's our goal,' and then [Mr. Fox] said, 'That's *your* goal. I don't think you can force your will on someone for your own mental health.'"

Dr. Martin and Dr. Hansen conferred about how likely it was that Mr. Fox would act on his suicidal thoughts, and agreed that it was simply important that he keep coming back. "We can't force him to do anything, just get him back here frequently until there's trust," Dr. Martin said. An hour later, however, when his shift was over, Karl was still talking about Mr. Fox, this time with Christine, the nurse practitioner. "I mean, 'Yes, it is good for me when you do better,'" he said, "'but it isn't *selfish*.'" Christine was consoling. "I think there's a lot of depression that is driving it," she said. Then as she left the office, she joked sarcastically, "You're so selfish," and Karl laughed unhappily.

I was not in the room where Karl met with Mr. Fox, so I cannot assess whether his recalcitrance reflected some sort of failure on Karl's part to see and reflect the man well. But the aftermath suggests that there was indeed some form of empathic rupture, although we cannot know exactly why. Clearly, to the doctors at least, Mr. Fox represented several challenges—how to get him to trust the practitioners, how to make sure he did not represent an actual suicide risk, how to talk to him so he would be motivated to take the drugs and protect himself and others. But one of the challenges that Karl returned to again and again that morning was the way Mr. Fox perceived Karl's efforts: as "selfish." The word was like an irritant to Karl, who devoted many hours to the care of clinic patients. While he surely derived other satisfactions from the work, his frustration with Mr. Fox suggested that he also looked to patients for some small understanding that he was not—at the very least—"selfish." The encounter was highly unusual, as most patients acted more like Mr. Shiflett, the man whose case opened this chapter: anxious, sometimes deferential, but often simply grateful for Karl's sympathetic, informative style. Only in the absence of this normal exchange can we see the value Karl gleaned from it, a moral identity that he could believe in.

After dignity and motivation, the third connecting thread spun out by this work—when it goes well—is understanding of self and other. And like the first two, these benefits come to both parties to the encounter, both seer and the seen. Emotional recognition delivers sometimes transformative effects that go beyond the instrumental gains of any particular job or occupation.

A particular kind of interactive service work, connective labor is the reflective work that people do to convey an emotional understanding of the other. People in all sorts of occupations use it to help them with their jobs, to teach, heal, and counsel, to manage, sell, or even control others. The capacity to see another person, to understand their perspective, allows these workers to offer a bespoke service whose tailoring makes them more effective at their given tasks. Often invisible, this work is part of that which makes it possible for physicians, therapists, funeral directors, and a whole host of others—from teachers to hairdressers to police detectives—to achieve valued goals. Each of these jobs comes with a particular set of outcomes that are their ostensible purpose, such as controlled blood pressure, a decent understanding of algebra, or a good haircut, and connective labor helps people accomplish them.

I started out this research with the goal of making this invisible work visible. That visibility, I thought, could help workers gain more acknowledgment and reward from school principals or insurance companies. As it happens, however, I found a much broader world behind this work: it turns out connective labor is more than just what's useful instrumentally, smoothing the information delivery of physicians or teachers. Instead, the work is valuable not simply because it helps people to do the ostensible tasks for which they were hired, but rather for the meanings it gives themselves and others.

Along the way, the experience of seeing and reflecting another person effectively generates a host of other goods, from purpose to dignity to understanding, in both workers and others. Indeed, viewing connective labor as solely that which makes it easier to teach or advise well is a bit like considering only visible light as the sum total of the universe. But as the physicists tell us, dark matter might be invisible but it causes gravitational effects that cannot be explained by simply looking at that which we can see.

When we allow ourselves to see the larger meanings that people create together, we understand that connective labor is not the outgrowth of an individual worker's skills or accomplishments—that may or may not be replaced or augmented by machines—but instead is a mutual achievement, collectively produced by humans in concert. The connective threads emanating from this work—the ties that define and bind worker and client together, the knots of social relationships and social meaning, the resonance—are spun from its core humanity.

The work beyond the ostensible tasks is vital work, and yet in many occupations it is under threat by systematization and automation. As employers try to extract greater efficiency or profit from connective labor, they impose new systems, data collection mandates, and technologies that shape its conduct and experience, for good and ill. The next chapter explores these trends with regard to automation and AI, considering how automation is being sold, what it portends for humane interpersonal work, and the unintended consequences it bears for what it means to be human.

3

THE AUTOMATION FRONTIER

One spring before the pandemic, I visited an experimental school in the San Francisco Bay Area, where—like a wave of other schools popping up that sought to "disrupt" conventional education—kids used computer programs for customized lessons in many subjects, from reading to math and more. A warren of rooms on the first floor of a busy corporate building, the school was tucked away under the sweeping fragrance of majestic eucalyptus trees arching over the parking lot, like a shady perfumed oasis amid the sunbaked lawns, busy roads, and Starbucks franchises dotting the Silicon Valley landscape.

That day, a handful of kids sat around kidney-shaped classroom desks, staring at computers with their headphones on, the quiet settling around them like dust motes. While I watched, an irrepressible little redheaded boy, looking about six years old, seemed to want to interact with others about whatever he was learning. He kept looking up from his screen to try and catch the eyes of the kids around him, while they bent their heads to their own tasks. One time he called out, "What's five minus two? Three, right?" Nobody replied. Later, apparently finishing a problem, he exclaimed "yes!" and pumped his arm in triumph, looking quickly around to see who had heard him. Again it looked as if no one had, except me.

Increasingly, schools are turning to technology to teach, a trend that started even before the coronavirus pandemic turned so many families into homeschoolers. The school I visited that spring took aim at the traditional classroom model: loosening up some of the age segregation to group kids according to their abilities, emphasizing project-based learning, and most of all, relying on software to deliver on one of the hottest buzzwords in education: "personalization." The students had spreadsheets that organized their day, which was broken up into independent study, or "goal time," and chunks devoted to different specialized apps, like Quill for grammar, or Tynker, a coding program (see figure 3.1 for one student's spreadsheet). Students also pursued "passion projects," which bore more than a passing resemblance to nearby Google's famous "20 percent time" practice of allowing its engineers one day a week to work on their own pursuits.[1]

Most of the kids I saw there seemed interested and engaged, although they could also sound like harried adults. Zoe, a student about thirteen years old, once told me she liked staying for extended day because it was quiet and she could get work done. "It's not that the classes are too long, [but] there's, like, seven hours of 'goal time' and fifteen hours' worth of work, so it's hard to, like, squeeze it all [in]," Zoe said. Her sister complained to her sometimes, but "I'm like, 'Oh, no, this is normal,'" she said. "'We're all running around like headless chickens. It's normal.'"

Zoe and the redheaded boy may have been learning from apps, but they were not entirely on their own. The school had added more and more time with adults since its founding a few years back, as the limitations of mechanized education became clear. But these were not conventional teacher roles: most of the adults on-site were of two kinds, either "content specialists" or "advisors." While content specialists taught brief small-group lessons on, say, a particular math concept, advisors sat in a central area with their computers open, and met with the kids one-on-one once a week for forty-five minutes, keeping track of their progress; they were the ones who asked the kids, "How's it going for you overall?"

After the redheaded boy had left the area, I listened while an eight-year-old Asian American girl named Kim met with her advisor, Shelley, a white woman. They gazed together at a computer screen that displayed an Excel sheet summarizing Kim's latest goals and accomplishments. With the girl sitting next to her, Shelley asked, "So, in programming, what do you want to do this term?" "Honestly, I'm not sure," Kim replied,

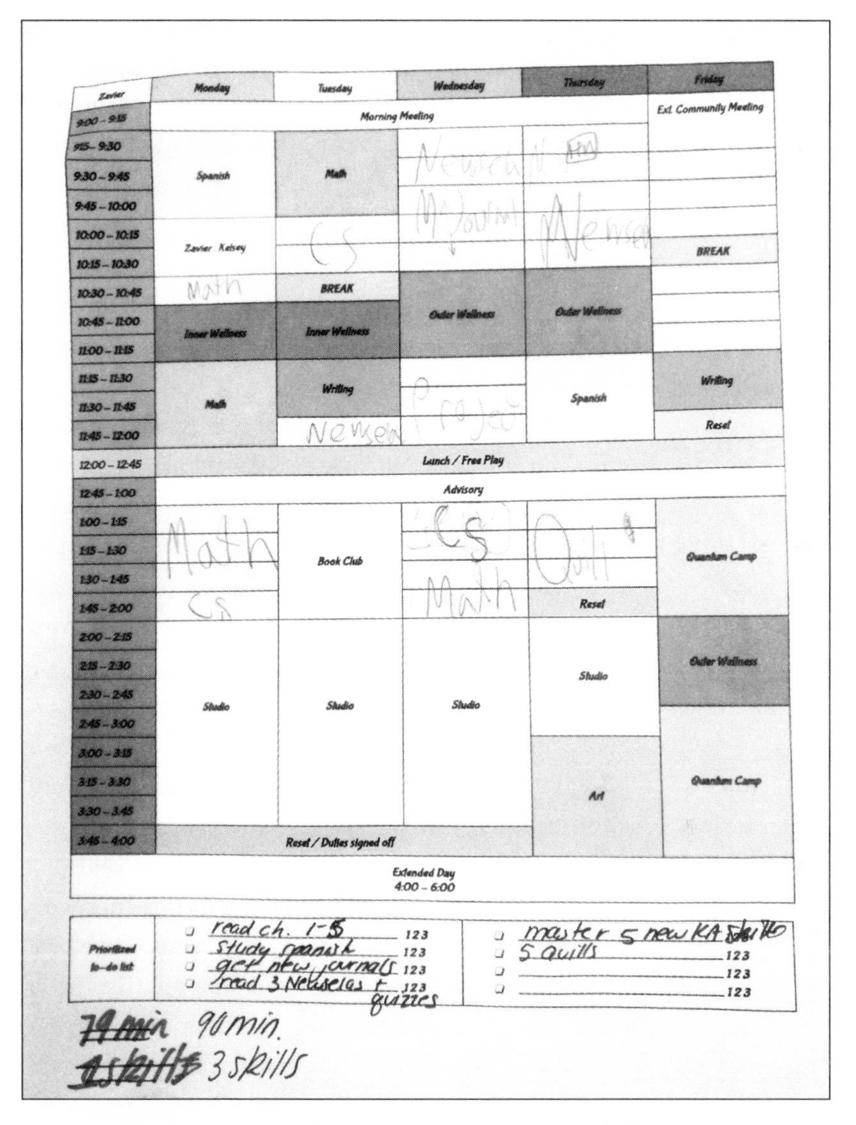

FIGURE 3.1. A spreadsheet with a student's schedule for the week.

thoughtfully playing with her long black ponytail. "Well, is there anything you're interested in?" Shelley asked. "You could combine it with your passion project. What's your passion project?" Kim mentioned a business she was starting (we were in Silicon Valley, after all) and how "it was based on an invention" that she called "a genius idea." Shelley listened patiently, murmuring little sounds of assent and encouragement. There was all the time in the world for them to figure out a customized plan for programming.

I visited that school several times that spring, attending classes, observing meetings, and interviewing teachers, students, parents, and administrators. Humming with a low-level buzz, the school seemed lively yet calm, the kids independent and learning-focused. Nonetheless, its experiment was a radical one, particularly for how it managed the connective labor of teaching. The school practiced a comprehensive decentering of the teacher, in which machines in the form of targeted apps for various subjects took over many core instructional responsibilities long shouldered by humans. But it also rearranged how people related to each other in school, dividing up parts of teaching—"content delivery" and relationship—that for centuries of schooling had been assigned to one person.

Standard schools asked teachers to be good at too many different tasks, the principal, Faye Owens, told me. "We ask them to be great mentors and coaches, but we also want them to be really good at explaining their content area, we also want them to be good at analyzing data, we also want them to be creative and project designers," she said. "For the most part, those skill sets aren't traditionally found in the same person."

The advisors, and the relationships they forged with the kids, were the emotional glue for students like the little red-haired boy who spent mornings with the apps. "The advisor is in charge of the relationships," Faye said. "They are someone to talk to, someone who cares about the kids and vice versa, someone who makes them feel like they belong." The content specialists, on the other hand, gave the kids targeted lessons beyond what the apps had to offer, focused on getting the kids to understand long division or what makes something a dystopian story. Just as Frederick Taylor broke down the bricklayer's craft into its component parts more than a century ago, the school disaggregated the teaching of particular content—happening every morning via laptops and headphones, as well as in the small-group lessons—from the connective labor that helped kids actually care about learning that content.

"So, the advisor really is meant to be someone who is skilled at mentoring children," Faye said. The content specialists? Approvingly, Faye quoted one as saying: "I know what I need to be good at, and that's explaining math to kids." And in the morning, the kids had machines to teach the rest.

The Silicon Valley school certainly touches on the issues of central concern to most writing on AI to date: job disruption (what happens to teachers if machines do the teaching?), privacy (who gets to see, use, and commodify the student's voluminous data?), and algorithmic bias (how does the software intersect with preexisting inequalities to sort kids?). Yet these matters divert our attention from an important conversation we are not yet having: about the promise and peril of automating relationship. This chapter charts the cultural reverberations of the drive to mechanize connective labor, in the ongoing drama of how we demarcate what falls on which side of the automation frontier.[2]

The social movement to introduce automation to connective labor is powered by Big Tech, which pours billions of dollars into AI and app start-ups in the field, in its relentless drive to colonize the human lifeworld. Its energy comes from the breathless (or breathlessly cautionary) media coverage about the use of AI in humane work, and the artists and philosophers who dream out loud about an AI future, but its mandate comes from the gleeful marriage of capitalism and technology in the twenty-first century. I spoke to and observed engineers and administrators from California to Boston to Japan who were working to bring such technologies to connective labor, from virtual therapists to virtual palliative care. They made claims of varying ambition and scope for AI, apps, and automation, ranging from "it's better than nothing," to "it's better than humans," to "we are better together." Most commonly, technologists argued that AI and apps could offer solutions to common problems in connective labor, to "scale up" its delivery or to take care of mundane tasks so that humans might be "freed up" for more meaningful work. But even if that prospect were likely, questions remain about who is going to be freed and what counts as meaningful.

The school I visited might seem very different from a conventional classroom, but its story is nonetheless a familiar one to anyone who follows the movement of tech into connective labor: innovators use technology to "disrupt" the way it is conventionally delivered. To its credit, the school took action when some of technology's social costs were laid bare, increasing the presence of human "advisors" to add "motivation" to their model; the school's abundant resources allowed it to surround students with caring adults in various guises. Yet in other contexts—less generously supported, more attuned to "efficiencies" or profit—technologists' claims to deliver connective labor "at scale" end up shrinking what counts as that labor. Ultimately, bringing AI and app technology to connective labor leads us to redefine the automation frontier, and along the way, to a cultural reckoning of what it means to be human.

The Spread of Machines in Connective Labor

The current state of mechanization in humane interpersonal work is a partial one. The COVID pandemic accelerated the introduction of virtual connective labor everywhere, making each of us experts in one form of mechanization: the remote interactions between people over computers or text messages. We all bore close witness to its pros and cons, seeing how it enables access for those who can't be in person, for example, or how some features like Zoom chat have coaxed forth new forms of participation, but how the lack of eye contact or gestures or casual conversation makes it harder for people to connect emotionally. Many of us lived through what researchers had already found: that therapists, teachers, and physicians can achieve some of their goals virtually, but that there are also significant costs; we know, for example, that compared to in-person classrooms, students suffer substantial learning loss in virtual environments.[3]

Other forms of automation include apps or programs that actually assist or replace the worker with machine-generated words or even robots, and while this is a wide range, the technology tends to share the same outcomes of limited benefits and certain costs. Watson, IBM's AI once lauded for triumphing on the game show *Jeopardy!*, failed to live up to its promise when it was deployed in medical contexts; cancer centers spent millions for Watson to provide treatment recommendations customized for patients, but the collaborations stumbled over data issues, with physicians frustrated by having to spend time on data and technology rather than caring for patients. A 2020 review concluded that "AI for primary care is an innovation that is in early stages of maturity, with few tools ready for widespread implementation." Meanwhile, bots that replace the human entirely have had some success in therapy, surprisingly—educating users in mental health issues, for example—but sometimes their answers can miss the mark. One 2019 review, for example, reported that chatbots sometimes responded to the cue "I am depressed" with "Maybe the weather is affecting you." In 2023, when an eating disorder helpline fired its human workers, they instead offered the services of a chatbot named Tessa, who doled out inappropriate advice such as weight loss tips.[4]

But these mixed findings have not stopped the explosion of the market for health-related apps, which is enormous and growing, including anything from the meditation app Headspace to the "diagnosis assistant" Sensely. Nonetheless, there is very little data available regarding their effectiveness; one meta-review described a sort of Wild West of marketplaces, cautioning that "at present, anyone can create and publish health and medical apps in the app stores without having to test them, and patients must experiment with apps by trial and error." The reviewers concluded that "the overall evidence of effectiveness was of very low quality."[5]

It is certainly likely that AI and automation will improve in these fields. The release of ChatGPT and its inheritors, which use generative machine learning based on billions of texts across the web, offered the most fluent responsiveness to date, sparking heated public conversations about plagiarism, job loss, and computer literacy, along with more than a few speculations about the inroads of AI into counseling, teaching, and other forms of connective labor. A group of researchers reported that the bot had passed the US medical licensing exam, but it also was prone to many reported factual errors delivered in the same confident tone. Another study found that the bot generated better responses to medical questions posted to Reddit, answers rated by medical practitioners as more empathetic than those posted by human clinicians. Future iterations of these models were slated to include not just text but also sound and images; designers appear to be less focused on improving the machine's relationship to truth, and more focused on making it able to handle different kinds of human expression.[6]

Despite the seemingly boundless future of AI, there are caveats to note. Lucy Suchman, the anthropologist and critic, has called artificial intelligence and automation "technological solutions looking for their problems." The state of the field is nowhere near the capacity to be able to provide connective labor in any real sense, she told me. Instead, in all contemporary examples, humans are very necessary as forms of support for the machines to work, a point underscored by all the "tweaks" innovators told me about making. Yet the hype around AI, and the susceptibility of the media to it, has propelled a kind of cultural imaginary that "perpetuated the idea of the suitability of these machines for this labor," she said. "It does that by erasing all the other forms of labor that would still be in place in order to make these technologies possible."[7]

In addition, while researchers and scientists have worked hard to create AI apps and agents that might be able to function in the connective labor space, they did so without fully understanding what they were even automating. Even some AI researchers themselves agreed. "There's a lot of hype," said Sue Carlson, an engineer in Massachusetts who has written more than a hundred papers on human-machine collaborations. "What does it mean to have a relationship? Most people working in the field don't ask those questions. They want to make the robot seem great for kids, make it enticing in certain ways, and then say it has emotional connections. Look a little deeper, and you'll see, 'Oh yeah, it's vaporware.' You need to look at it with a jaundiced eye, look not at what they say they have they built, but what have they tested in interactions with human beings."

Geoffrey Janney, an award-winning researcher of emotional AI working in California, echoed Sue when he told me that engineers were not using a robust conception of relationship. "A lot of us who are dabbling in this are doing these really superficial aspects of relationship building. [We're making agents that are] nodding and saying 'Uh-huh' or [throwing] in a bit of language, but there's not really a physics behind it. There's this veneer of sociality." Laughing, he said, "It's like lipstick on a pig."

Thus, AI and apps in connective labor have so far demonstrated limited benefits, feature a continued need for invisible human labor, and rely on a remarkably superficial version of its raw material. Yet the automation of connective labor proceeds apace. What is its appeal? What do AI and apps promise us that human laborers do not?

Three Kinds of "Better"

Advocates voice several rationales for what these technologies have to offer. Some think that connective labor is uniquely human in its conduct and experience, and that any replacement is bound to be inferior. Artificial intelligence and other software in this space are bound to be worse, they think, but at least they are "better than nothing." Others argue that the technologies can improve upon some of the characteristic flaws of people's connective labor, that it is instead "better than humans." Still others think, in the received wisdom of Silicon Valley, that AI and apps will not replace human labor but instead augment it: thus technology and humans are "better together." Gleaned from interviews with engineers, researchers, and administrators, these positions inform the kind of problems the technologists can even see, those they view as important, and those they tackle as solvable.

Better than nothing. The argument that apps and automated connective labor are "better than nothing" depends on the observation that there is stark inequality of access to empathic, perceptive teachers and doctors, for example. The contemporary degradation of connective labor—in which physicians and therapists work in increasingly hurried conditions that encourage rushed, dismissive, or even scripted interactions—is directly linked to these arguments. "It's where we think we can have the most impact," said Timothy Bickmore, a Northeastern researcher who had developed an AI couples' counselor, exercise coach, and a host of other applications. "We try to find these areas where there's either no service provided or, you know, the service that's there is—doesn't meet the needs of the individual, so we're trying to fill a niche. There's a low-literacy population or a particularly disadvantaged population that our current healthcare system is not adequately helping, and we think by, you know, automating, we can greatly improve the care that they're getting. And so hopefully, the AI is better than nothing."[8]

A few years ago, Bickmore created a "virtual nurse," an AI program to help low-income patients at Boston Medical Center with discharge procedures, hoping that it would enable them to understand sometimes long and complicated instructions. The virtual nurse, whom coders dubbed "Louise" or "Elizabeth," was no more than an animated figure on a screen, the mechanized voice asking viewers if they were Red Sox fans before going over the after-hospital care plan. Bickmore was surprised, however, when 74 percent of the patients said they liked getting their discharge information from the virtual nurse better than a human one. "I prefer Louise, she's better than a doctor, she explains more, and doctors are always in a hurry," one patient told Bickmore. The virtual nurse gave the patients more than just information; she also gave them time. The software program offered the patients, many of whom were identified as having low health literacy, the extra minutes they needed to digest complex directions.[9]

"Providers spend an average of seven minutes with patients at discharge," Bickmore said. "And it's clear from our studies that especially for low-literacy patients, [they're] going to need somewhere like an hour." The unspoken message here was that the busy clinicians were offering disadvantaged patients a weak facsimile of connective labor, akin to "nothing." With the virtual nurse, patients could proceed at their own pace, as opposed to the quick interaction they were likely to get with doctors—who wanted to give good care but were also sometimes palpably counting the minutes, plagued by the long list of tasks lying ahead.

The "virtual nurse" was not a panacea, however, and Bickmore considered that important enough to say so. While he thought it worked well to provide information about discharge to needy patients, he was careful to limit the program's utility to those exact scenarios, and scornful about researchers who made broader claims. "I've seen, over the years, demo after demo of [researchers who claim,] 'Here's this animated nurse that you can just talk to about your health,'" he said. "And that's a recipe to start killing people, as far as I'm concerned."

"It's—You can't possibly understand everything a person's going to say," he said. "We don't have that ability for the foreseeable future, and people [patients] can start talking off-topic about things that your system's not designed to handle, or start using, you know, metaphors or 'I'm going to off myself' or, you know, anything." "Better than nothing" arguments have an inherent modesty to them; without humility about what your program can do, automation threatens to "start killing people."

Penelope Mason had a similarly modest view. A therapist who was starting a business to bring teletherapy to college campuses, Penelope was open to therapy apps, but only as a potentially helpful tool for people who did not have severe problems. "In a lot of ways the mental health apps are 'app-i-fied' workbooks," Penelope said. "Those totally have space in the market, especially if you have no other access to any other resources. Probably if you're psychologically minded, you really can pick up a self-help book and get some insight there, depending on the complexity of what has caused the person to be stuck and what they're stuck on."

Apps could help, she thought, but only for those people without "complexity." "The reality is for most people, there's a parent who is alcoholic, or complex relationships, or they have a biological vulnerability, and that complexity makes it hard for them to see just through their own observations, just through apps. Based on how the technology is working right now, [the capacity to] observe and mirror back accurately and give hope and motivation? We don't have apps that have that ability yet."

The apps might not be ideal, but these developers are not asking how to hit the ideal. Bickmore had developed a "relational agent"—a tablet that patients would take home—to serve as a "palliative care consultant." As in the case of the "virtual nurse," he told me that the existing system was leaving too many people underserved. In the United States, he said, palliative care is given much too late, and many patients could benefit from services at a much earlier stage. His lab had written a module that involved "discussions" (with a machine) about spiritual beliefs and religious background. I asked how patients seemed to be accepting the automated counseling, and Bickmore told me they had just finished a pilot study just to see whether people would tolerate having an end-of-life discussion with a tablet. "And by and large, no issue," he said. "We did ask them, 'Would you have rather talked to a person?' and they said yes. But 'Do you think this is important to talk about your spiritual background at end of life?' And everybody, the majority, said, 'Yes, this is very, very important to do.'"

By separating out the question "Is this important?" from the question of "Who would do this work?," researchers make discursive space for the "better than nothing" argument, starting with the notion that there could even be such a thing as palliative care counseling without a person to deliver it. Unsurprised by the fact that people prefer to talk to other people, the questioners instead disassemble connective labor into parts, disengaging the end result (the counseling) from how it came to be (the clinician) and for whom. "Better than nothing" arguments set aside the whole question of how it might actually be best to provide connective labor as unnecessary, unrealistic, or irrelevant. Most important, "better than nothing" proponents assume that a major contemporary problem is how to "scale up" connective labor, necessary because labor costs are fixed and high and thus human service is increasingly, and inevitably, a luxury good. Proponents of AI and apps often seem to focus very narrowly on the interaction, instead of the larger backdrop shaping it, such as why human practitioners have such an impossibly long lists of tasks to start out with. The busy clinicians at Boston Medical Center may have come in second place to Bickmore's machines, but if the doctors were the ones imposing industrial "time-discipline," as the historian E. P. Thompson once called it, they did so under orders, bent to the dictates of the hospital and its schedule.[10]

Helena Edwards worked at a firm developing an app that offered monitoring, support, and advice for diabetes patients. While people already received good care when they were very sick (because they were in hospitals) or very wealthy (because they could hire their own nursing help), she said, the app was crucial for less-advantaged people whose problems were not very severe. "There is a huge amount of the population that just isn't getting any medical help at all," she said. "A lot of that middle population needs pretty basic care." This "middle population," however, was not actually in the middle of any single dimension, but instead, it seemed, they were people who were not very sick and not very affluent. And while advantaged people with mild diabetes received their guidance from human beings, the cost of labor became more relevant for low-income patients.

"There aren't enough medical professionals, not even close, to handle the number of people who need them," Helena said. Insurance companies had set up nurse hotlines to fill "this gap," she said. "These nurses essentially just get burned out because they're saying the same very basic things over and over and over and over again, and that's the piece that we're trying to automate on a scale." Of course, this kind of burnout reflects the degradation of nursing, and might be addressed by designing their jobs to have more variety and depth in patient care, not just by replacing them with a machine. But when I asked whether employers could hire more people, she shook her head. "I just don't know how you would do that because you have to pay people's salaries. It seems like you need something like AI to be very scalable for certain applications there."

Such access-driven arguments are not only reserved for health care. In 2020, even before the pandemic, thousands of young American children, including nearly half of Utah's four-year-olds, were enrolled in virtual preschool, online programs that use animation and songs to teach pre-reading and other skills. "We simply don't have the money to provide a quality pre-K experience to every child in my state, even though I absolutely agree that a face-to-face, high-quality pre-K is the best option," Craig Horn, a North Carolina legislator, told the *New York Times*. "But when it's not an option for the child, I refuse to ignore that child." David Cardenas, the mayor of a rural California town that is home to mostly Latinx farmworker families said: "With this program, those kids who were left out in years past, now they're going to be included."[11]

This is a "better than nothing" rationale, and it is pervasive and compelling—who could argue with giving low-income people access to therapy or preschool or palliative care counseling they might not get now? Not the clients themselves. Some parents of preschoolers told journalists that the online programs made their kids "smarter and more ready to learn."

Many consumers of therapy apps and virtual preschool are apparently ready to embrace such opportunities rather than hold out for affordable, accessible, human-driven programs that are unlikely to materialize.

Furthermore, humans give the benefit of the doubt to machines. Researchers have repeatedly found that if programs or robots signal understanding or relationality, people will approach them as if they had feelings or empathy. One of Timothy Bickmore's earliest experiments was for a companion-type "agent" on a computer that went home with elderly adults. "They really liked the social dimensions of the interaction, and even though they were clear that it was a computer, they talked, they used a lot of language, as if it were interacting with a person and were looking forward to their next chat," he said. "They would like to chat all day with this thing. I mean, these are largely adults who were living alone at home." As MIT psychologist Sherry Turkle points out, "roboticists have learned those few triggers that help us fool ourselves. We don't need much. We are ready to enter the romance."[12]

Yet the assumptions underlying these "better than nothing" arguments are themselves particular ways of seeing the world that are hardly inevitable. First, the arguments presume that relationships—an interactive, mutual accomplishment—can be reduced to emotions, a point I discuss further later. Second, these advocates presume that the only outcome that matters, the only goal we need to pay attention to, is the learning or healing or advising—the ostensible task served by connective labor, as discussed in the previous chapter—and not any other gains that stem from seeing the other.

Most important, however, current staffing levels and compromises about quality are not fixed or inexorable, while the only way to relieve the pressure on busy clinicians and strapped education systems is not necessarily to automate the work. As AI computer scientists and ethicists Amanda and Noel Sharkey note, "'[The] robots are better than nothing' argument could lead to a more widespread use of the technology in situations where there is a shortage of funding, and where what is actually needed is more staff and better regulation."[13]

Practitioners I spoke to agreed. Carrie Koppel, a therapist at a VA hospital, questioned whether apps or agents could ever offer the nonverbal acuity that she considered crucial to good therapy. But even if they could, she viewed their development as a political decision. "Even if machines [could] pick up nuances and facial expression and that kind of stuff, why are we doing that though? So that people can make money in tech, so that big, huge industry can continue to blossom? Why do we have to do that? So, that would be my question."

Better than humans. For some, automated connective labor is not just "better than nothing"; it actually improves upon what humans have to offer. To them, machines solve problems such as standardization, stigma, shame, unpredictability, and inauthenticity, and they do so with customization, perceived privacy, and logic. Machines also provide not just better service to clients but can help workers by doing unappealing or dangerous jobs, much like the robots the US Army deploys for bomb disposal.

The bedrock for many "better than humans" proponents is the notion that humans are, in the words of one engineer, "not particularly special." Researchers and administrators argue that human beings—their emotions, their relations—are knowable and reducible, and thus replicable, allowing engineers to design "relational agents" or apps to tackle such disparate problems as how to make children on hospital wards more comfortable or help seniors feel more connected or, in kiosks in current use at the US border, to assess whether immigrants are lying.[14]

Gregor Siemens was an entrepreneur who had started several highly successful data analytics firms with more than $100 million in venture capital; his latest effort was a company that automated the sales process. It began when he noticed a basic inequity in tech firms, he said. "Sales reps are the highest-paid individuals for the least productive ecosystem of the company, the most unreliable performers." In the software industry, sales and marketing cost "five or six times" what firms spent on actual product development, he said, and all because of what he considered a misguided mystique surrounding salespeople and the sales relationship.

"All of them think that they are the unique, only person that can do this, and sales is like magic. That only they know, and only their super charismatic approach to the customer is what makes the magic happen," Gregor said. "But being kind of a data guy, what I did is a lot of analysis about their behavior, predicting if a deal will close, predicting how long it will take to onboard those calls. And I was way more accurate predicting if a deal will close or not." Ultimately, Gregor said, "we identified a sixty-five-step process to sell a product." The experience of doing so convinced him that salespeople benefited from a reputation they did not deserve.

"Sales is not that 'white magic,' charismatic sales guy," Gregor said. "I feel that the opposite is true. Sales is just a repeatable process. It's just delivering the right information at the right time to the right person." In other words, the connective labor—the listening, the seeing, the witnessing—that salespeople engage in to enlist the loyalties of the customer is "white magic," only serving to obfuscate what Gregor decided really matters to closing a deal: following a sixty-five-step process that he ultimately had programmed into an app. "I think a lot of people are not realizing how replaceable they are by the next generation of technology," he said.

Gregor's company countered the notion that humans were particularly unique or difficult for researchers to reverse engineer; in his view his app solved the problem of human unpredictability, further mystified by the emphasis in sales on charisma. But to my surprise, after we had talked a while about his app and its impact on the economy, he told me he was worried about the automation of interpersonal work. "The human things can't come from a machine," he said. "Interactions, respect, empathy. It's very dangerous. I'm concerned." Many salespeople would say these particular "human things" are integral to their process, of course. Yet even if we believe that sales did not involve any of those dimensions and thus was no loss to automate, Gregor's proposition—that humans were effectively modular, interchangeable, and not very mysterious, with behavior reducible to sixty-five steps—resembled the arguments I heard for the mechanization of connective labor in not just sales but all kinds of work.[15]

If you consider humans as not particularly special, and furthermore quite limited in their capacities, then it is not a big leap to think automation can do what they cannot. This conviction undergirds the experiment that is the Silicon Valley school whose example opened this chapter. That school is actually one of several in the San Francisco Bay Area aiming to use technology to "disrupt" education; most of them private schools, they generally serve a fairly privileged population for whom "better than nothing" arguments are not going to be appealing. Nonetheless, a number of these schools choose to deploy apps for much of the instructional day instead of teachers in classrooms. For these schools, the apps allow the children's learning to be pegged to their individual ability level, allegedly better than any human teacher—short of one-on-one tutoring—can manage.

Tabitha was the principal of a school that used a very similar model to the one I visited. In an interview, she told me the apps allowed valuable customization of learning "for students to have choice and voice, and you can do that through technology on some level, right? Based on what assignments they're getting, what level they're getting it at," she said. "So, if you look at personalization, [it's] 'I know that you're at this reading level, it's better for you to do this activity first thing in the morning, so you don't forget, and you need to be quiet alone.' Right? So, that's space, that's place, that's level, there's all different kinds of ways of looking at that." Technology allows the school to customize the subject, the level, and the learning environment for each child; finally, advocates say, AI promises to take on the problem of mass education.

Dorothe Pfaffner, the mother of an eight-year-old in the school I visited, said this kind of customization let kids advance much faster. "Some of them, like my son for example, he's just like any other kid, he's better at some stuff than others, but he's breezing ahead in Khan Academy," Dorothe said. "I mean, he's doing like fourth-grade math now and he's supposedly in third grade, so he's like way ahead, at least a grade level ahead. So, I don't really think he needs the [human teaching in that] because he's breezing through it."

The opposite of such customized learning, the negative case these proponents are solving, is the menace of overly standardized public education. To some degree this menace is surely exaggerated, since many contemporary human teachers use "activity centers," groupings by competence, and other pedagogical tactics to enable them to differentiate instruction for their students. But perhaps more significant, beyond what kids absorb of a given academic subject, learning in groups offers some benefits—the experience of being faster and slower than peers in different subjects allows for a more nuanced view of merit; the experience of instruction not perfectly personalized can teach resilience and humility, particularly to more-advantaged students, for whom childhood is increasingly customized; the experience of group learning generates a collective culture that teaches that sometimes other people have as good or better ideas, and that what we share together is sometimes more essential than what an individual can achieve. To be sure, each of the schools attempting to "disrupt" education includes some group-learning experiences. Yet when the most important question is how fast a student can go, standardization becomes an acute problem.[16]

Customization is not the only way in which mechanized connective labor surpasses humans, according to these advocates. Machines can handle more data, but they also have demonstrably more endurance than human workers, who need sleep and burn out on repetition. Ananya Rajendran, who worked at a large tech multinational firm, noted that the firm was developing technologies for a concierge application, she said. "We can imagine a hotel, like Hilton, wanting to build a concierge conversational agent, embodied in a robot that stands in the concierge area, maybe during off hours, to help their visitors with different kinds of questions they may have," she said. The leap from "available during off hours" to uncomplaining or nonunionized labor was not made explicit, but Ananya's comments underscore the threat to labor encapsulated in many "better than humans" arguments.

Some of these apps and programs aim at work that humans find distasteful or boring. The diabetes app that Helena Edwards worked on was designed to replace nurses in call-in centers who endured repetitive and draining shifts. Similarly, Geoffrey Janney, the AI researcher, said teaching a person with brain injury how to eat again was particularly difficult. "It is traumatizing both for patient and caregiver, because they have to do it over and over again, and it's kind of embarrassing for the patient, and for the caregiver it's very frustrating. That is an ideal setting" to replace human with machine, he said.

Yet "better than humans" arguments tend to go even farther than claiming that humans need relief from bad jobs, that they are not particularly special, or that they are not as good at tasks like managing patients' individual needs or staying up late. Humans also judge other humans, and human judgment is sometimes wrong, delivered harshly, or crippling in its impact. Engineers and administrators who advance "better than humans" arguments thus take on not just human frailty but human wisdom.

"In some domains like addiction and drug use and abuse, people love to talk to a hotline bot because they are more willing to express their problem, [because they are afraid of] being judged," said Ananya. It was the same in education, she pointed out. "If you have a question about a very simple concept that you haven't [understood], you'd rather ask somebody who wouldn't judge you."

Osamu Kimura was a Japanese roboticist who worked to bring AI into language tutoring. "You know the Japanese person always want to talk in English, but we have a different, you know, some grammar and it's difficult to learn. You know, the Japanese person is hesitant and embarrassed to talk in English to the actual person," Osamu said. "So that's our idea. Make robot teachers, English teachers in the home," where language learners could make mistakes unobserved by another human.

People widely consider bots or apps as a judgment-free zone. They are routinely more honest with computers than with humans; this surprising fact is one researchers keep looking for and finding again and again. Adults will disclose more to a computer about their sexual practices before they give blood. They'll tell a computer more about their financial troubles. Children are more willing to disclose bullying to a computer than to a person. These findings recur even though computers are not very private, not very anonymous, and do not actually refrain from assessment, as anyone subject to insurance rates or bank loan terms as computed by algorithm might report.[17]

The persistence of this belief is more than just an interesting or paradoxical observation about human misperceptions; honest confession is consequential. It actually matters for improving the accuracy of diagnosis of learning or health problems for which a person might need treatment. Teachers need to know how little students understand the material, for example; therapists need to know that a client has an eating disorder or has had thoughts of suicide, although therapists often told me that they cared more about how they might come to that knowledge rather than about knowing any particular fact. But if the fear of judgment silences some patients or clients or students, practitioners worry about that silence.

The perceived lack of judgment in apps and automation also works to lure in people who might not seek help in the first place. Veterans are notoriously reluctant to seek mental health assistance; the military has funded numerous studies examining whether they would accept AI therapy. "Can you have a simple chat box that would interact, for instance, with veterans that come back with all sorts of psychological problems?" said Peter Almond, a computer science researcher. "Psychotherapy is totally stigmatized, so they just will not go. Knowing that there is a robot on the other end, might they then open up, right? So I think they are really important applications."

It is not, of course, that computers do not judge at all, but rather that their judgments do not feel as personal or, ironically, as biased. Some technologists confessed they shared these views. Helena, who worked on the diabetes app, said she herself was suspicious of people whose job it was to offer support, compared to the app, whose support would come only if you actually deserved it, say, by posting better glucose scores or following better health regimens. "I almost feel like the bot is a little bit more genuine because it's coming from a place of logic, whereas the person is paid to say something and you know that they're just going to say 'good job' no matter what," said Helena. "[The app] seems less complicated." Automation improves upon humans, then, when the humans are particularly scripted.

At first blush, the lack of judgment seems like a strong selling point for apps and relational agents. As we shall see in chapter 7, misrecognition, negative judgments, and shame are serious risks for people in human labor. Yet practitioners argue that if clients and consumers are able to be more honest with computers, it is simply because they do not want to be judged in an unsafe environment, and it simply reveals how bad the connective labor was that they had had to date. It is not that people prefer to "be seen" by computers rather than people; it is that they prefer them to judgmental or unsafe listeners, as opposed to reflective and empathic ones.

Even for those who think computers are mostly better than humans, the potential of judgment-free automation has a cost: that of removing the power of accountability, of aspiration, of the client wanting to do better. When I asked Peter Almond, the engineer, what he thought humans still had to offer in this work, he said: "an audience that matters." In his view, robots would someday do most everything humans could do—in education, for example, that included grading papers and answering questions about the material. He still wasn't sure, however, if one could "project enough humanness onto a robot that you want to make it proud of you."

As in the case of "better than nothing" proponents, however, most of the problems that "better than humans" advocates see as important and solvable are ones that lend themselves to technological solutions, are created by specific (and changeable) labor practices, and are themselves revealing of priorities, such as the individual attainment of particular (usually high-achieving) students. Indeed, many of the problems researchers are tackling likely stem from how the work is being organized, as misrecognition and judgment become more likely when workers lack the time or resources to go slow and take appropriate care with their witnessing, while inauthenticity is the price we pay for scriptedness. The working conditions of contemporary connective labor turn out to be an indispensable backdrop to the conversation about its automation.

Better together. Finally, some argue that AI simply augments humans' efforts—in other words, that AI and humans are "better together." Researchers and administrators make these arguments (1) when what is at stake is of an overwhelming volume: of knowledge to master, people to coordinate, or work to accomplish; and (2) when they distinguish between thought and feeling, with the latter reserved for humans. The "better together" argument is ascendant among scientists and AI researchers, possibly because it makes their work seem less threatening, in that it will not decimate jobs but instead just reconfigure them.

Particularly in medicine, researchers and administrators were decidedly enthusiastic about AI's capacity to help a profession drowning in must-dos. Sarah Jahani was a researcher working on a training app for nursing resource managers, who directed nurse workflow on a given hospital unit. The app was aimed at substantially reducing the cognitive load such nurses shouldered, predicting issues like room turnover and some patient needs, and making suggestions for what the system should do next. The job is "harder than an air traffic controller, and they do it without any decision support, like they do it with pen and paper," Sarah said. "They might have like twelve to twenty direct reports, nurses that they're tasking and other resources that they're managing. That's at the edges of human capability."

Many primary care physicians I spoke to may have been ambivalent about using AI in their own clinical practice, but actually embraced the idea of such programs for other physicians, such as specialists who needed to sift through accumulated medical knowledge. "Like, [for] a cancer patient who has a specific stage of a specific histology, of a specific tumor with specific genetic markers, like, I can see how AI would be extremely powerful," said Nathan, who had worked as a physician at the VA. "Because it's like, it would be impossible, because it's just getting more and more fractionated, and it's getting too complex. There's too many subgroups of lung cancer, there's too many. So in that context, information is highly, highly valuable."

On the other hand, he said, the technology is far less apt for primary care. "In primary care settings, 95 percent of the time it [the medical situation] is obvious, right? It's like, 'You need to lower your blood pressure, you need to control your blah, blah, blah, your sugar, your weight,'" he said. "The *content* has some complexity to it, I mean, but the essence of it is not. It's entirely whether the person does anything. I mean, if you look through a thousand general medicine notes at the VA, you would see the same ten problems, and they wouldn't surprise you, probably: high blood pressure, diabetes, high cholesterol, coronary disease, smoking, weight, COPD, enlarged prostate—that's it. Literally." Under those circumstances, finding and treating the proverbial "zebra," the complex unique disease or disorder, is not the issue, and technology becomes far less useful. "Then," he said, "the relationship becomes everything."

But AI developers are not content to develop technology that is only for sifting through reams of medical studies; they are taking aim at the emotional tasks as well. I sat in on a Boston conference held for those working in the socioemotional field of AI, and watched one presenter report on her latest research on the pediatric ward of a local hospital. "We're bringing 'social robots' into the units," she said, where nurses are normally trained to engage children in emotional support, to distract them from being in a stressful environment. "They [the nurses] are way understaffed," she said, reporting that they attend to twice the number of child patients they can handle. "They can't be with every child every time, and the idea [for the AI] is extending their services." Her lab was conducting experiments in which they analyzed the spoken dialogues and facial expressions of child patients in conversation with a robot, versus a plush toy, versus an avatar. The nurses were open to the prospect of using the technology in the unit, she said. The possibility of helping to "keep a constant record of how a child is doing is very interesting to them."

What's common to all these arguments is the language of "freeing up" humans for other, often more meaningful, work. An influential series of reports by the consulting firm McKinsey, for example, argued that it was far more likely that AI will change jobs rather than eradicate them, as well as create new ones that we cannot predict. "Capabilities such as creativity and sensing emotions are core to the human experience and also difficult to automate," one report notes, but "the amount of time that workers spend on activities requiring these capabilities appears to be surprisingly low," with less than 4 percent of US work requiring creativity, and less than 30 percent requiring emotion sensitivity. By taking away other tasks, the report continued, AI might free up workers for more meaningful work.[18]

Underlying the language of "freeing up" is the notion that some tasks are valuable for humans to undertake and others are not. Researchers try to ferret out what work might be considered rote and thus optimal for automation. "There are things that machines are starting to do that are presumed to be not rote, these so-called creative tasks," Geoffrey Janney, the AI researcher, said. "Is writing a news story, is that rote? Some people think it is, some people think it's not," but programs were in use right now to write up news items, in particular for business and sports. "You've got companies that are playing with [automating] a computerized intake nurse, which people would have thought was not rote, but on some level it [the intake nurse's job] is just asking a bunch of questions."

While it seems clear that automation will generate jobs and industries that we cannot foresee, however, the "free up" language seems strikingly optimistic to me. It is surely possible that with the cost savings of automating work formerly performed by humans, employers could keep the same number of employees and instead imbue human jobs with newly meaningful tasks. We can see this at work in the experimental schools, where teachers cheered their new freedom to concentrate on offering one-on-one help to those who needed it. But given the history of US employers driving costs down to offer cheaper products—opting for the "low" road versus the "high" road of, say, Germany and other high-wage, high-quality producers—the experimental school was probably the exception rather than the rule. Instead, employers will likely cut the people they can. If history is any guide, in the US, we're less apt to free up workers than free up their parking spaces.

Gregor, whose firm was producing an app to automate sales, was not afraid to say that this was the outcome he anticipated. Sales staff were very overpaid, he thought. "There's a huge difference in the productivity and the pay," he said. "And therefore, I think technology naturally will correct that. And I see myself just being a little wheel that—" "A part of the correction," I finished for him. "And if you follow this through," he continued, "then what will happen is the first company [to adopt his app] will use less sales reps and make more sales, and significant higher profits. But then, you know, if we believe in the free market, that will lower the price and it will have a positive impact on the country." "Better than humans" proponents did not always shy away from the implications of their work for job dislocation.

Some "better together" arguments rested not on the premise that humans needed help to manage unknowable volumes of knowledge, but that feeling and thinking were distinct and separable, and that if computers were thinking, humans were for feeling. Tara Singh, a medical school administrator, illustrated this belief when she described how the coming of AI would change the practice of medicine. "The whole goal

FIGURE 3.2. "Reassuring," by the xkcd comic.
Source: https://xkcd.com/1263/.

of AI is to take the physicians' cognitive skills and offload them to machines that don't get tired," she said. "The humans will have to do the interpersonal stuff that no machine will do." The end result could be to reinject meaning into the physicians' work, she said. If robots take on the diagnostics, physicians could become more about being present with the patient. "Medical practitioners talk about wanting to help people rather than being robots," she said, and concentrating on the emotional tie with the patient, "the interpersonal stuff that no machine will do," could be their chance.

Of course, some engineers working on socioemotional AI refused to cede even this much ground. Instead, they maintained that AI usefully served in emotional jobs—with apps programmed to express empathy to people, for example. Indeed, the discursive ground was littered with claims about what AI would never be able to do—white-collar work, creative work, coding itself—that were later abandoned as the automation frontier plowed right over them (see figure 3.2). Yet if AI and automation could do even this intensely personal, emotional work, what was left for people?

To true believers—those engineers who are pretty sure AI can do much of anything that humans can—the crumb they are willing to offer the human worker is "motivation." Humans may not be particularly special, but they might still be special to other humans. "Humans help humans feel motivated," said Penelope, the therapist and entrepreneur. "It's not really skill development—it's hope."

At the experimental school, they were explicit about relying on humans as motivation. At first the school depended much more on programs like Khan Academy, said Faye Owens, the principal. "As we have gone on in time, we've learned that that relationship is more important," she said. The school had started out with a lot more computer time, she confessed, but kept expanding the advisor role to enlist the children's hearts. "I mean, I think every edit we've made to our school model since we've opened has been to give more and more time to the students to have one-on-one with the teacher and to allow them to develop that bond." Apparently, "personalization" needed the personal touch to make it stick.

Yet the vision of human connective labor as simply motivation is quite a thin one, and wherever it held sway, I saw engineers and administrators underestimating, in my view, just how much practitioners were doing. One afternoon at the experimental school offered an example.

Vivian was a dynamic Asian American veteran teacher who had been hired as an advisor for older grades. One day, I watched her meet with Adam, a white boy who looked about twelve. "I'm looking at your reflection on page fifteen," Vivian said, as they both stared into their computers. "Did you answer the how? Pretend you are explaining to someone who really doesn't know the answer. How does that happen? What would need to happen?" Adam mumbled something into his screen that I didn't quite catch, but Vivian picked it up immediately. "That makes it easier, true, but how would it give me the result that I want? If I really want this specific result, what is it about the process that can get me there?"

Later, with Anna, a thirteen-year-old Asian American girl with dyed purple hair, Vivian said, "In your reflection, I'm wondering how having that knowledge allows you to attract attention. But it doesn't say anything about informed choices. What do you now have as an artist, what you didn't have before?" "More variation, more control, more selection," Anna answered, and the two of them continued discussing what it meant to have more understanding as an artist.

Vivian was working here to bring students to the next level of learning, asking not just what they knew but how they had acquired that knowledge, how they felt about knowing it, how they knew they had mastered it. These kinds of meta questions were exactly what kids needed to answer to be able to really learn, according to modern pedagogy. "Truly, if a kid is going to learn, they need to have an internal barometer of whether they know the information or don't know the information," Burt Jester, who founded a school in the East Bay, told me in an interview. "And they need to have an internal barometer for what an aha moment is. And that, I think, takes an intimate relationship with another human being to be able to develop those barometers." Jester thought that witnessing—where a student is seen by another human, who helps the child reflect upon and narrate the story of what they are learning and how—was the core of great teaching.

The advisors at the experimental school built those barometers every day. They were not just "in charge of the relationships" but also worked—by combining thinking and feeling—to contextualize what the students were learning. Administrators likely knew they were doing this—they had designed these jobs after all—but their vocabulary to describe it was quite limited; when they described the "tweaks to their model," it was about calling on the advisors to ratchet up student motivation. Yet much more than the "audience who matters," to recall the engineer Peter Almond's summation for what humans were for, the advisors made rich, profound meaning with the students in their charge.

When school administrators split the work of teaching into "content delivery" and "motivation," they were participating in a legacy of specialization at work that resembled the way industrialists broke the back of skilled labor. The scholar Harry Braverman argued that modern capitalism invented "de-skilling," dividing artisanal craftwork into jobs that could be done by cheaper labor with fewer credentials and less training. Similarly, jobs like those of the advisors, which involved more feeling and less content specialization, are traditionally considered less skilled, and usually receive less pay. While the content specialists and advisors had the same credentials at Faye's school, one might imagine how a different district, with more financial constraints, might reinterpret the school's model to save money—replace content specialists with apps, and replace advisors with feelers who come with less training and could command less pay. Indeed, journalists repeatedly report stories of public schools that rely on Khan Academy videos to teach math without the advisors to make it stick.[19]

Each kind of "better than" argument sketches a vision of what AI and automation have to offer—as reluctant stand-ins, triumphant replacements, or willing partners to the human workers—and each paves the way for the widespread adoption of machines in connective labor. Yet this adoption takes place within an existing ecology of modern capitalism and modern bureaucracy, where human connective labor is squeezed by twin imperatives of profit and austerity, erecting a social architecture that constrains their witnessing. As we shall see in chapter 5, for some, hiring a human is already a luxury, as marginalized populations currently receive connective labor that is scripted, surveilled, and sped up.

Given that existing ecology, "better than nothing" arguments hold sway in the public sphere, convincing lawmakers and administrators to adopt AI virtual preschools and virtual nurses. On the part of affluent users, on the other hand, "better than humans" arguments have a certain cachet—busy people talk a lot about preferring automated services to avoid having to waste time interacting with others—but the convenience and status of having personal services delivered personally continues to offer a meaningful appeal. With the trajectory of automation shaped and determined by these inequities, we are hurtling toward a future where they are amplified: one in which less-advantaged workers provide the connective labor in person for the wealthy, while receiving automated services for themselves.

The Unintended Consequences of the Automation Frontier

While engineers and administrators varied in just what kind of "better" AI and automation was, they all advocated deploying automated connective labor to solve common problems such as access, judgment, and standardization. Along the way, many attempted to divide human-to-human interactions into heavily symbolic and meaningful on the one hand, and "mundane" or "rote" on the other. Researchers and others used this kind of discourse to draw some cultural lines, to render some work as more and less open to machines. Yet their efforts led to three unintended consequences: first, the erasure of connective labor; second, the explosion of all things data; and third, the transformation of what it means to be human. Such are the implications we all face of crossing the automation frontier.

Invisible witnessing. As we have seen, particularly among those who considered AI as an exciting reservoir of untapped potential, researchers and administrators seemed to view humans as "motivators" at best, there to boost client attachment by eliciting their emotions. Many apps that aim at behavioral health, such as social anxiety or depression treatment, relied on this view: offering mechanized connective labor, with humans positioned here and there in the process merely as cheerleaders and monitors. Yet people rarely acquiesce entirely to their own underestimation, and connective labor sometimes ended up seeping in unrecognized anyway.

Veronica Agostini was a young white woman with a fresh BA in psychology when she applied for and got a job as a "coach" at a new online start-up trying to cure social anxiety. The company had developed an app that incorporated the principles of cognitive behavioral therapy (CBT) in a twelve-week online program, with coaches like Veronica to interact with the clients. "It wasn't like the coach was therapy, or anything like that. It was purely like, 'How can we get people to engage with the app?'" Veronica said. They may not have been doing "therapy," but as Veronica described her work, it sounded increasingly familiar. "Basically what happened is you talk to someone one time, and then, you know, if they were interested and, like, you did a good job on the first call, you would talk to them every week for sometimes, like, twelve weeks. I talked to some people for, like, months," she recalled. She described one client, for example, a middle-aged man in sales who said he had generalized anxiety.

> And he was like, "I'm new to this field and I'm just so nervous about it, I don't know what to do. Um, and I still make my quota at the end of every month, but it's just so anxiety-inducing. I need some way to deal with this. And I've never talked with a therapist before or

> anybody—like, I've never shared anything, like, about this with anyone." And he went through the program, and he was, like, a stickler about it, like he was, like, very engaged, on—Like, always on time, always picked up the phone when I called, which is not true of most people, um, and really took advantage of it. And he started, like, just blowing his sales quota out of the water. He was just, like, doing so well they, like, gave him a promotion, he got a new sales territory, like, it was awesome.

After the man had completed half the program, with these rather spectacular results, he stopped showing up, and Veronica said she was worried. "I was like, "Oh no. Like, 'Hope everything's OK,'" she said. But she also said it wasn't that unusual for people to disappear. "That would happen sometimes; people were doing well and they were just like, 'I don't need this,'" she noted. But she managed to talk to him, and he told her, "Well, I've been doing really well at my job. Like, I don't need help anymore. But I started this new relationship, and I'm really nervous about it, I just have all this anxiety." So he ended up coming back to the program and completing it. "That was a really good one," she said. "He stands out in my mind."

Some aspects of the job made it clear that it was not therapy. Like the anxious salesman, when clients mastered their problem, often they would just drop the app, behavior more akin to closing a self-help book than it was to the sometimes-lengthy experience of ceasing a therapy relationship. Furthermore, Veronica said, there were rules in place to keep the coaches from acting as mental health professionals. "So you were allowed to reflect what was in the program, but you could never give advice. You obviously could never make a diagnosis. Um, you couldn't, you know, tell them what to do or anything like that. Like, your job was getting them to talk about what's going well for them, how they've been able to fit it into their life, um, what, um, you know, what's working for them and what's not. 'How are you feeling about what you just did? Here's what's coming up. What do you think about that?' Those types of questions."[20]

But the rules dividing their work from therapy seemed more for the coaches than for the clients, some of whom were happy to think of the coaches as private counselors. "Lots of people would just want to talk, and they wouldn't be going to a therapist," Veronica said. "Yeah, they can't afford therapy or they don't have the adequate services wherever they live. So this was like, 'A hundred bucks a month?' Like, 'I can scrape that together,' like, 'Why not?'" Other aspects of the work echoed therapy as well. Coaches listened while clients shared profoundly intimate details; they were reflecting and witnessing as they tried to help the clients get better, and they suffered from burnout. "I think in the beginning I—It really was hard," Veronica said. "It was so hard to set boundaries and I would think about the clients after I went home. It's like, there was no training on how to deal with that."

The impact of this work was palpable, she said. "[Coaches] would be, like, really upset during the day and it was just like the leadership team was kind of deaf to all that. And I think in some ways they had a pretty callous attitude about things, where, um, it was sort of like, 'Well, it's at-will employment, so you can quit if you want to.' Which is just like, 'No, people are here because they care.'"

At one point, the firm hired a social worker to come in and train the coaches in "motivational interviewing," a systematic approach to reflective witnessing that highlights the client's role in defining and enacting goals. The training changed the job dramatically, enabling her to help clients more and making the job much more sustainable.

Still, Veronica left after four years, not just because of the lack of support, but because the company started increasing the client load to untenable levels. "In order to make it work financially, you will need to give the coaches as big a workload as possible," she said. "I mean, I was literally talking to seven or eight people every single day. There were days when you would talk to twelve people." While the load was daunting, Veronica said she found the work moving, humbling, and powerful. "You genuinely had an impact on people and you talk to them every day, and you saw how things were for them," she recalled. The work filled her with purpose, and she hoped to one day train as a therapist. "I loved talking to all these people. I loved feeling like I had an impact."

Yet despite both the profound meaning of the work, and the emotional wallop it delivered to the coaches, Veronica's own language joins in minimizing her effect. She "loved feeling like I had an impact," but quickly followed that with: "Even though it was really them that was doing—You know, they were doing all the work, you know, I wasn't really doing anything. I was just cheering them on and helping them work through some hard things sometimes." Thus the firm maintained that Veronica was not doing therapy—and benefited from that claim by not having to train, hire, or pay for skilled workers, while extracting the value that their witnessing helped create. The clients came and went as if they were not doing therapy, treating Veronica as an extension of the app. Even Veronica asserted that she was not doing therapy. Their collective insistence rendered invisible the work she actually did to witness the clients' needs and struggles, and the price she paid for it in the emotional residue their trauma left behind.

The notion that humans motivate other humans is the operating consensus among engineers and researchers working in the automation of connective labor, and they are not wrong. Yet as the example of Veronica's experience with the social anxiety app makes clear, this assumption also underestimates what is produced when someone sees the other. Like the folk belief that separates emotions and thinking, impoverishing them both, the notion that humans' best role is merely as "an audience that matters" blinds us to the other kinds of work we do.

Without a robust understanding of what connective labor is, it becomes much easier to render it invisible in the contestation over the "automation frontier." The debate over AI is also a negotiation about what human activity is and is not available for automation, which work might be considered rote, routine, or unimportant, and which might simply be too much for humans to do well. But technologists, aided by the invisibility of connective labor, cultivate a sense of inevitability that lends their efforts a certain momentum.[21]

Olivia Nadal was a psychiatric nurse who traveled to work on short-term contracts; her job often involved evaluating someone else's conduct, she said. She mused aloud about the prospect of AI in nursing. "Maybe a lot of part of the job could be taken over [by AI], but a big part of what I have to do is assess the person and I have to observe them, their behavior. And I don't . . . Maybe the thing could follow them around and like report on what they're doing. Yeah, I don't know," she said, doubtfully. "I would like to think, and I think it's important, especially that aspect of talking to someone face-to-face. They're going through crises. I don't know if people are going to want to talk to robots exactly. But maybe everything else about my job could be replaced." Her tentative phrasing—"I would like to think"—suggested that even if she "[thought] it [was] important" for a human to do the job, she felt a sense of helplessness in the face of the heralded AI revolution. "But maybe nothing is really that important that it can't be taken over by it," she said, defeatedly. "And that's probably the truth." The automation frontier had moved again.

The expansion of data needs. Automation made data expand like a sourdough starter. For example, we have seen how the advisors at the experimental school were "in charge of the relationships," the heart of the school for students who spent many hours learning from apps. At the same time, however, the advisors were also responsible for tracking the students' individualized learning plans in all their subjects—what their goals were for this week, this month, this term, and whether they were meeting them. All this tracking was an entirely predictable result of the school's customization: because each child received an individualized education, dictated by the apps every morning, the school had to spend a lot of time assessing and monitoring each child's current level of understanding and near-term goals. If the school did not, the students might spend their time repeating lessons they already knew because they were more comfortable there, or they might accelerate beyond their actual grasp of the material. What that tracking looked like, however, was a lot of data entry in practice, and kids and their advisors ended up spending their one-on-one time peering together into parallel computers, staring at shared Excel sheets, typing.

In a session typical of the ones I observed, the advisor Vivian typed into her computer, as Jackson, a very quiet Chinese American boy of about thirteen, sat next to her looking at his. In their meeting, they spent a lot of time talking about "deliverables"—business-school speak for a measurable outcome, but in this case a way to signify and record Jackson's progress.

VIVIAN: You did six hours of IXL time, let's put that in there. Is there a deliverable for language?

JACKSON: Just go to class, reading comprehension, and write essays.

VIVIAN: Write essays, how many?

JACKSON: Three.

VIVIAN (*excitedly*): Three essays! That's a deliverable. Did you put in your Labster score? What was that in?

JACKSON: Animal genetics. I got a 79.

VIVIAN: Let's stick that in.

At another session, Shelley and Nelson, a white boy who seemed to be about eleven, were going through his activities together, and Nelson was complaining about the coding program. "I could use it, but it doesn't teach you. It expects you to code on your own," he said. Shelley's response focused in on a different issue, however: how to monitor progress. "Measuring it becomes a lot harder," she told him. "We need something that's a little more measurable."

The school faced a problem—the primacy of data—that might initially seem incidental to their hoped-for disruption of traditional education. While schools everywhere in the US collect assessments and manage student data, particularly in the aftermath of the 2002 No Child Left Behind Act, in this Silicon Valley school, as in the modern primary care clinic, data was unusually visible, its negotiation a constant presence—particularly jarring since (just like in a primary care clinic) the people in charge of collecting it were often the same as those ostensibly responsible for the "relationship."

When I mentioned to Faye, the principal, that the advisors seemed to spend a lot of time on data entry with the kids, she said the school was working on that, and that they had instituted some new guidelines for the advisors: start off the meeting with your computer closed, or sometimes take a "walking meeting." She reported telling the advisors: "Sometimes it's not about the computer, sometimes it's just 'what's going on with your friends.'" She also sent out a spreadsheet of tips for questions that advisors could ask the students. "Who have you been hanging out with this week?" it suggests. "Anything you wish would be

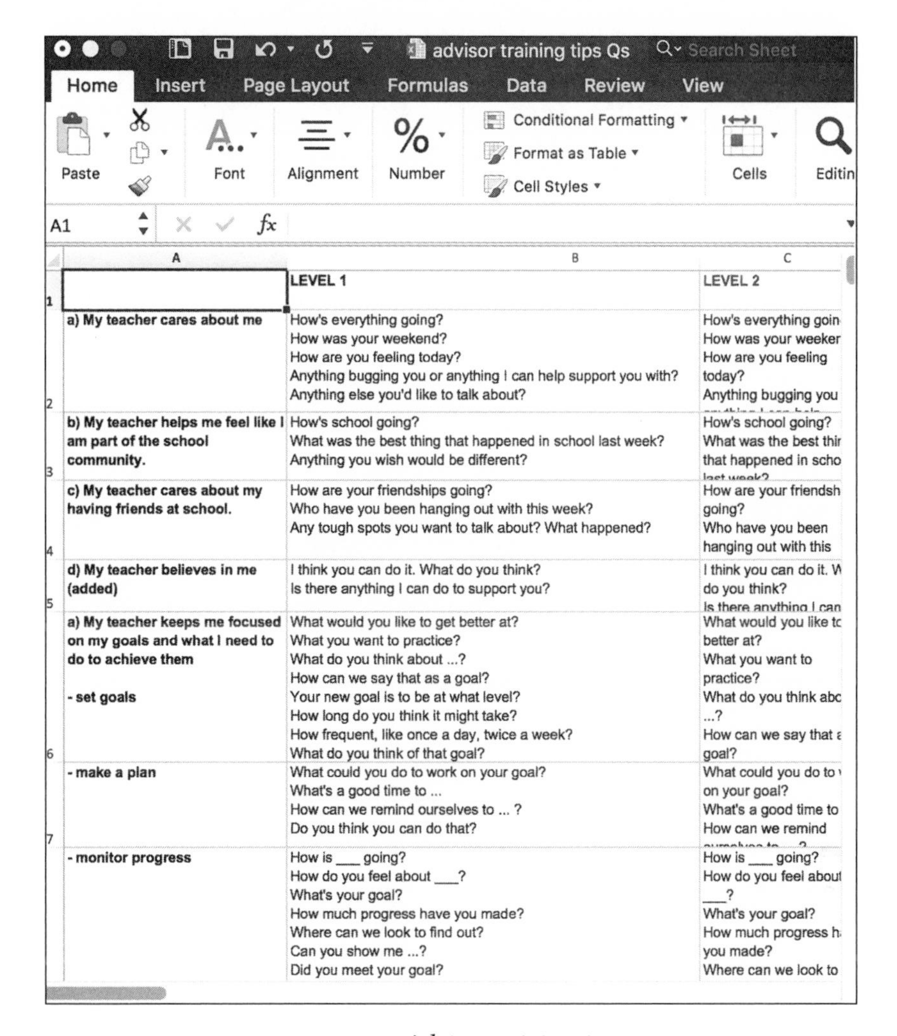

	A	B	C
1		LEVEL 1	LEVEL 2
2	**a) My teacher cares about me**	How's everything going? How was your weekend? How are you feeling today? Anything bugging you or anything I can help support you with? Anything else you'd like to talk about?	How's everything goin How was your weeker How are you feeling today? Anything bugging you
3	**b) My teacher helps me feel like I am part of the school community.**	How's school going? What was the best thing that happened in school last week? Anything you wish would be different?	How's school going? What was the best thir that happened in scho
4	**c) My teacher cares about my having friends at school.**	How are your friendships going? Who have you been hanging out with this week? Any tough spots you want to talk about? What happened?	How are your friendsh going? Who have you been hanging out with this
5	**d) My teacher believes in me (added)**	I think you can do it. What do you think? Is there anything I can do to support you?	I think you can do it. W do you think?
6	**a) My teacher keeps me focused on my goals and what I need to do to achieve them** **- set goals**	What would you like to get better at? What you want to practice? What do you think about ...? How can we say that as a goal? Your new goal is to be at what level? How long do you think it might take? How frequent, like once a day, twice a week? What do you think of that goal?	What would you like tc better at? What you want to practice? What do you think abc ...? How can we say that ɛ goal?
7	**- make a plan**	What could you do to work on your goal? What's a good time to ... How can we remind ourselves to ... ? Do you think you can do that?	What could you do to on your goal? What's a good time to How can we remind
	- monitor progress	How is ___ going? How do you feel about ___? What's your goal? How much progress have you made? Where can we look to find out? Can you show me ...? Did you meet your goal?	How is ___ going? How do you feel about ___? What's your goal? How much progress h you made? Where can we look to

FIGURE 3.3. Advisor training tips.

different about school?" (See figure 3.3 for an example.) While the school was clearly taking action, the guidelines seemed to suggest that the advisors were erring on the side of data entry because they wanted to. Yet the sheer prominence of data was neither happenstance nor due to any particular predilection on the part of the advisors for collecting and analyzing it; instead, it was a systematic outgrowth of the school's emphasis on customization.

The expansion of data affected even the students, who aspired to "mastery"—the top of the school's internal ranking. In the middle of Nelson's meeting with Shelley, for example, she stepped away for a moment, and he looked over at his spreadsheet on her computer, checking out the different colors she had assigned to the subjects indicating his level of understanding.

NELSON: Hey, wait, why did I get a yellow?
SHELLEY: Because I think you are still working through it.
NELSON: But I finished it.
SHELLEY: But remember you need to give specific examples?
NELSON: But it didn't say that. I answered everything in there.
SHELLEY: You get a green if it's done in multiple different ways.
NELSON: I did a storyline, and then made some edits.
SHELLEY: Wait and see what Lindsay writes. Plus, when it's a new style of writing, it is not usual to be a master of it right away.
NELSON: I want to be a master.
SHELLEY: I know you want to be a master of it right away—just hold your horses.

Data monitoring was one way in which adults maintained control over children's progress, allowing the school to direct the children's advancement, but also requiring an enormous investment of resources. We can see that the second unintended consequence here—the expansion of data primacy—applied not just to advisors but also to students, who spent substantial time scrutinizing, considering, and even arguing about their own data.

The primary impact of the expansion of data was in squeezing the connective labor by shrinking the amount of time available for it. Furthermore, it also signaled the colonization of relationship by a paradigm of measurement, bringing connective labor under the sway of those who would render it more systematic. Even AI engineers—at least those who conceived of it as "better than nothing"—were not sure about that move. "When I look at people who are kind of trying to make the relationship systematic and boil it down, they have trouble . . . it seems to be more than the sum of its parts," said Timothy Bickmore, the researcher who had created the virtual nurse, palliative care agent, and couples' counselor. He recommended retaining some of the mystery of relationships. "The things that work better seem to be like when we make time for it maybe, but don't really go into the specifics of exactly what has to happen."

The transformation of the human. Finally, the mechanization of connective labor cast a certain shadow upon human workers. The advent of AI, apps, and automation in connective labor seemed to create a new task for them: to prove they are not robots.[22] The specter of being mistaken for a robot was real, particularly in highly systematized or text-based work. Paul Giang had a long history of working in sales, but after a layoff in 2015 he started working for Uber, TaskRabbit, and a host of other gig economy outlets. If there was a line dividing robots from humans in their work, he could feel that line passing over him repeatedly. Sometimes his customers perceived him as a machine, he said. "[They'll say,] 'OK, yeah, just put it over there,' and then I drop off the stuff, and they just tap [the button]. I think they see it as automation. They see you as like just a system. I have friends that tell me like, 'Yeah, you're essentially working as a vending machine.'"

With the social anxiety app, Veronica said that at first her job was to conduct an initial phone call with the client to assess whether they would be a good fit for the program. Subsequently, she was supposed to respond via emails whenever the client completed a task on the online app, but she quickly realized the emails were not effective, and started texting. People were more responsive, but she started getting a new kind of question, she said. "A lot of people were like, 'Are you a robot?,' basically." I asked her how she handled that—did she try to convey that she was human? Yes, she said. "I think I basically just tried to small talk with them. Um, so I would just ask them, like, how their week was going or how they were feeling about getting started with the program. And then if they responded, like, I would try to respond with something that seemed, like, more personalized. It wouldn't be, like, just a generic text message. Ask another question, maybe share a little bit about myself if it was appropriate." Working within the context of a social anxiety app, whose service delivery involved automated assignments modeled after

CBT, the specter of robotics was prominent for its human workers and clients. In fact, in other apps Veronica might indeed have been a robot.

Others distinguished themselves from machines through their capacity to decode hidden parts of humans. Even if AI elicited more secrets, it would not be able to understand or manage the secrets it was told, said Grace, a white therapist in private practice. As an example, she said she had a client who lied frequently. "And so—and you know, we have a societal judgment around lying, but once I established, 'Well, this is a weird pattern that I don't usually see,' then I went into, 'What's useful about it to this person?'" The answer was that lying was the way her client asserted himself against others around him. It was how he "made sure he has his own mind," Grace said. "Because he merges with everyone. And so, when he lies, it's when he's feeling like there's no him." Then she paused. "See a computer do that." Grace was proud of her empathic insight into a mysterious and shame-ridden pattern, and considered it proof that she was different from—and better than—"a computer."

A third way that people apparently experience humanness is through "human error." This was especially evident for connective labor practitioners working in highly systematized environments, where mistakes are like flickers of authenticity. Faculty who starred in online courses featuring thousands of students sometimes left in mistakes for just that reason, said Sebastian, an online course specialist. Mistakes were "opportunities to be authentic or communicate things. Or being humble, sort of the humility of making mistakes. Or not trying to be the ideal self in the space. I think they do contribute to a connection, even at scale."

As a signal of humanity, of course, making mistakes was not available to all practitioners—physicians often talked about them with shame or dread, and workers from less-advantaged backgrounds did not feel like they had the same leeway to make errors and still be considered professional—but for many therapists, teachers, and others, it had surprising effects. Mistakes seemed to convey to clients or students that they were getting a second look, that they were worth seeing correctly, and that what was doing the seeing was a human being and not a robot. Workers reported the powerful capacity of human error to cut through a prevailing sense of scriptedness, a sense that seemed just one step away from mechanization.

For some engineers and researchers, the line between human and robot was an amorphous one. Many were enthusiastic about using AI to extend human capacities, in a future of cyborg-like meshing of human and machines. "Well, the boundary between the human and robots someday [will] disappear," said Ichiro Haruguchi, a Japanese roboticist. "The robot is going to be more humanlike and the human is going to be more robot-like, and then someday the boundary disappears. So, look at the paralympic player—they use a prosthetic arm in the race. Do you think we need to have a flesh body to be a human or not? In current society, we recognize a flesh body is not a requirement to be human. So probably it's a matter of more inside stuff, like mental, right? But what is 'mental'?"

But despite this sanguine prediction, to be mistaken for a robot posed a new sort of existential challenge for Veronica and other workers in the field of automated connective labor, whose task became not simply connecting to others because she is a human, but demonstrating her own humanness because her own connective labor—in its scripted, counted, disembodied form—was not enough to do so: in essence, proving her own humanness to clients accustomed to machines.

The erasure of connective labor, the expansion of data primacy, and the new demands to prove humanness—these are all consequences of the relentless movement of AI, apps, and robotics into interpersonal emotional work. At the automation frontier, we are all drawing lines around work, again and again, to define what matters, what is meaningful, what is human.

The introduction of automation and AI into connective labor is not some distant prospect, in a dystopian vision or a futurist's dream. Instead, it is ongoing, ushered in by a massive investment on the part of Big Tech and its researchers, the media's perennial fascination with the human-robotics divide, and consumers. It also benefits from the American love affair with technology, the national faith that tech will solve our way out of the challenges that stand before us, the conviction that technology is inevitable. Paul, the salesman-turned-insecure-gig-worker, talked at some length about his uncertainties about what to do next. Then he looked hopeful. "I think that maybe an app in the future might help gig economy workers decide what they're good at," he said. AI career counseling—that was the solution.

Many of the "better than" arguments take aim at those moments where humans have failed other humans with connective labor that is inaccessible or unsafe. They seem to assume that current staffing levels and compromises about quality are fixed, however, and that only by automating the work can we relieve the pressure on the overtaxed public sector, strapped healthcare systems, or busy teachers. In essence, then, technologists paradoxically combine an optimism about employers using AI to "free workers up" for more meaningful tasks with a pessimism about contemporary politics and the need to adopt automation to solve labor scarcity that will not be going away. Even more important, automated programs compete for public spending for human connective labor; in Indiana in 2017, for example, the state spent $4 million for preschool but put $1 million of that toward an online program.

When I look back on my springtime visits to the experimental school, its students looked joyful and self-directed. Yet it seemed likely that the happy climate was due in part to the presence of plenty of caring adults around to make for a warmer experience than that suggested by the quiet morning scene of laptops and headphones. What if a much poorer school system adopted the model, and saved money by cutting out such expensive labor? The outlook of exporting this model to other schools was not entirely rosy. I could imagine the kids in a cut-rate version being more like the little redheaded boy I saw that morning, looking around in vain for witnesses to see and share in his discovery about math. When we integrate automation into our connective labor, much depends on whether we rely upon it to do technologically what we refuse to do ourselves politically: redress inequality of access to caring, supportive reflecting work.

The solutions we choose to pursue through automation also affect more than the problems we target. As engineers automate connective labor, they have quite reasonably focused on the ostensible tasks that workers are trying to achieve, such as giving medical advice, therapy, or spiritual counsel, and they have made strides toward producing programs that achieve some sort of limited benefits. Along the way, technologists encourage clients to think they can be protected from the risks of judgment. Yet the turn toward mechanization relies on a particularly thin vision of what humans do for each other, and how they do it. When we talk to practitioners, they offer a fuller reckoning, pointing to practices that challenge their automation, the characteristics that make it particularly human work. The next chapter explores the mystery of how people do connective labor as an artisanal craft, and the uniquely human practices that enable them to find their way to each other.

4

HOW TO BE A HUMAN

Connective Labor as Artisanal Practice

When Wanda Coombs, an African American therapist in the Bay Area, described how she knew whether she was doing her job well or poorly, she said she relied on her own body for the signals. We sat on couches in her living room, in a modest house with bars on the windows, talking together while the mechanical whine of passing trains nearby rose and fell periodically. Her preschool-aged daughter had just come home from daycare, and climbed all over Wanda while we talked, pulling on her shirt, touching her face, but Wanda was unruffled. The signals she pays attention to are unmistakable and physical, she said.

> You feel it. In both situations, you feel it . . . There comes a point in session where I start to tingle, like the room gets . . . When I'm with a client and the room gets slightly fuzzy and I'm tingling, and I just feel like this energy. Yeah, I feel it, and that's when I know that I'm at like the meat of their stuff, like the good stuff, like we're getting somewhere here.
>
> And when things are wrong or off, you feel it too. It feels like a stopping, like a halting. Something stops and something halts and there's kind of, like, silence or a dead air, and that's when I'm just like, "What just happened here?"

Wanda reads the air for clues about how the session is going with her client. And while not all people who do interpersonal work talk about "energy" or fuzziness, it is not unusual for them to rely on their own capacity to absorb and act on nonverbal sensations. Its physicality is certainly one reason why interpersonal work is sometimes called "the last human job."[1]

Wanda never intended to become a therapist. As a very small child, she suffered from recurrent, violent, and terrifying nightmares from about four to ten years old, and she never found the therapists she saw then very helpful; she eventually just grew out of it, she believed. She had planned to be an OB-GYN, but did not have the grades, and eventually backed her way into studying psychology as a college student, whereupon she realized that she actually loved the field. Even now, however, she pursued a style of therapy—collaborative, intuitive, explicit about inequality—diametrically opposed to the kind she had as a youngster. "We sit in a really powerful seat, and I like to share some of that power," she said.

Wanda used her body as a vital tool in the craft of connective labor, to sense the tingling or the halting. The very act meant listening not just to what a client said but also to the emotional currents beneath it, and that kind of listening tapped into the nonverbal field of communication pulsing beneath their words. Emotions figured prominently in her practice: she interrogated her own feelings to ascertain what she understood about others, and also to figure out just what she should convey to others about what she saw; she also actively cultivated a caring self that enlisted her emotional allegiance to the other as an ally and supporter.

Furthermore, her work was deeply collaborative, a dialogue in continual making, which meant that it was also unpredictable, depending in part on what the client brought into the room. She may have collected all sorts of formalized therapeutic tools in her training, with words like "gestalt," "narrative," "postmodern," and "psychodynamic" dotting her language, but in the end it was the interaction with the client that directed the encounter. "Now, I pretty much have, like, what I do, but then it's also that I don't actually know what happens. I just kind of see where things take me and I've [been] very transparent with clients about that. Like, I don't know what the end of the session is going to look like.

I don't know what five minutes from now is going to look like. Like, we're just going to be in the moment and see where this takes us." That very unpredictability required her flexibility; she always had to be ready to seize any opportunity to bring the other forward in some way.

Sensing the other, feeling their way, adjusting, creating a mutual understanding, responding in the moment—this is how humans do this work. To be sure, there was considerable variation among the people I spoke to, with some even scoffing, for example, at the very improvisational quality that Wanda considered fundamental. Yet despite the variety, it became clear that overall, connective labor was a bodily activity, an emotional, collaborative, spontaneous kind of meeting of the minds, a creative effort resembling the fleeting artistry of a jazz band riffing or a choir rehearsal. These practices were part of what made this work uniquely human, and uniquely difficult to tame by algorithm.

And while all the talk of "tingling" and "energy" and reading the air might sound like some sort of New Age anti-scientism, it gains support from an unlikely corner: that cutting-edge field dubbed "second-person neuroscience," where researchers argue that what happens when people encounter each other "is not entirely reducible to knowledge 'inside' any two individual[s] but exists 'between' them." As one influential review put it, how we know other minds is the "dark matter" of neuroscience.[2] This chapter sheds some light on that "dark matter," reporting on how people do connective labor in the course of making a living. In a world of checklists, manuals, and AI apps, these practices illustrate how people wrestle with the messy chaos of human interaction, how they conjure the wild magic of seeing and being seen.

The Human Practice of Seeing the Other

Other writers have tried to nail down how teachers connect while they teach or how doctors do it while practicing medicine, frequently in the form of checklists or other attempts to itemize best practices for busy people. In contrast, this chapter focuses on those practices that are at the heart of connective labor and yet not easily standardized—methods that are common across many occupations yet evade measurement, challenge time limits, or resist commodification. These are the ways of doing connective labor that make it particularly human work.[3]

Most of the material here comes from interviews with practitioners, which means that we are sometimes left wondering about the other person's perspective. Where possible, I include observations and other information that helps us consider the "other side." But ultimately, interviews are the best way to reckon with how people think about their work, the techniques and strategies they use, and the lessons they learn along the way—in other words, about connective labor as an artisanal craft.

From therapists to teachers to physicians to a host of others, people describe a human endeavor that largely coalesces around five distinct practices: (1) using the body, (2) reading and deploying emotions, (3) collaborating, (4) managing spontaneity, and (5) making (and reacting to) mistakes. And while these practices are core to how practitioners do the work, they also raise some common tensions—about expertise, judgment, and vulnerability—that people must then endure, manage, or resolve.

The Body as Instrument

The bodies of connective labor workers were a vital instrument for their work. People used their bodies to set up a ritual space for the encounter, to get a sense of themselves and the other in the moment, to reach out to others through physical touch, and to manage and convey power. Workers and their clients, patients, and students found a human connection in this very physicality.

Many people use certain rituals to center themselves and others, to carve out space for the moment of meeting the other within the press of other demands, even to consecrate their interaction. A crucial part of establishing this ritual space was the use of the body. Even such basic habits as hand-washing could be done as a mindful ritual, said Andy, a primary care physician. "You know, maybe there's a little mindfulness that needs to happen just before you go into the room, collecting your thoughts," he said. "And the greeting, I think, is always important. You know, acknowledging everyone in the room. And the washing of the hands is sort of in a sense sacred." By adopting small rituals, connective labor workers created a small, protected space around their work. The rituals served to slow time down and separate out an interaction from what came before or after, enabling people to attend to an event or experience and make particular sense of it. Rituals like these create a mutual focus and sustain a shared mood, according to sociologist Randall Collins, and through these means they bind people to each other.[4]

Hank, a hospital chaplain, told me that the first thing he does is "stop myself and pause before I go in a room, and say, 'What am I believing right now the most?,' scanning my body and my feelings. Do the same thing I'm asking maybe clients to do down the road. 'OK, now can you be present?'" Others, like Wanda, use their body to read the air, to ascertain how an encounter was going, with either their unconscious sensory apparatus picking up on nonverbal signals the client was emitting, or, as Wanda believed, some sort of emotional energy transmitted between people doing close, intimate work. When it was going well, she said, "And the room gets hot too, right? Like, we're talking about energy. It's like it gets hot. Like there's something that's happening in that moment that is not something—but it's definitely on that, like, other realm level where it's like there's something that's happening here. I don't know what it is. I'm not trying to question it—we're just going to go here."

Still, while Wanda may have been more open to the metaphysical than most, many workers told me that they used their body to focus on nonverbal cues, a "vibe in the room," even their instincts. "I really try to attend to my intuition and also just the feel in the room and the feel inside me," said Carrie Koppel, a white therapist in a VA hospital. "I see it as just being—It's a very dynamic interchange between the patient and the therapist. If I'm just really tuned in and centered in a session I can just really be creative and pull things out. Like, maybe I'll do a role-play, you know, all of a sudden, I'll be like, 'Wait, we have to do this right now.'" When I pressed her to explain where the sense came from, she demurred. "It's something I can't describe. How do you know when it's the right time to kiss somebody for the first time? How do you know when it's the right time to intervene when your kids are having a fight with another kid? You know, how do you know when the right moment is? How do you know when to say goodbye for the last time when someone's dying? I mean, it's in here. *(She gestured to her heart.)* It's in here."

It was not just therapists who used their bodies in this way. Ken Margolis, a middle school teacher, said he continually relied on "a Spidey sense," an awareness of what was happening in the classroom—was one corner too quiet?—or to a particular kid. "Like when the kid comes in, they're not bubbly anymore," he said. "I mean, you just read—you read the face or the body language, the face, the aberration from the normal, all those."

Other practitioners talked about how they had learned to develop their "Spidey sense." Ellis Gable, a funeral director in south-central Virginia, called it "good listening," and he said that when he was learning his job, he could see how crucial it was. "I think the most important thing that I probably ever learned was by watching people listen, because, you know, you can hear but not listen. But you can listen and not hear."

"The nonverbal cues actually say a lot," said Charlie Jiang, a primary care physician. "It's what helps me anticipate what to ask, what to say next. Where to stop and just shut up and listen." More of his colleagues need to do that, he thought. "Nowadays with computers you see a lot of physicians typing in front of a computer. I'm like, that time would be better spent if you sat, not typing in front of a computer and just reading their nonverbal cues. A lot of the diagnosis comes right there. You can tell, you can sense it."

Some practitioners used their bodies to make physical contact in their everyday work. When a job required that contact—for example, massage therapy—it led to some intense interactions, workers said. "Once you start touching people, a lot of barriers come down and, yes, the conversations have gotten into deeply personal stuff," said Harrison Thomas, a hairdresser. "Deaths of family members. Troubles of family members. Evictions of family members. Abuses. Whatever's in their lives, it sometimes comes out," he said. "If they want to divulge this information and tell you about it, you listen. You absorb it. It is in a way an intimacy that you develop with people because you have to be in their personal space."

Birdie Mueller, the nurse practitioner who talked about washing the homeless man's feet, said she considered physical touch an important part of the way she helped heal. "Most patients hug me when they're saying goodbye," she said. "And I think I have, like, magical powers, because if you speak in calm terms and you rub and kind of speak nice calm thoughts and distract them, I usually can bring their blood pressure down. It's, like, powerful or magical."

A tall woman with long blond hair and a brilliant smile, she relied on her charisma and her touch to bring patients comfort. "And I am aware that I am a presence, I am a strong presence, it's a big presence, I am aware that does feel—that's a lot of watts in your direction. That intensity is strong. I'm very excited about whatever thing that they're excited about, and then physical touch, I think a lot of our patients, particularly those who have their families in Mexico, they're not touched. A hug or an arm, or 'you're doing great,' they just want reassurance. How much of that [touch] is what they need, you know."

The availability of touch could differ considerably even within the same occupation. While for some, physical touch would be deeply inappropriate at work, in other jobs its use varied depending on local norms and expectations, and the unevenness led to some confusing uncertainties, bringing up issues of consent. One day, Wilson Parker, a white faculty member at an elite education school, brought three student teachers and me to visit a classroom in a juvenile detention facility in California. At the end of our visit we talked about the power of touch, and how schools differed in how welcome touch was. Imogene, a tall white woman with stylish square black glasses, said she had worked in Chicago and Memphis in mostly African American schools, and that Memphis was open to touching but Chicago was not; her placement now was in a place where the principal welcomes hugs. "It's a very huggy culture," she said. Juan, a man in his late twenties with an air of calm readiness about him, told us that his placement was in an almost all-Latino school, and that because Latino culture was more open to physical touch, he was thus able to use it to reach kids. Touch was powerful, but it was also subject to varying cultural regimes and organizational rules.

Bodies may have been central to the work of connective labor, but they were also suffused with other meanings that could get in the way of the connections workers sought to forge. A sturdy white man who clearly spent time at the gym, Wilson Parker used role-play and theater in his classroom to make bodily power visible to his student teachers. He said he asked them regularly: "How is it that you come up to a child to talk to them about their writing?" Do you stand over them and look over their shoulder? Or do you get on your knee next to their desk, and talk with them face-to-face about the ideas on the page?" Wilson tried to get the budding teachers to understand that students won't hear their feedback if they are intimidated. "This is something that I model, this part of my lesson that I do, that I do a whole routine like a theater stage, right?" he told me. "And I do egregious things that obviously show that my body is scary, and you know, what is this big white man doing standing over whoever. And, you know, their eyes just bug out of their heads." By standing over them in a mock-threatening way, Wilson worked to make the impact of bodies visible to the student teachers. "This is a simple thing you can do with your body, to give feedback. If you don't know you're supposed to do that, then what you're going to say is 'OK, bring your paper up to the front and I'll grade it.' And the student's not going to read what you write on the paper. So, how is that helping them?"

He was trying to show the teachers how bodies can communicate threat and welcome, which affects students' capacity to learn. "Teachers need to understand that part of their daily work is to foster those relationships and practice strategies that will allow them to develop those over time," he said. "I mean simple things that some people don't know how to do, aren't very good at, like eye contact. You know, you can call it charisma, or you can call it, you know, some innate ability to do something, but you can also practice how you use your body." Bodies were tools workers used both to sense and convey messages, but unless they were made aware of it, many messages ended up being implicit or unintentional.

The Language of Emotions

While people used their bodies to do their work via physical touch or a "Spidey sense," they also relied on their emotions as a sort of divining rod, a guide to detecting and managing sentiment in the other person. Emotions were central to the practice of connective labor, at once as tool, signal, and practice: a means of feeling out the other, but also a way to establish trust and forge bonds, a key to unlocking change and a conduit of support.[5]

Some people thought it was specifically their capacity to empathize with others that helped them know what was going on. For teachers, empathy was not only useful for connecting to kids; it also helped them know where to expect trouble, said Bert, a middle school principal and teacher. "So, if I'm really empathizing with my students in my classroom, I can feel, to some degree, when something bad might come up or when a kid is going to go off the rails because we're talking about a divorce in this book and I know that his parents just got divorced," he said. "Whatever it is, like, if I'm really empathizing with the kids, then my intuition will be more highly tuned to that."

Therapists called this "attunement," meaning listening closely not just to what someone said but to the emergent emotional undercurrent that was the clinician's raw material—the straw they spun into gold. They were particularly articulate about what it involved. "The feeling is the antenna, the thought is also the antenna because there's content going on the entire time between us, and yeah, it's a constant loop," said Grace Bailey, a therapist in private practice. "It's a constant loop of content coming in, checking with the feeling, checking with the client's feeling—and not just, like, bodily but verbally 'How are you feeling?' And taking in all of that information, realizing what I'm being pulled to do, and then thinking about 'Is that the right thing to do?' And I don't think you can think about that without a feeling. That's a gut feeling."[6]

Even those clinicians who reached for a more "scientific" approach acknowledged the importance of this emotional sense-making. Gerard Juliano was a therapist with the VA who said he was "just adamantly against" the notion that therapy was "intuition, [with] counselors flying by the seat of their pants and following the patient and having no sense of where treatment is going." His litany seemed targeted to the ways Wanda described her work, even though they did not know each other; my sense that they were speaking to one another was compounded by the fact that I interviewed him the day after Wanda. Despite his critique of the intuitive, however, Gerard said he had had an unusual capacity for empathy ever since his father died when he was just six years old.

> I'm really able, kind of automatically, to put myself in other people's shoes, try to see things from other people's perspectives and understand where it is where someone else is coming [from]. And in doing that, I think it made me someone that people would turn to with a problem or a dilemma or when there was a crisis. And I think I got a sense of reward from that. To me it's always been a real privilege or honor for someone to self-disclose to me; that means a lot. And also, I think it was the way that I learned to be useful, like I was able to help people by listening nonjudgmentally.

In my experience, Gerard was a bit unusual for this combination of being at once an early onset empath and someone who rejected an improvisational approach, who instead talked regularly about "evidence-based approaches" and "objectivity." Yet despite his distrust of "seat-of-the-pants" counselors, even he admitted that therapy relied on an emotional intuition. "That's where the art is. The art is in knowing what it is that the patient is conveying. In what they're saying or what their body language is saying that's important versus not important. And not always knowing, that's the intuition piece. You don't always know why what it is what they conveyed was important or not important. You just feel that it's important. And the knowledge sometimes has to catch up." While Gerard's therapy had a science to it—the "evidence-based" methods, the "manualized" approaches—its artistry lay in feeling something before "the knowledge" caught up.

Emotions were not just a tool to sense what was going on for another person; they were also a means of establishing trust and forging bonds with people who felt adequately seen. One day, I witnessed Russell Gray, the master therapist, talking to the students he was training about how clinicians sometimes had to establish a connection with different family members in the same room at the same time. "The most serious problem is that there's some conflict in the room, or at least differing perspectives, and they're kind of saying, 'Choose a side,'" he said. First, he collected the students' ideas of how to respond in group therapy to a teenager's statement: "When my mom micromanages my life, it drives me crazy." You could ask for an example, one student suggested. You could ask them how it makes them feel, another chimed in, and validate their emotion even if it is not true. After a few others, Russell stepped in. "Everything you are saying makes sense, but for establishing rapport, you have to be able to understand them. That's going to establish you as the center," he said. "The rapport piece is all going to be empathy."

But not all empathy was equivalent, Russell said. "There's sort-of-accurate [empathy] and there's sloppy empathy. You will get nailed if you say, 'That must be so tough having your mom micromanage you.' You will be dead, you will get hostile glares," he said, while the students laughed. "But if you're doing it right, you can go from [pretending to speak to the daughter] 'You're really upset how things are going with your mom, you're really worried' and [pretending to speak to the mother] 'you feel like you can't just let go, and you don't like it that your daughter's pushing back on your basic rules,'" he said. "If you've said it right, you can say both of those things and they can't say you're wrong." A student spoke up to add: "You can validate the emotion without validating their behavior."

"And you can go farther," Russell said. Again speaking as if to the mythical daughter, he volunteered: "'You really don't like it when Mom is involved with your life, and you don't want to have a bad relationship with your mom,'" then, switching to the mother, he said, "and 'you don't want to be pushing your daughter away.'" He summed up: "I can get to positive intent." That only works after a while, however. "Until they've got some trust in you, you won't get very far. But one of the things that is so often the case is: as much as everyone is talking, no one feels listened to, so the extent to which you can [show that you] listen to them, that's pouring stuff into their tanks." In this training session—characteristically dotted with bits of real wisdom like "as much as everyone is talking, no one feels listened to"—Russell was teaching the students how to use emotions as a means of both seeing their clients and helping them see others.

In many jobs, providers talked about how the power of seeing the other lay primarily in naming a hidden emotion, which then drained it of its magnetic force, particularly useful when it was getting in the way or distorting someone's vision. Greta Kendrick, a pediatrician, said that to her surprise early on in her career, much of her practice involved reassuring overly anxious mothers, and that she had learned to do so by naming their emotions explicitly when they gave off certain signals. "It's often the ones who keep coming back and keep coming back that actually—that's the signal in many cases of particularly moms, women, who have a significant mental health problem in the moment. You know, they have postpartum depression or an anxiety disorder. [It's] their body language or they're hyperventilating in the room with me or, you know, all these things that we can kind of sense about people," Greta said. She would make sure their child was not seriously ill, and then turn to the mother. "And what's left is I need to sit down and say, 'I see lots of patients who have this very same thing and I've told you about it many times, and I'm not worried about your kid. Let's talk about you. What I notice is that you're worrying more than the average women who tell me about that. How are you?'" Empathy—and naming the emotion they could recognize—worked like a magic key turning in the lock, opening them up to change. In these conversations, Greta said, "some of them start crying."[7]

Emotions served as a tool for these workers, helping them ascertain the truth and elicit trust, but for many, sentiment also came with its own mandate: to care for their charges. Wanda considered caring for her clients her first task, she told me. "There's some people I care [for] immediately, right? Especially the folks with marginalized identities or—especially in the beginning—if they have a strong trauma history." She told the story of one client, a trans woman, who had spent years going to summer camps as a teenager that involved "conversion therapy," and were affiliated with the "ex-gay" Christian organization Exodus International. "That's so traumatiz—I can't—I immediately felt my heart go out to her and was—She had difficult relationships with her family and she needs someone who cares, and I really want to be that person who cares," Wanda said. "And I think we can offer—There's always that to give and I think that's part of my job is to give that. Not just do therapy but also to care."

Many workers talked about caring for their charges, about liking or loving them, giving them unconditional positive regard or support—even, in some cases, when they had to work at it. "First and foremost, the person has to look like they love kids," said Bert, the principal, as he described how he hired teachers. "You can kind of tell it early on, if a kid says a joke and like they don't really laugh, and they don't see it as, at least, clever. You know, it's in the little things, like how they might remember a kid's name."

I sat in on a class of future school psychologists when the instructor, Linda Wallace, urged them to "love" the future students in their schools. "Here's the problem with parenting adolescents: you have to have their permission," she told the class. "They vote with their feet, and if they don't feel like you have investment, they won't let you. You can ground them, et cetera, but they won't let you parent them. It's the same for teaching. The kids won't let them teach them, if they feel the teacher doesn't have any [love] for them."

Other practitioners agreed, and said it was so important that they had to work on their own emotions to get to the right feeling if it was not forthcoming. Sarah Merced, who worked at the VA, described how she ginned up her positive regard. "I've cared for them all, honestly, I really have. Some of them I liked much more than I liked others, and so in those cases what I try to do is I try to find something likable and, or, what's been helpful for me is to imagine somebody as a child and to kind of tap into that vulnerable kid that didn't have something, because most of the time that's what's going on when I'm having a hard time connecting."[8]

Mandatory caring did not work for everybody, however. Carrie, a VA therapist, said her patient population included people who had been so traumatized that she thought it was unrealistic to expect staff there to like every patient. "There's a high, high level of complex PTSD here, meaning childhood trauma and then military trauma, so, I mean, we deal with a lot of people who are very nasty. I mean, we have people here that are abusive to us," she said. This disagreement actually echoes a debate ongoing among care-work researchers. Two schools of critique have emerged—from both critical race scholars and disability rights activists—to argue that care-work research is too focused on how much and whether care workers feel for their clients. When the "rhetoric of love" cloaks work relations, as in the case of domestic workers and the families who employ them, it may make those relations more meaningful, but critics argue that care workers pay a heavy price for that language. Love rhetoric, these critics point out, obscures the skill involved in care work, which then justifies the exclusion of this work from labor regulations; it conceals how care is a collective responsibility; and it hides the steeliness of intersectional inequalities of gender, race, and class beneath a velvet blanket of warm sentiment.[9]

Still, many I spoke to—including workers of color like Wanda—found the notion of caring for the other inherently meaningful, even crucial, for their work. I ended up thinking that the politics of emotions meant not that we should deny their relevance for some people's work, but that we should instead develop a more complicated vocabulary for the uses of sentiment, including as a tool that people use to sense and reflect the other, regardless of how they feel. People can still see the other and do it well, even when they themselves—like Carrie—are more guarded; the mutuality of connective labor does not have to be symmetrical, and sometimes what workers get out of it is a sense of their own efficacy, rather than a sense of their authentic self being known. Discrepancies in care highlight this variation, but while emotions were a conduit for care for some, they were a means of attunement for most.

Collaborating in the Give-and-Take

The collaborative element of connective labor—the way it involved an encounter between two or more people—was another crucial piece of it as a human practice. For Wanda, for example, cooperation was at the heart of her approach, the way she distinguished herself from the therapist she saw as a young child, and the style she most admired in her mentors.

> I really—I'll go places that my clients want to go and I won't go places that they don't want to go. And if I am going to challenge or nudge, I'll let them know. So, I'll say, "You know, I hear that this is some place

> that you don't want to go right now. I just want to let you know that eventually it will be important for us to talk about this." And then maybe if we touch, touch, touch and there's just like "No, no, no." Then as the relationship develops at some point, I'll say, "You know, we've talked about this before and I'm really curious about this now. I'm wondering if we can explore it a little bit more."

Collaboration involved giving the other a voice in their experience and treating them with respect. Workers talked about embracing a certain flexibility and making space for others to affect what transpired. Yet because students or clients or patients had to—in some way—assent to their own mirroring, to the image that was being reflected their way, collaboration also ended up challenging workers' expertise, when clients simply did not agree with them.

This truth was brought home when I observed a supervision session that Ethan Riordan, a clinical psychologist, held for several student trainees. Two of them, Kelsey and Rupert, were seeing a fairly difficult couple in marital counseling. While six of us sat in a small room facing a large TV, they showed a video of the most recent session, with the man and the woman on chairs facing them, and the woman began speaking. "This is an area that's been an ongoing challenge for us," she said on tape. "He'll say, 'You don't want to get in shape.' I will say, 'Yes, I do want to get in shape. I'm struggling to figure out how to put it in my schedule.'" On the video, Kelsey is heard saying, "Sounds like being busy does different things to you." The man replied, "No."

The students watching the video laughed. "He does that all the time," Kelsey said. "That's hilarious," said Sean, another student. The man continued. "You put a lot of busyness on your plate," he said, speaking to the woman. At another point, the woman raised a point of disagreement about their child. "People have different ways of showing caring, love, commitment," we heard Kelsey saying on the tape. "It sounds like there are some differences." Commenting to the man, she said, "You're seeing that growing your kids' independence is showing them caring, while for you," she said to the woman, "helping them out is showing them caring." The man said flatly, "Thank you for this summary."

The students laughed again. "Can you get him to perspective-take?" Sean offered. The video continued, with the man saying to his wife, "You've been doing this for fourteen years making their goddamn lunches." At that point, Ethan, the class instructor, paused the video, and looked at Kelsey, the student trainee. "If I were in your spot, I would have looked at her and said, 'What do you think he just said?' Sometimes a summary stops the underlying message of what he is trying to say. What he was trying to say was: 'You're full of shit.' These two can handle being exposed to one another's emotions, and I would try to get those emotions out there so they don't just get caught in their cycle. Don't stay too polite with their summaries." The man refused to be drawn into the process by Kelsey's attempts to reflect his point of view, his refusal a potent illustration that witnessing required a willing collaborator.[10]

The Spontaneity of Connection

Precisely because connective labor was collaborative, involving humans in an encounter with other humans, it introduced an element of spontaneity that meant interactions were never entirely predictable. The work was more than just delivering a soliloquy to a rapt audience and being done with it; instead, practitioners had to be ever ready for an opportunity to engage the other, to take advantage of what educators called "a teachable moment." This volatility made connective labor akin to a live performance, subject to the synergistic waves of the people present in that time and space.

For Wanda, the spontaneity was part of the mystery of communing with the other. "I still don't know how sometimes we get to some of the places that we get in sessions, like, I don't know how," she said. "And then, like, they're in it too and we're just in this space where I can't describe what happens, but it's happening. And so if I say something and they say, 'Can you repeat that?,' I can't, because I have no idea what I just said because I'm just so here with you right now. I'm just in the moment." The unpredictability made trust particularly salient, because both practitioner and client were going forward without entirely knowing the path or outcome. For Wanda, her improvisational style meant that it was vital that she establish trust early on with the client, and she did so by trying to help them toward some sort of epiphany in the first session. "They have to feel the change," she said. "Like there has to be a moment where there's a shift, or there's a change and it's like, 'Oh.' It's like an 'Oh,' moment. It's like 'I never thought of it like that. That makes so much sense.' And it can be about a very small thing. It builds trust. Yeah, it's like, 'Oh, she knows what she's doing.'"

In addition, however, the spontaneity of connection required that the practitioner trust in the process as well. "Like I see something here and this might be really meaningful, but I have to trust that it's going to come up when it needs to come up," Wanda said. All this reassurance indicated that the unpredictability was a bit destabilizing for both worker and client. The uncertainty also posed particular challenges for the systems—the checklists, the manuals, the randomized controlled trials—with which administrators and engineers attempted to nail down some of the volatility of this work. But for Wanda, attempts to make it more predictable—by, say, restricting reimbursement for only the more "manualized" approaches such as cognitive behavioral therapy—reflected a core misunderstanding of what success was and how she achieved it. She did not take insurance, she said, because it required standardizing the unpredictable. "You can't put time limits and constraints and those sorts of things on healing," she said. "It just doesn't work like that."

Wanda may have been on one end of a spectrum of belief, someone who considered connective labor an artisanal craft out of reach of many systems. Yet even those who were more open to making the work more predictable and standardized spoke of its core spontaneity. A particularly vivid example was recounted to me by Alan Krupner, the chief hospital chaplain, and the one who required Erin and the other trainees to update three databases after every spiritual visit, leading to all the clicking and typing we saw her do in the introduction. A stocky white man with a jovial air, he sat with me in a busy café and talked about a striking memory when he had improvised to meet the moment.

It was the worst day of his working life, but it was also the day he did his best work, Alan said. He was helping a pediatric transplant team in a Cincinnati hospital when a toddler girl was scheduled for a transplant of her small intestine. "She was fourteen, sixteen, eighteen months old," he explained. "Sitting up on the gurney blowing kisses to everybody as she went back to the operating room. Young parents." But halfway through the transplant, something went terribly wrong, and the little girl died on the operating table. "It was late at night. They arranged to bring her body out into the pre-op area where the patients wait to go into surgery so [the parents] could spend time with her. And the patient's father, I think it was his mother who took me aside and said, 'He's beside himself and he's completely paralyzed. Can you all give him something so we can get him home?'" The young father was sitting, frozen with grief, unable to move, "barely able to breathe." But since he wasn't a patient, they would have had to admit him to treat him, which would have been "completely counterproductive to anything going on," Alan recalled. "We needed to do something. He can't get up. He can't stand."

So Alan went into action, but what he did was "not even particularly religious," he said. "So I got down on my knees in front of him and just put my arms around him and just started talking to him in kind of a voice like I'm talking to you right now, so I focused him now with my voice. 'Listen to my voice.' And he was hyperventilating. I said, 'I want you to listen very carefully and do what I say. I want you to take a slow breath in, hold it for just a moment, and let it out.'" He could hear the father's breath begin to alter a little bit, so he knew the man could hear him. "And I would keep saying that, and interspersed in that said, 'This is the worst day that you have ever had to face, and I pray to God it's the worst one you'll ever face. There's no way around it, and I am so sorry. Take a big breath in, slowly hold it, and let it back out. I have no idea what this must be like for you.' And I kept going back to the breathing. 'And you will find a way to move beyond this day.' Breathing in, breathing out." He crouched there, breathing with the man, coaxing him along. "It felt like it was twenty minutes," he said. "It was probably closer to two."

Finally, the father's breathing returned to a more regular pace and he was able to stand up and walk out of the room and go home. Alan found himself utterly shaken, devastated, and exhausted, but proud too. "That was a day when I felt like—I don't know what happened after that, but I was able to help him start a journey toward dealing with this loss." Kneeling before the father, Alan felt that he had been able to touch the man in part because he responded to the moment, reaching deep within himself to find his own vulnerability, taking a risk so that he could connect to the man in his despair. "It was very intimate. I had to . . . and I made myself . . . I mean, he was a young Black man. At the time I was a forty-something white dude. We didn't know each other. And I don't know what his thing was about physical intimacy, what he would be able to hear and not hear. And what I was . . . He could just as easily have jumped up and took a swing at me. But to share with him the depth of sorrow that I sensed and that I felt for him, worked."

The nurses who beheld this tableau were amazed, he said. "When the family left, [the nurses] turned to me and said, 'I have no idea how you do this day after day,'" he remembered. "That's always been the thing that I hear." But this particular memory—crouching and leaning into the father after his daughter's tragic death, coaxing the man's breathing while narrating his grief—has stayed with him. "This happened probably twelve, thirteen years ago, and it still affects me," Alan said. "[It] haunts me."

Alan was proud of his work that day, but he also viewed that kind of succor as unsustainable. "It was so dependent on the intimacy of the relationship and the vulnerability of the chaplain in order to open the door," he said. Normally, he favored a combination of empathy, curiosity, and respect that did not require such abject defenselessness. "It's kind of hard to be in a caring profession and not bring in some sense of vulnerability, but if that's your ace card that you always play, that's a problem. It puts the caregiver at an unacceptable risk," Alan said. "I have to do that and do that. Again and again. Not just from time to time, but regularly." Still, even as he did not recommend that sort of response as a regular practice, that day his work embodied how practitioners responded to the needs of the moment, when few moments were predictable, illustrating how he answered to the utter spontaneity of connection.

Making and Managing Mistakes

Finally, as the saying goes, it is only human to make mistakes, and for some connective labor workers, errors were opportunities to paradoxically draw closer to the other. When Wanda sensed the "halting" that told her something was off, she used that feeling as a sign that she needed to check back in with her client and to correct the path. "And I'll say it, you know, 'I felt like maybe there was a shift here. Does that feel like that resonates for you? Or does that not feel like that resonates for you?' and invite them to come in to provide clarity." She viewed those moments as occasions to shift the power back to the client, she said. "Sometimes it's like 'I just want to say that I see you and that I respect what's happening, and so if you're feeling uncomfortable I want you to know that you can stop right now and you can take a break right now, or we can maybe shift and come back to this later if you want,'" she said. "So, a lot of permission-giving around taking control and power in the room, because ultimately that's control and power in their healing, and eventually—I mean, I'm not going to be in their lives forever. So, you [the client] need to be able to have that control and power."

Such mistakes could be valuable moments, agreed Grace, the therapist. Sometimes "I didn't hit it, I was off, I saw them in a way that they can't possibly see themselves or that they're ashamed about," she said. "So usually in the next session, I can say, 'that thing,' or, like, mostly I don't even have to go back to it. Mostly it presents itself again and I can say, 'You know, I'm thinking about that time, and here's where I think I missed you.'" She recounted a moment when she misunderstood what a client who was involved in a polyamorous relationship meant by commitment, and the rupture and repair she had to do there. Those moments of rectifying mistakes can prove redemptive for the therapist and pivotal for the client, she said. "That's when I get the most growth."

Therapists were clearly open to the potential of these ruptures to actually be positive for the relationship—some told me they were taught to look for and manage ruptures, while research has identified the potential of such moments. After a number of therapists brought them up on their own, I started to ask all the people I interviewed about connection mistakes: What counted as a mistake? How did they handle it? What effects did it seem to have on their relationships?[11]

To be clear, mistakes were not by themselves redemptive. (By mistakes, I refer only to those involving the relationship, such as misreading someone—not, say, medical or grading errors.) Indeed, many told me about mistakes they had made and not corrected, that they regretted, that haunted them even years later. When rectified effectively, however, mistakes were often powerful moments of communion with the other, and not just for therapists.

Majeeda Nur was a middle school administrator who said she could count on one hand the times she made relational mistakes of the kind I was asking about. One time, it was a math class she was teaching in middle school.

> Something was happening and the student either wasn't paying attention or something was going on, and I got upset and I called him out on it. But I did it in a way that, in retrospect, I felt like it could've made him feel singled out and hurt his feelings, you know? Really hurt his feelings. And so, the next day, because I reflect on what I do in the classroom, I was in front of the class and I said, "I said something in a way that I think may have been unintentionally hurtful and I want to apologize to Miguel. I want to apologize to Miguel in front of everyone that I made a mistake and that I should be more thoughtful in how I use my words, and I hope he accepts my apology."

What was the boy's reaction?, I asked. "He accepted the apology and it seemed to make a difference," she said, musingly. "It was interesting because it seemed like he was surprised that I cared enough to even notice that it had bothered him. It opened up our relationship. I just remember him being very receptive as we moved forward with the math and he and I being able to interact in the classroom, the collaboration with groups and whatnot, it just seemed to make a difference. It's like I did see him, I did acknowledge him." The moment—making a mistake, apologizing for it in public—was clearly an intense one for Majeeda, enough so that years later, she still remembered the event keenly. Yet in the aftermath, she concluded, rectifying the mistake made a difference for the student. We do not have Miguel's version of this story, of course, so we cannot know to what extent the mistake actually felt redeemed for him. For Majeeda, however, the lesson she derived from the searing experience was that correcting a recognition mistake helped convey to the other person that they were seen.

Just even recognizing that there was a problem sometimes worked to help the other person feel witnessed, people told me. A few years ago, Sarah Merced was seeing a patient at the VA, a white woman who had experienced sexual trauma in the military, and at the end of the third or fourth week, the woman left the session with a comment that she might not be able to return, about how "she might get busy."

"I was like, something was just kind of off," Sarah recalled. "Like, it didn't feel like the same. It just didn't feel right." So Sarah called her before the next week and told her about her sense that something was wrong. "I think I said something like, you know, 'The session felt different today. I'm wondering if maybe I missed something, or didn't hear something right, or something. I think it could be helpful to talk about that if you're able to come in again.'" That moment ended up being pivotal, Sarah said. "She ended up coming in again and we were able to talk, and at that point the relationship really shifted where she ended up being one of the most consistent people and she ended up making tons of progress that year."

As the treatment came to an end, Sarah asked the woman what she thought had worked for her. "And she was like, you know, 'There was this point where you noticed that I wasn't happy with whatever you did.' She was like, 'The fact that you even like noticed that was a big deal.'" Sarah divined that it was actually the rupture itself, and her attempt to rectify it, that had helped the relationship. "That's happened in other situations too," she said.

Her client's low expectations seemed to come from a lifetime of not mattering enough, of enduring misrecognitions from other practitioners who did not stop to notice. Sarah thought the woman's background made it particularly powerful when Sarah stopped to correct what had gone wrong. "Yeah, I think this was someone who hadn't had her needs seen before for much of her life, and so she was already used to and expected people to just not know or care to notice what was happening for her. And so the experience of actually having someone who was, like, attuned enough to notice that there was something off, and then bring it up, I think was very powerful for her." Certain people are going to be more susceptible to the experience of seeing someone take the time to fix a recognition mistake, she said, "those people that come from invalidating environments, homes, like—Yeah, just having had multiple experiences where they weren't heard or seen." Part of that heritage for this woman included the time she had spent in mass institutions—mass schooling, the military—that treated everyone the same. Rectifying mistakes might matter to many, but it had a special resonance for those whose contexts made it scarce.

In contrast, most physicians generally rejected the idea that mistakes were helpful, a response surely brought about in part by their acute liability, life-and-death responsibility, and a culture of perfectionism. When they told me about mistakes, they sounded more like confessions, moments about which they felt guilt or shame about wrongdoing, as we shall see in chapter 7. But a few gave me examples of redemption and repair: a medical resident who responded to a hospital patient's anger so earnestly that he convinced her to take him back as a caregiver, a primary care physician and patient who argued when the doctor mistook some lab errors for treatment noncompliance, but who emerged from the conflict "closer than ever." Even for physicians, relational mistakes—when they were acknowledged and addressed—could be new opportunities for people to connect.

Tensions of Human Practice

Reading and conveying emotions, collaborating, managing spontaneity—these aspects of connective labor were common across many different kinds of occupations, even as no two practitioners might deploy them in the same way. As a group, these components posed particular challenges to those who would make the work more systematic by introducing elements of standardization, such as forms, checklists, or manuals, not to mention AI or automation. Yet even as these humanistic aspects were core dimensions of the practice of connective labor, they also highlighted tensions that were inherent in its conduct—tensions between client-centeredness and expertise, between safety and judgment, and between vulnerability and sustainability. These tensions are most obvious when we look closely at collaboration.

The Tension between Client-Centeredness and Expertise

The emphasis on collaboration brought to a head the tension between the clients' active participation or direction and workers' expert knowledge, often accumulated at great expense and over many years. What happens if the patient doesn't "collaborate" with what the expert knows? Yet what is collaboration if it doesn't allow an active role for the patient? I witnessed an example of this tension one day in the HIV clinic.[12]

Christine, the nurse practitioner, was seeing a longtime patient, William, a patrician white man in his late seventies wearing a neat sweater and a big college ring. When she walked into the exam room, he stood up and they hugged, and William joked with Christine about seeing her sister at a casino in Reno. "My siblings are night owls but not me," Christine replied, smiling, while she examined him. They talked about a colonoscopy and some dental implants, and he made jokes while she asked him questions, but the tenor of the visit changed when he mentioned that he had been experiencing some shortness of breath lately; sometimes when he was doing something, he had to sit down, and it lasted for a few minutes. It was a lung issue, not a heart issue, he said, and he gestured to his chest while asking her about the TB treatment he had a few years back after some exposure. But TB was coughing blood, wasting away, and night sweats, Christine said, none of which he had. Instead, she wanted him to go to a cardiologist right away. But William resisted the idea, partly because he didn't want to offend Dr. Montrose, his primary care physician whom he would see in a few weeks.

Christine often used silences to good effect in her appointments, pausing after making a suggestion or an observation; it felt like a subtle way of nudging patients to her point of view while giving them space to give an opinion. She also had a habit of glancing down or to the side when relaying particularly difficult news, as if to turn down the intensity volume, to lighten the impact of her words. But that day in the clinic, her silences did not work their magic with William.

They discussed a heart attack he had in 2006, and the stents put in, and gently disagreed about whether it was Dr. Montrose or Christine who admitted him then. She remembered doing it, she said. I did not quite understand why the point mattered—were they arguing about whether or not Dr. Montrose is fully involved? Or whether Christine had a history of being worried for good cause?—although it was clear that both of them, in their low-key way, thought it was important.

William told Christine that Dr. Montrose's father had recently died at ninety, to which she replied, "It's always hard when you lose a parent." Sitting in the exam room trying hard to be unobtrusive, I wondered to myself about why they were talking about the grief of the (absent) primary care physician, although William's sympathy and affection for Dr. Montrose was obvious.

Christine urged William to allow her to refer him to cardiology; he repeated a few times that he would wait to see what Dr. Montrose wanted to do. His appointment with Dr. Montrose was about ten days away, he said, but that felt too long for Christine, and she asked again, rephrasing it. Eventually he said it out loud, "I don't want to offend Dr. Montrose." Christine replied softly, "Well, I can make the appointment and you can cancel it depending on what you and Dr. Montrose say."

"I respect Dr. Montrose's expertise," William said, but he must have thought that sounded disrespectful to Christine, who as a nurse practitioner was lower down the medical hierarchy than an MD, because he then added, smiling, "Not that I don't respect yours."

Christine gave up then, temporarily, and we left the exam room then to collect the attending physician for his sign-off on Christine's care, as was customary at the clinic; on the way Christine told me she was just not sure about Dr. Montrose. She definitely remembered being the one who admitted William to the hospital in 2006, and that he'd had all these irregularities in the EKG, and that Dr. Montrose hadn't done anything. In the tiny side office where the attendings sat in front of their computers, Christine presented William's case to Dr. Lewis, explaining about cardiology and Dr. Montrose. After sorting out which Dr. Montrose—"Oh, that's Rob. He's a good doctor," Dr. Lewis said—they both entered the exam room, where Dr. Lewis underscored Christine's point. "We need to get you to see someone for that shortness of breath," he told William. William repeated that his appointment with Dr. Montrose was soon, but when he conceded that it was not next week but the week after, Dr. Lewis said, "Well, that's getting a little far. You really should see someone." He suggested that William call up Dr. Montrose's office and ask for an earlier appointment, telling him they save time for the more acute appointments. "Otherwise you're doing great, good work," Dr. Lewis said, and left the room.

"I'm not going to die here, am I?" William asked Christine in the wake of Dr. Lewis's departure. He joked, "They need me to make money in Reno."

"We want you to stick around," Christine smiled. Then she said, looking down momentarily, "Having HIV can make you more at risk, plus the heart attack and other things."

William stood up. "I'll mention it to Dr. Montrose when I see him on the 25th," he said. He gave Christine another hug. After he had gone, Christine told me "he was acting weird," and said she would call in the appointment to cardiology anyway. "They can cancel it; it will get him seen sooner," she said. I asked her why she hadn't engaged William on the science of it, telling him that shortness of breath meant heart, not lung, trouble. "Well, I just saw him digging in his heels," she said. "Some patients just don't respond to that."

The scene illustrates how Christine follows a collaborative model of medicine, in which she makes her opinions known—William needed to see a cardiologist as soon as possible—but uses questions, jokes, silences, and even carefully calibrated eye contact to help patients relax, exchange information, and express their own views. The incident also reveals the limits of that collaboration, however, both because Christine decided to make the appointment anyway, over his objections, but also because she could not make him go. Clearly, the model fosters some inherent tensions between expertise and cooperation; they could not just tell William they were making the appointment, and William did not have to listen to them if he wanted to prioritize his relationship with the grieving Dr. Montrose over his own health. I left the clinic not even knowing whether William was going to get to the cardiologist in time, a state of ignorance shared by Christine and Dr. Lewis and further reflective of the challenges of collaboration. But by the end of the session, watching these seasoned practitioners express their strong united opinion about the danger he was in, I ended up feeling all the urgency that William sought, in his deference to Dr. Montrose, to dismiss.

The Tension between Safety and Judgment

Collaboration also amplified another tension inherent in connective labor: that between safety and judgment. As we shall see in chapter 7, when people allow someone to get close to them and to hear their struggles and triumphs, they open themselves up to being quite vulnerable. Both workers and others worried about the risk of judgment involved in that vulnerability, and workers talked about trying to make it safe for the others to be honest about their problems. At the same time, however, workers' jobs often involved weighing in with some sort of judgment based on their accumulated expertise—you need to get more exercise, you need to study more to understand the lesson. The knife edge of shame lurked beneath these interactions, and practitioners were constantly evaluating when they should step in.

Tiptoeing around a client in matters of shame was Wanda's métier, and she thought it had led to a recent major success. Battling an eating disorder and shrouded in humiliation, one client had never wanted to talk about when and how much they ate. For a full year they met and talked about other things, with Wanda every once in a while testing the door. She played out how she acted in the face of the client's reluctance: "And so 'I understand you don't want to share that. That was really great that you gave me that boundary because I know how hard it is for you to be assertive. Maybe we can talk about it at some other time where you feel more comfortable.'" But in the last month or two, she said, "they told me how much they ate."

I asked her why that felt like such a major success. "One, they don't tell anyone how much they eat; they are very secretive about their eating and they have a lot of shame when they eat in public and they prefer not to have anyone watch them eat. Their shame response when I would initially ask how much they had eaten or what it meant to binge, was so high that they would completely shut down." Just being able to talk about it was an enormous step, and one the client was able to take, Wanda said, because the client was in charge of when that took place.

What helped her to be able to take that step?, I asked. "Safety," Wanda said firmly, and then continued, "because we processed how they were able to share that and it was about safety. And I think safety with some of the things that torture clients is so important. It is an ever-evolving thing." The capacity to wait for the client to lead was what had led to this success, she said, and was the primary reason why she rejected insurance. Pretending to be an insurer, she said: "'If you want me to pay you, I need to be able to see your notes, your notes need to say this, and then I need to be able to see this, this, and this happen in this amount of time.'" She rejected these stipulations out of hand. "No. No. Like I could never say—If I did insurance, like, with that client that recently told me about how they ate, that [took] a year [before they told me], right? That's not an immediate result. It's an amazing result, but it didn't happen in any specific way. I could have never had [a] way to get that result quicker because it wouldn't have happened, because it had to happen in the way that I can't explain that it happened." The insurance requirements would have meant she could not have waited for the client to feel safe, and would have precluded their most recent success, Wanda said.

But waiting for the client to feel ready to disclose what was shameful had its costs for these expert practitioners. Sadie Aronson was a primary care physician who considered herself a collaborative doctor. She sometimes had mixed feelings about it, however, which came out in some unusually convoluted syntax in our interview. "It's not my philosophy of care to give—like, be directive in the sense of, like, authoritative, right, because that's just not my—or, yeah. I'm just not—That's not my style," she said.

Her style was challenged, however, in the case of a patient whose husband was violently abusive, and who found herself unable to leave. "I think for her, it was just like, 'This isn't working,'" Sadie said. "'Whatever we've been doing is not working.' Like, she has all the information. She knows the statistics, she knows that, like, intellectually she knows where this is going. 'Yeah, of course you want this to work. But has anybody ever told you that it's not going to? And have you ever actually sat with that and let that marinate for a little while?'"

Sadie struggled with whether or not to tell her longtime patient to leave her abusive husband. Maybe it was like telling a smoker to quit, she mused, trying out an idea that would justify a more direct approach. "Because so often, you know, it's like the smokers who say their doctor never told them to stop smoking, right? Or how they say, like, you know, with smoking cessation, one of the most powerful interventions is having your doctor tell you to stop. And it's like, yeah, I don't want her to keep going having nobody told her . . . that this isn't a good idea. Or that this isn't going to work out. Like hope is a valuable—hope is valuable in most all scenarios, but in this one it's working against her and it's hurting her."

With this patient facing the threat of violence in her own home, the stakes could not be higher, Sadie thought. "Like, 'This is going to lead to your demise, and I get it,'" she said. "'I see the psychological patterning. I know what your brain is doing. I know that you're in love with him and there's this part of you that is going to always forgive him and want to be with him. But that needs to go—otherwise you're going to die. He's going to kill you.'" Our interview was starting to seem like a role-play for Sadie to practice saying more directive things. Normally a collaborative practitioner, Sadie clearly wanted to convince herself that her patient needed to hear her judgment.

Proponents of collaboration, however, maintain that if people don't want to comply, all the direction in the world will not make them do so. Gunika Navani was a primary care physician who said she tried to make space for what patients thought of her advice. "Yeah, it's usually not direct feedback, but you can get that sense of maybe they're not buying into the suggestion or maybe they look a little skeptical. And

I'll just say, 'Tell me what your concerns are,' or 'Are you comfortable with what the plan is that we're talking about?' and then just wait for a moment and see what the response is," she said. She grinned: "Usually it starts out, 'Well, I read on the internet . . .' But at least it's an opportunity to address the information that they're receiving from other sources, because if it's not addressed then they may not necessarily move forward with what I think we've agreed to. If we haven't really agreed then it may not happen."

Workers' witnessing often reflects a particular agenda, most commonly to hold clients accountable to a particular standard that comes from their professional expertise—to get them to drink less or exercise more or hold boundaries with intimate others. They take that agenda seriously, viewing it as their professional judgment, the value they add, the reason for their even holding the job. Doctors try to get women to leave their abusive husbands, teachers try to get a student to address where they lag behind, couples' therapists strive to get a cheating spouse to be honest about behavior and intentions. Yet the fact that they are not just seeing, but also seeing with judgment, invokes a specter of shame that haunts both practitioners and clients, a specter that motivates many calls for automation, as we saw in chapter 3. Collaboration gives rise to fundamental tensions between the client's experiential knowledge and the practitioner's expert judgment.

The Tension between Vulnerability and Sustainability

Collaboration also brought up another tension that workers struggled to manage: that between being open to the other, and setting limits for their own protection. Being available to the other person was necessary to empathize with them, many thought, and yet it also meant in part becoming vulnerable to them. At the same time, as we saw in the chaplain Alan's example of comforting the bereaved father that sorrowful day, they feared the costs of that vulnerability.[13]

As we might expect, Wanda was very open. "And as a relationship grows—I really care about my clients and I don't really have that thick wall of a boundary that I imagine with other types of therapy," she said.

Exhibit A of her thin boundary was that the next day, she was having one of her clients over for tea—the one who had recently shared about her eating habits. The client had been distraught over a recent election, and came up with the idea of bringing tea to share with Wanda. "They were like 'I just want to,' like they want to bring me something," Wanda said. "I'm just like, 'Yeah, sure, bring some tea, but we're going to share in it together.' This is my thinking that we're going to share in this together, we're going to be in this healing moment together and I really want to make it therapeutic, and they're eating in front of me, right? So it's not just that there is this kind thing and they're doing caretaking, which is part of their stuff, but they're also eating in front of me," Wanda explained. "I said I was going to bring some treats and they were OK [with that]. So, this is going be hugely—" "A big moment," I said. "Oh, yeah," Wanda said.

Wanda considered the tea an opportunity, but she also noted that it would probably be frowned upon by some other therapists. "I'm not really good at following the rules, I don't think they tell—I'm pretty sure they tell you that you need to have professional boundaries and that includes maybe not having tea with your client because they're really distraught over the election. There are the things that in training that you officially learn, and these are not things that you officially learn." In fact, Wanda maintains, with the exception of one "really good professor who was so empathetic," she did not think she learned how to connect to others in school. Instead, she said, "I feel like they teach you about boundaries in the therapeutic room more than anything else."

Teachers also wondered about how much vulnerability was possible, and how much was too much. Wilson, the teacher trainer, said being a good teacher relied on being open.

> Some of it does have to do with the degree to which you're able to be open and vulnerable to the people that you're working with, or you're able to be a human being they can connect with. And that when you have five classes of thirty-three kids, you need to be that person for all of them, most of the time. And you need to know that that's one of the things you're supposed to be doing. And you need to have a

vision for what that looks like while you're doing it. And it's not going to be the same for everybody.

Yet how much vulnerability was possible with a hundred sixty-five students? "I think for me, the level that I'm involved in my students' lives and what I want to do, what I'm doing for them is probably not sustainable," said Pamela Moore, an African American teacher at a low-income school in Oakland.

Furthermore, what if one of them was a student who was overly dependent? Conrad, a middle school teacher, said it was a problem, but a rare one. "You might have a student who might think that because you're asking questions and you connect, they want to connect with you more than what's appropriate," he said. "If it becomes a problem, you just have to be professional, you have to know when something goes a little too far when a student is trying to connect with you more than what should be happening. It is a problem because we don't have a lot of training on that."

Even in training settings, the guidance was not very exact. I observed a graduate class for young school counselors for one semester, and one day a guest lecturer came, someone who worked in the field. A student asked him: "How do you draw the line with a student who's too dependent on you?" He recommended that they do so carefully, in the beginning. "You guys will each have your own personal boundaries," he said, "but one of the things I always found helpful, is that if I'm spending more time thinking about the client's situation than the client is, that's a sign I need to step back." A wave of agreement washed over the room; some students bent their heads to take notes. "If you're feeling overwhelmed and consumed by the student, that's a concern and that's when I would call in a supervisor," he continued. "That can get tricky. You guys might run into that. And I don't know if there's a hard-and-fast way to resolve that."

Despite the imprecision, however, most workers avowed that boundaries were crucial. "For psychiatry and all these other fields where you're really intensely involved in people's lives is—The emotional boundary is actually a problem," said Rebecca Rooney, a primary care physician.

"Because if you are empathetic and you feel the pain, or if you're sympathetic and you feel sorry for them, you are somehow drawn into their drama and you're feeling it. Then you walk away not really feeling like your job is meaningful, and you're having to protect yourself the whole time." Her solution was "compassion," which she defined as a sort of empathy without overinvolvement. "If you walk into the situation like, 'I'm going to have compassion for what I hear,' and you're able to convey warmth in that visit, you feel better, and the patient feels better," she said. "It's a win-win."

Without boundaries, people could easily get overwhelmed and depressed about their lack of impact, Rebecca said; boundaries actually helped workers stay oriented toward the other person by protecting them from overinvolvement. "So, that's where the joy of practice is, when you feel like you have meaning, you can have impact, and you can be involved with people's lives, and you can be engaged."

The tension between being open and being closed seemed like a constant, inherent in jobs where people had to both work to sense the feeling in another, and yet retain the capacity to not absorb every feeling they could sense in the other. Achieving that balance was the tricky part, the solution to a problem embedded in the collaborative element of connective labor.

Five core components of connective labor make it an artisanal practice, a humanistic enterprise at its heart: (1) using the body as an instrument, (2) reading and deploying emotions, (3) collaborating, (4) responding to spontaneity, and (5) making and managing mistakes. While there was variation across fields and practitioners, for many—who work in vastly different jobs but that all involve some form of connective labor—these characteristics feel fundamental to their efforts. The work of connecting to another human being draws on our very humanity.

These components also kindle some significant tensions—inherent conflicts between participation and expertise, safety and judgment, vulnerability and boundaries—that these workers find they must navigate.

The tensions arise expressly because connective labor is a mutually constituted event, a reciprocal, momentary performance captured by the well-worn teacher's adage that "education is more like lighting a fire than filling a pail." If connective labor were just about filling a pail, these workers would never have to worry about how to center the other, how to help them feel safe, how much to let them in; it is in its very interactivity that connective labor is both powerful and unpredictable.

These elements also pose distinct challenges to their systematization, as each involves aspects that are either difficult to standardize (collaborating, responding to spontaneity, managing mistakes) or that specifically rely on the humanness of the practitioner (reading and deploying emotions, using the body as an instrument). That said, as we saw in chapter 3, there are ongoing efforts to standardize and automate exactly these areas, from emotion to collaboration. There are even engineers who are programming AI agents to make "mistakes," guided by some research that shows that it leads their human interlocutors to feel a stronger alliance with the machines. The automation frontier is in constant contestation.[14]

This chapter has explored the ways in which workers in many different fields practice their own humanity in the doing of connective labor. The next three chapters, however, reveal what happens when we take the humanity of the last human job for granted.

5

THE SOCIAL ARCHITECTURE OF CONNECTIVE LABOR

One Thanksgiving weekend about thirty years ago, Ruthie Carlson, just a few years out of college, started volunteering in a hospital. "They were short-staffed, and I was taking water in to patients. They were desperate for conversation," recalled Ruthie, a tall white woman with an earnestness periodically lightened by a flashing smile. "There was a little old lady that had fallen and broken her hip visiting her family. Just incredibly lonely and totally embarrassed that all this had happened, and I sat down and talked with her." Their brief conversation was like a bolt of clarity for Ruthie, who was thinking about going into medicine. She had just come from volunteering on the orthopedics floor, where "medical care was not well managed by surgeons," she said. "It was very upsetting to me because I thought they could do a lot better." After her conversation with the elderly woman, she knew she had to make sure she was in a position to connect with patients. "I realized that the key piece—that makes the difference, that's what's so needed in healing—is the connection with people," she said.

When she entered medical school, she was determined to end up in primary care for a rural community. "I went in knowing what was needed was direct contact with patients from the ground up," said Ruthie, who peppers her language with words like "gal" and "folks," and describes herself as "not a big-city person." The first two years of medical school are spent almost entirely in the classroom, however, and the experience almost crushed her plans. "I really questioned and doubted what I was doing," she said. Finally, in the third year, the students were assigned patients to talk to. "Then it was OK. Then it made sense what I was doing and why I was doing it. As crazy as things got, if I could go and spend time with a patient, it was OK. I could go without sleep, I could do whatever."

Ruthie told me her story with some urgency, leaning in over coffee together in a busy café, narrating her pursuit of "that connection with people" over the course of her medical career. She most enjoyed geriatric patients, whom she found saucy and inspiring. "I just loved my little buddies with A-fib telling off their doctor in the ER that they were going home, they didn't need a pacemaker and they could stuff it up their ear. I love the patient like that. That was my patient. It was like, 'Good for you, gal, go for it.'" She considered pediatrics, but ultimately decided against it because of her search for "real conversations." "You don't get the connections with the peds patients that you do with your older patients," she said. "You can play with them and have fun with them, but you can't have the real-life conversations that I want, do we do this or that [intervention], and what really counts and why," she said. "You know, those real conversations, that's what gives meaning to what you do in medicine."

Ruthie and I had already known each other for years from a local sports group in Virginia, but I knew I had to interview her for this book once she told me she was leaving her current position to start her own practice in "direct primary care." Often called "concierge medicine" (mostly when referring to physicians who charge much higher fees than Ruthie), direct primary care asks patients to pay for membership in a practice, monthly sums for the privilege of seeing a physician undistracted by having to get insurance authorizations or the press of twenty sequential fifteen-minute appointments. But Ruthie had started out her career dedicated to the cause of rural medicine, bringing primary care to people in need in the hinterlands, those with few resources and little access to doctors. How did she end up founding a members-only practice of physicians for hire?

Over the course of several conversations, Ruthie told me of her path from medical school and residency, to a rural clinic serving the mountain people of Appalachia, to a tony practice in Virginia's horse country, and finally to her entrepreneurial experiment in direct primary care. I heard her story as a mix of tragedy and hope, of sorrow and determination, but it also highlighted a crucial point: that connecting depends on more than just the dedication or skill of an individual worker or even the receptivity of a given client. Instead, Ruthie's story shows us that organizations like hospitals and schools have a substantial impact on whether people can effectively see and be seen there, through their social architecture. And while there are undoubtedly an infinite variety of these architectures, Ruthie managed in her career to pass through three typical ones: mission-driven, corporate, and retainer for hire.

In this chapter, we follow Ruthie and others like her as they work in these different settings, getting a sense for what it is like to be there and the kind of compromises these contexts ask of workers. Along the way, we will find a new explanation for burnout and a new understanding of how context shapes connection. Ruthie's trajectory illustrates the kind of tough choices that people who provide connective labor often have to make, not just in medicine but also in classrooms, offices, and stores: choosing between work that is sustaining and sustainable, between whom you see and how you see them, between service and servitude.

The Mission in the Mountains: Serving the Underserved

Ruthie entered medical school determined to go into primary care in a rural setting. She was not shy about her plans, which she maintains were not met with enthusiasm, as most top-tier schools focused on training their students to be specialists. "Definitely, at that time it was not looked highly upon, shall we say," she said. "They'd say, 'You're smart, why are you thinking that way?'" She graduated undeterred, and finished her residency in family practice, which she prized because of its intensely pragmatic relevance. She recalled a mentor who had spent ten years in primary care in Fairbanks, Alaska, "in the middle of nowhere," flying his patients in. "He came and taught us and he was like, 'This is what you're *supposed* to do, but here's what you *can* do. *This* is where the real heart of it is.'"

After Ruthie left residency training, she got a job in the Southeast, joining a group practice of six partners that served the mountain people of Appalachia. Posted there in a medical desert, she found the work profoundly rewarding. There was an almost total dearth of specialists, "two cardiologists for the whole county, no nephrology, no psych, no derm," she said. "You're it. You kind of had to do it all."

The people were initially suspicious of newcomers, but once they accepted her, she said, they were uncommonly generous and loyal. One day, a forty-five-year-old man came to see her in clinic, "a down-to-earth kind of camper, all about just hunting and fishing and whatever. You'd see him every once in a while, with just a shoulder injury or various things." But that afternoon, he had a lump in his groin, and when the biopsy came back from the lab, it turned out to be cancerous. "Very quickly I got to know the family," Ruthie said. "I mean, he was young and he was kind of taking care of his mom, and it progressed pretty rapidly, so I had to do home visits. He couldn't come into the clinic." While she was visiting, she noticed how pale the man's father was, and after coaxing him to the office, discovered that he had rectal cancer. Soon she was heading to the house every week. "I was accepted, and you become part of the family," she said. "It was a unique experience."

"Very quickly I'd go there and we'd have dinner together. I don't know how to explain it. They kind of looked at me as immediate family because I was there once a week, so of course I would go to dinner. They just assumed that I would want to do this," she recalled. "And I did. They were great. It was expected that I would come for the visit after work, and so I would stay and have dinner, and whatever. That was just the way it was done."

She grew ever closer to them. The family noticed that she was trying to learn how to ride a motorcycle, and decided to step in. "They realized I had taken a motorcycle safety class and they were like, 'Doc, you need some help,'" she said, laughing at the memory. "Because I was not good at riding one. The son took me under his wing and got a guy on the street to work on my motorcycle so they could get the gears better. I was stalling [out] and they were like, 'We need to fix this for you,'" she remembered.

Her original patient, the forty-five-year-old hunter/caregiver, ended up dying from his cancer, and at the funeral she sat with the family. "They made a point that they did want me up there sitting with them," she said. "[Mine] was a revered, respected role that really meant a lot to the people." They had let her doctor them, and in so doing, enveloped her in a social world that extended beyond the doctoring. "Even to this day, I still talk to the family," she said. "This is fifteen years out, we still call each other at Christmastime. I still talk to the older brother, and [when] his mom fell and broke her hip, they consulted me about where to go and who to trust."

These relationships were a bit unusual, but practicing in the rural community meant that Ruthie was widely embraced. "I mean, you couldn't help but stumble over these patients if you went to the grocery store or you went to get your eyes examined," she said. "You go anywhere, there are your patients. I go to the bank to cash a check and the bank teller's putting her foot up on the thing and showing me the ingrown toenail we did last week."

And the population was needy but inspiring. "I had a fourteen-year-old who delivered a baby at home in the toilet. On her own and she hadn't told her parents," Ruthie recalled. Ruthie kept track of the girl, who was older now. "She is one of the most amazing women who's going back to college, is going to be a nurse, and she is going to show that place what she can [do]." The community was full of people enduring amid some tough conditions.

Ultimately, Ruthie said, she found the experience of providing primary care in the rural area intensely fulfilling. "I loved it because they needed—I wanted to be in an area where patients really appreciated and respected and needed our care. They were very accepting of us." Serving as a physician in a tight-knit rural community meant forging deep relationships that mattered to the people involved, as they mattered to her. "It's just the way it was," Ruthie said. "But it was rewarding, it reminded me of why I was doing it. It was worthwhile."

The Flip Side of Mission-Driven Social Architecture

Valued by the community and driven by her own sense of calling, nonetheless after seven years Ruthie left the mountain. She had moved there dedicated to serving a people who had so little, and she was gratified and sustained by the work and the relationships she made. But ultimately the clinic ran her ragged. "I enjoyed it, I was passionate about it. But it was very concerning. There weren't enough of us," Ruthie remembered. "There was never enough."

The circumstances in which people do their connective work matters. Researchers dispute that empathy is wholly an individual trait, available in some people and not others; instead scholars have found that empathy varies within the same person, sometimes even within the same day, and that it can be enlisted by particular cues and experiences. Indeed, those cues and experiences are wrapped up in how the work is organized.[1]

Ruthie's mountain practice reflects what we might call a mission-driven social architecture. In these kinds of organizations, workers serve people with often acute needs who are disadvantaged by their poverty, race, national origin, or other social categories mapped onto the topography of inequality. Doctors, teachers, and others in these places can derive profound meaning and purpose from their work. But mission-driven social architecture also typically features extreme workloads, with extraordinary numbers of patients or students, in environments of severe resource scarcity, with not enough time or space to serve them well. The organizations often espouse the value of public service, but workers without enough support sometimes feel like the stated values are hollow because they are not backed up by adequate resources, as practices cut corners—for example, packing students into classrooms and patients into clinic schedules—to be able to meet demand within austerity conditions often dictated by the state.

There are a number of dimensions that make up a social architecture. Material resources like time and space are its fundamental building blocks, the frames and crossbeams of how an organization shapes the connective labor within it. Resources that matter include the time an organization allows for a given interaction, the ratios of worker to client, the cognitive load workers face, the extent of technology and data use—any of the material conditions that enable or impede people's connection. In my research, I also found two other dimensions to social architecture that matter for connective labor, including its relational design—how people are put in relation to each other, such as the leaders, mentors, and peer groups that act as sounding boards for feedback and support. Finally, a social architecture has a "connective culture," or a shared set of beliefs and practices that affect how people are able to conduct relationships. These dimensions often combine in patterned ways to create unique strengths and risks; in the mission-driven context, the architecture's strength was the profound meaning it offered its practitioners, but its risks included the potential for rushed and alienated connective labor, and for workers, the sense of never being enough.

While each industry has its own unique history, the fact that workers who serve the poor in health care, education, housing, and other social services inhabit similar kinds of institutions suggests that there is something in common here: that under late modern capitalism as represented in the US, basic needs are often framed not as public rights but instead as private goods. Strapped public schools and community clinics are staffed by people who might disagree with that equation, but who cannot overturn such a pervasive cultural frame on their own. As one critic wrote, the community health clinic exemplifies this paradox, "address[ing] only the health needs of the disaffected, using crumbs tossed from the groaning table of medical corporations." Operating with inadequate support from state or local sources, mission-driven social architecture often relies on individual heroes who endure in unsustainable jobs as long as they still find their service to the truly needy sustaining.[2]

In addition to Ruthie's periodic home visits, daytime clinic hours extended well into the night, the physicians regularly took shifts on call in the hospital, and physicians visited nursing homes every week. "Sleep was good," Ruthie said flatly. "It was difficult if you'd been up all night and try to go to work the next day." The hospital attempted to hire hospitalists to take the nighttime shifts, but the patients rejected their advice as from strangers. "[The patients] would come in to [see] me, and say, 'Who does he think he is, telling me to take this med?' They weren't going to take it from anyone that wasn't their doctor," Ruthie said. "They would just call us the next day. 'The doc said I needed this. I told him I'd call you.'"

The sustaining relationships became unsustainable when a crisis arose in the local hospital—a whistleblower had documented some errors, the administration did not respond to a federal inquiry, and the institution was shut down temporarily—leading local physicians to band together to try and rescue the hospital, which was the only one for many miles. The extended emergency broke Ruthie down.

> I mean, it's a stressful job to begin with. I can't remember how long it went. It was exhausting. We were seeing patients nonstop in the office, but then we would [also] meet at seven in the morning at the hospital or six or eight at night, so it was on both ends of the day. I was friends with the chief of staff, and I sort of took it upon myself that my job was to keep her going. I literally would go to her house and [administer medications] so that she could attend the meetings and get things back up and running. I was doing home visits and trying to do my [pediatric visits].

The experience was utterly depleting. "It was very personal and very miserable," she said. "I made it through that year, but I think about that time, I just remember saying, 'I can't stay here, I'm just exhausted.' [Yet I d]idn't know what else to do because my patients needed me." When she snapped at a new nurse, she realized it was time to leave. But, she said again, "it was very personal. It probably took something of that magnitude to get me out of there. I was really attached."

Ruthie had arrived in Appalachia fresh from years of training, full of dedication and eager to finally put what she learned to work. Yet seven years later, she was leaving, discouraged, exhausted, and burnt out. The hospital's mismanagement crisis contributed, to be sure, pointing to relational design problems—in particular a lack of leadership and a depleted cadre of peers and mentors—but these issues acted as more of a catalyst to her departure. Rather, the depth of needs she faced there, and the intense inadequacy of resources available to enable her to meet them, meant that she was always on, always responding, always doctoring.

"It's Just Like They're Singing Their Siren Song"

Ruthie's rural medicine experience echoed that of other people I spoke to who worked in mission-driven environments, practitioners stretched beyond capacity within a resource-poor environment. Everywhere, committed professionals inspired to work with underserved people confront this kind of overload: overwhelming demand, strained schedules, and little support, with stories equal parts inspiring and daunting.[3]

To some degree, Pamela Moore, a Black middle school teacher at a public school in Oakland, sounded just like Ruthie—driven, dedicated, and facing significant overload. Only two years into her teaching career, Pamela told me that many of the children in her classroom were "going through a lot." "I have students who are homeless," she said. "I have several students who have lived in group homes or are currently living in group homes. I have students who have parents incarcerated."

While working there was traumatic, she also had a strong sense of calling. "I want my students to have the same opportunities as my own children," she said. She was starting next month by setting up a camping trip for her impoverished students. When I asked her why camping, she talked about her mission. "You know what? That's what school is for. Why is it [just] kids who have parents who live in a different zip code [who get all the experiences]? We take them ice-skating. We have an ice-skating field trip. We're going to go camping. I want to take as many kids [as I can]. I go camping with my own family. It's so important for kids to get out away from cities." She said she wanted to teach every child, no matter how difficult. "I wanted to teach every kid who came in my classroom, even if you were the most insane kid ever," she said. "I want to teach you."

Her mission was doubly felt because she was an African American teacher charged with educating kids of color. "The reality of teaching in a high-poverty, high-needs environment, for a teacher of color, who is teaching children of color, is we can't just teach," she said. "But because of your identity and what you represent—I'm African American, I'm one of those folks who is tasked with supporting African American girls at our school. We are disciplinarians, we are counselors, whether we are paid to do that or not, that is our obligation to our community. You know that going in, and it makes our jobs unsustainable."

Overload is a problem with many causes. On the one hand, needs have certainly expanded in a host of connective labor occupations, as an aging population requires more medical care, for example, and mental health diagnoses multiply. In recognition of these trends and their impact at home, Arlie Hochschild calls this a "care deficit," arguing that people are getting needier just as those on hand to care for them get busier—women, mostly, who serve as America's safety net in the absence of one, as sociologist Jessica Calarco argues.[4]

The supply of connective laborers has not expanded commensurately, however, and that reflects a different set of trends. Thanks in part to the convergence of capitalism's downward pressure on labor costs to increase profits and the shrinking of the state under neoliberalism, emphases on efficiency pervade the public and private sectors, leading employers and managers to see how far they can stretch a single worker.

Economists suggest the problem is that humane interpersonal work is ever more expensive, compared to the declining price of most everything else. Dubbing this a "cost disease," they say it stems from uneven productivity gains, because personal, face-to-face services resist the moves people might make to increase their output, such as, say, increasing class sizes. "With little room for genuine productivity improvements, and with the general level of real wages rising all the time, personal services are condemned to grow ever more expensive (relative to other items)," writes Alan Blinder, a Princeton economist and a one-time member of the Council of Economic Advisers. He recommends that we "train . . . more workers in personal services" and educate children in "people skills." While the jobs may resist moves that increase productivity, however, a relentless pressure to find any possible savings leads organizations to continually push workers to their limit.[5]

Other teachers echoed Pamela's sense of strain. Vivian Vaughan, a teacher at the Silicon Valley school I visited, had worked for two decades previously as a public school English teacher. "I had a hundred fifty kids full time, and I remember teachers saying, 'I don't even know some of the names in my classes,'" she said. "I have more of an advantage, I think, being the English teacher, because I'm a teacher that's reading your business and your mind and your thoughts and your feelings all the time [in writing assignments], but in other classes where it's just blackboard answers and doing worksheets, you probably don't know everybody's name. I mean, you have a hundred fifty, and some teachers get a hundred fifty kids each semester, so you run through about three hundred, four hundred kids a year. You just—you're just kind of a factory worker at that point." Just the pressure of having too many students with not enough time got in the way of being able to connect to them.

Public school teachers, public defenders, social workers—people who worked in a mission-driven social architecture all faced the same combination: the needy populations, the negligible resources, the patchwork support. It came with anguish and a sense of the absurd. For many, the degree of need they faced was overwhelming. "My patients, it's just like they're singing their siren song to whoever will listen because no one will take care of them," said Jenna, a county hospital pediatrician. "They're used to being—not getting needs met, and they're just desperate."

She described a "typical day," objecting to the "fiction" that all her appointments should last fifteen minutes: "Because some people arrive and they're complete—You know, it's like they are a family who moved here from Mongolia two weeks ago and I'm using a phone interpreter in Mongolian, and the kid has Hirschsprung's disease and had a surgery that the parents can't describe and is on a couple of medications that they don't remember the name of. And there's a Mongolian immunization record that my nurse is going to have to try and interpret."

Often, she could tell that patients and their families just needed her to be present for them, but she also dreaded what it would do to her schedule. "Sometimes a family will need the relationship," she said. "That's the whole thing. You know, kids who are having horrible behavior problems or other symptoms—sleep problems, skin picking, constipation—or a foster kid who has been removed from some horrible situation. I feel like the most important thing I can do is just be with them, and that's going to be the most helpful intervention, then I will do that. But it screws up the rest of my day," she said, adding mournfully, "I hate feeling that way."

Like Pamela's students, Jenna's patients may indeed have been desperate. Connective labor is rationed in mission-driven environments, so that practitioners often do not have the time or the bandwidth to do it well. When time is so limited, the capacity of workers to provide ample connective labor is restricted as well, increasing their propensity to make mistakes or to choose not to hear the complexities their clients face.

Like many who worked in mission-driven contexts, Jenna talked about feeling guilty for the steps she took to make the job more sustainable. Too often, she said, "I don't invite people to open up because I don't have time. And [yet] it's so helpful and cathartic [to do so]. And that is such a disservice to the patients. Because I see that [they] want so much more from me than I can give them." She confessed sadly: "It's a dynamic. There's not an invitation on my end. My hand is on the doorknob, I'm typing, I'm like, 'Let's get you the meds and get you out the door because I have a ton of other patients to see this morning.' And I have to decide how much of myself to be in the room with the story, or the trauma, or how much I can give in the interactions, and it is a conscious decision. You know, everyone deserves as much time as they need, and that's what would really help people is to have that time, but it's not profitable."[6]

The hand-on-the-doorknob tactic was common, said Nathan Weiss, who had left medicine to become a therapist. "There's depths of signals being sent, but if you're too busy, and if you unconsciously don't want to get it, because if you get it, you're going to be late for the rest of the day, those things are, they're very powerful. They just—Everybody does it," he said. "If you're [the patient] overwhelmed," he said, "they [the physicians] don't want to hear it."

Overload also robbed practitioners of the necessary time to reflect. Simon Jarret was a white physician working in a community clinic in the San Francisco Bay Area. As the son of a social worker and an engineer, he felt almost destined for medicine as a natural combination of the legacy they had bequeathed to him. "There's a valuing people and helping people and wanting to be a teacher and provider to people in need, and bringing science to that," he said.

At the community clinic, however, the crush of patients affected Simon's capacity to care for them well. He was good at reading the other, he said, but doing it all day is exhausting, racing on what he called "a fifteen-minute, you know, hamster wheel."

> So there's always a sense of there's not enough time to do a good job of my job. I do not have enough time to do what I really want to do, which is to help this patient manage their diabetes, which is to think really deeply. I don't have enough time to think really deeply about this patient's—the one unique interesting medical thing that I've seen.
>
> Most of the stuff that I do, I'm really good if it's rote. If I have something that is unique and I want to think about it harder, I just don't have any of me left. And then time is taken up by all these millions of little extra things that we have to do now. So I think it's a sense of overwhelm. I think it's just volume of stuff and the sense that a lot of that stuff is not really aiding in the care of the patient.

Simon was plagued by a sense that the time compression was making him fail his patients.

The "hamster wheel" inhibited the capacity for the clinician to experience their gut instinct, the crucial clinical sense that often underlies interpretive work, said Nathan, the ex-physician and therapist.

> Usually that stuff comes when there's a little bit of space, often, not always. But you know, you're not like "blah blah blah blah" and you get hit by it. It usually happens when there's a little bit of space.
>
> So, there's just so much less space. If I have twenty minutes [for] you and I'm like "What's important for you?" I should probably tell you your blood pressure and then we'll see what you want. If I have to worry about your sugar, and your cholesterol and your blood pressure, and teaching you about this and that, right? Then

> there's no disposable time. Either way, there needs to be some rest. There's no rest.

Clinical intuition would not come knocking, Nathan warned, but instead had to be invited by practitioners with time to hear it.

Faced with immense need but constrained by the mission-driven social architecture from meeting it, workers found both meaning and burden in whom they were serving. Their rosters stacked with complicated cases, they were often unable to take the time they needed to treat, teach, or even simply see the other well. They felt guilty for those moments when the enormous strain peeked through—the hand on the doorknob, not knowing some students' names. At stake was not just their capacity to see the other but also their likelihood of burnout.[7]

Burnout as Impeded Relationship

Scholars concur about what burnout looks like—emotional exhaustion, feelings of inefficacy or inadequacy, and depersonalization and cynicism—but there's some disagreement about its causes. Research on burnout suggests that it stems from high workloads, lack of control, and insufficient social support, and that it is linked to emotion work (e.g., ginning up feeling for others as part of your job) and emotional dissonance (the degree of distance between what you feel and what you have to fake that you feel). Most burnout research considers relationships as depleting, suggesting for example that treatment involves taking a break (from vacations to weekends to just taking a full lunch hour).[8]

More recently, however, scholars have explored the notion that relationships can actually be sustaining. Connecting to clients can be restorative when they require less "regulation" (a measure of how much the worker has to do to overcome emotional dissonance) and when they tap into personal sources of strength, particularly "perceived prosocial impact, positive affect and self-affirmation." These resources resemble the sources of purpose we saw in chapter 2: the good feelings ("positive affect") that workers get from witnessing vulnerability and connecting to individuals, their sense of having an impact, and the recognition, gratitude, and status that can contribute to self-affirmation.[9]

Given the meaning people derive from connective labor, then, we may need to recast what we think we know about burnout: perhaps it is not the experience of too much relating but too little. Burnout may instead be caused by the experience of *impeded* relationship, as what happens when factors—organizational, structural, personal—make it impossible for relationships to serve as sources of purpose or dignity. The conventional story of burnout relies on a metaphor of workers as buckets of compassion that spring holes, from which their compassion drains away. Perhaps a better metaphor would be to instead think of workers as soil, and relationships as rain—sometimes the rain is torrential or toxic, sometimes the soil is too dry to absorb it as sustenance. The assumption is not that we don't need rain, but that we need the working conditions that enable the rain to be restorative.

When Ruthie left the mountain, she was suffering from burnout, even amid the rich, profound connections she had forged there. They held a going-away party for her in a local church, and hundreds of people came. "My dad said, 'You know, it's a privilege to realize you made that kind of connection to people in their lives, because most people wait until the end to know that they [did that],'" she recalled. "It didn't feel like that, it felt like my gut was being wrenched out." She paused, adding wistfully, "But it was a beautiful tribute."

Corporate Social Architecture

In addition to mission-driven social architecture, connective labor practitioners often have two other choices for where and how to do their work: in consumer-oriented, corporate-like environments focused on efficiency, and in personal service for clients who could pay extra for it. At stake were the relationships they were able to create with their patients and clients, whether sustaining, sustainable, or something else.[10]

After Ruthie left the mountain, she returned to her parents' home in a small southeastern city, unsure and anxious. "I didn't know where to go or what to do, I just had to get out of that situation," she said. But when a local primary care practice found out that she was nearby, they offered her a job. She took it, thinking she could use the time to figure out what to do next. She knew she would not be serving her rural folk, and understood that both her employer and her patients were wealthier and savvier, a far cry from the hunters and campers offering her motorcycling advice in Appalachia. She quickly found out, however, that modern medicine in a corporate environment shaped the relationships she was able to forge, and that for her the new model was not much of an improvement.[11]

The first thing she noticed was that the other physicians in the practice were not allies, covering for each other, watching each other's backs, sharing an enormous unmet burden, but instead more like fellow employees. "My second week of practice, all my partners left on vacation for Christmas [leaving her to cover the practice because of her lack of seniority]. I didn't know anything. I just barely knew my way around the building. I had worked like three or four days and they were gone. Totally gone. All of them." She got through the experience, but for her it was a harbinger of what was to come.

The new practice was a culture shock for Ruthie. Her partners were worried about their productivity numbers, patient satisfaction grades, and patient care scores, she said, and did not want to spend time in meetings to generate a sense of team or a commitment to shared care standards. "They didn't even meet monthly to talk as a group. They were employees. They got their paycheck. They went home. They were done." At her old practice, doctors would not even leave for lunch without asking whether colleagues wanted something. "You just didn't do that. It was a family. We worked together," she said. In contrast, in the city, "here people just disappeared. My nurse would go without asking me," she said. "So, coming here was just kind of—It was a shock."

The patients were different as well. In the city, the patients were more educated, more demanding, and less appreciative of her time. Often, they came to their appointments having looked up their condition beforehand, and were not very interested in what she had to say about it. "They're very entitled," she observed. "They come in with their newspaper articles and they tell me what they need. 'I have Lyme's [disease], this is why, I just need you to give me a prescription, thank you very much.' It's like there's not any respect for what I might have to offer."[12]

The combination was utterly disheartening, with a dash of vertigo, as she vaulted from a 1950s version of the small-town GP to a 2020s version of consumer-driven corporate medicine. "The calling that I had been so passionate about became a job when I came back, that's probably the biggest difference," she said. "It suddenly was a paycheck and it was no longer about what I really—about why I went into medicine. It's sad. It's really sad." Her dissatisfaction crept up upon her like a disquieting shadow. It wasn't as urgent as it had felt in Appalachia, she said. "I mean it was nowhere near as bad. We had resources here. If I didn't see them [her patients], there were other docs. They weren't going to die in the hills." But something was definitely wrong. "Basically, I didn't have a connection with my partners here, and I didn't have sustaining connections with my patients," she said.

Ruthie's second act offers us a good look at another kind of organization, one with a "corporate" social architecture. Organizations with corporate social architecture have more resources than the community clinics and public school classrooms. They often serve a more-advantaged population of clients or students. They might have more professionals on staff, and the offices might be more spacious or well appointed. Yet paradoxically, the expansive resources do not necessarily mean that workers have more time, because the more-advantaged population is also more demanding, articulating their needs with more urgency, requiring that they receive customized attention.[13]

Vivian was the teacher who had left a public school after twenty years of teaching English to work at a Silicon Valley private school, which presented new challenges in how to handle the parents. "That is an ongoing—that's a tricky line—in that you do have parents who have certain expectations [of how much their child could achieve], and you have to, in some way, find respect and honor that, but at the same time you know student limitations, and so somewhere in there, you have to find like a middle. It is a private school. You're paying a large sum of money, I get that." She said she held three parent-teacher conferences a year, plus three parent coffees, plus open houses, plus a weekly meeting with a parent liaison, plus the school sent out a weekly newsletter. "Yeah, the parent part's tricky sometimes."

The corporate social architecture often involves a consumerist ethos, featuring practices such as collecting data on (and responding to) client "satisfaction," for example. Market pressures of efficiency and profit also come into play in these contexts as well, as we shall see in the next chapter. These aspects infuse the "connective culture" dimension of social architecture, by which I mean the shared values, beliefs, and practices of a given setting—including ideas about the main point of the work (e.g., public service versus customer satisfaction), what workers owe each other, and what counts as going above and beyond.[14]

Corporate-like environments can be found in therapy practices, private/independent and affluent public schools, and hospitals; corporate social architecture is a feature of not just for-profit organizations but also those that have simply adopted corporate-friendly priorities of efficiency, data analytics, and customer satisfaction. The crux of the difference between the mission-driven and the corporate social architecture seemed to be whether the service provided was a public right or a private good.

While some corporate entities had ample staffing, the introduction of marketized values did not always protect practitioners from time pressures, but meant that those pressures came in under the name of profit rather than poverty. Therapists who worked at public clinics and private start-ups sounded like each other, for example, with those at the former telling me they did not have time to add "another bipolar group" despite the demand, while those at the latter were talking about the burden of seeing many patients in one day and "not having control over your workflow."

Social architecture thus shapes how it feels to practice connective labor, with effects particularly obvious to those who have experienced both kinds of environments. Veronica Agostini was the "coach" for the start-up that was chronicled in chapter 3, offering online therapy through an app; at one point, to make money, the company pivoted from serving anyone who signed up to contracting with businesses who would offer the app as an employee benefit. The new business model changed what it was like to work there, Veronica said. "It just felt like those people already have access to lots of resources. It's not to say that they don't need help or that, you know, some people are more worthy of getting help than others, but to me it just felt like kind of cheapening the mission," she said. The only exception was when they were developing new programs, so they would offer them for free to beta testers. "Um, and then you would get people who actually, like, needed the help," she remembered. "Um, and we had a partnership with, like, a disability insurance. And *those*—lots of *those* people genuinely, like, needed help." Whom you were serving affected how it felt to serve them.

When Melia Santos wanted to test out optometry as a possible career, she went to work for an optometrist in a big-box store one summer. The environment encouraged the doctor to view his patients as just a source of income, she thought.

> The way he treated staff was OK, but the way he treated patients really stood out to me. He would talk to us, he talked to me about how he saw each patient as a dollar sign. And he was upset that he wasn't a millionaire at this point and that he should have been in his mind. He was like, "If I want a CD [a compact disc] or if I want something, I'll just see a patient walk in and I'll say, 'There's my CD.'" He just didn't show me that he cared much about each individual patient, and he was just trying to churn.

His approach to patients appalled her and sent her looking for meaning; today she worked as an optometrist in a mission-driven community clinic, and talked passionately about how people deserved quality care no matter what they could afford.

Corporate Social Architecture in the Platform Economy

While corporate social architecture prevails in many organizations that prioritize efficiency, productivity, and consumer satisfaction, it is also dominant in the burgeoning industry of connective labor delivered by platforms in the gig economy. As tech scholar Julia Ticona has observed,

Care.com and Urban Sitter have created pools of care workers comprising millions of people—far greater than Uber or Lyft, although those organizations are often thought to epitomize platform economy "disruptions." While gig care workers often toil in individual homes that might come with their own social architecture, the platforms themselves shape what it is like to secure that work as well as perform it. The dominance of customer "ratings" and reviews—which apply to the worker and not to the family or client—as well as algorithms that determine their search ranking based on their active participation in the site (e.g., how recently they have logged in, or how quickly they respond), coupled with the precariousness of the work itself, all serve to emphasize connective labor as a consumer service, in which care is counted, shaped, and surveilled by technology.[15]

"The thing is I have to get in those websites every day, sometimes two or three times a day. Otherwise nobody will contact you, just see your profile," said Graciela, a nanny. "So you have to offer your services to the families. So you have to go every day. Otherwise no one will ever [contact you]."

Gigs were often short-term arrangements, which had an impact on the kind of relationships the workers forged. Hattie Arnold had worked on many different gig platforms as a caterer, mover, and babysitter. "I think a lot of the time people who are doing gig work are viewed as disposable," she said. "So it's like you're used to a high turnover, and so then it's, you know, there's less likelihood to be interacting with them at that deeper level or to want to connect, because you're like, 'It's just going to be somebody different next time, so why should we even form that?'"

Even when the relationships lasted longer, however, the platform shaped the work and the workers in particular ways. The story of Veronica Agostini, the "online coach," helps us see how. As she explained the start-up: "Basically, someone had taken, like, a cognitive behavioral therapy workbook and just put it on the website. The whole idea was that, like, they would get the therapy from the app, and then the coach would be there to just be someone who's, like, checking in on them and, like, encouraging them to go on." Veronica's experience illustrates what we can say generally about connective labor in new economy jobs. First, the work is still connective labor, even though it is radically mediated by these technologies and platforms. She told me, "I think a lot of what many people got out of it was just having someone there who was, like, nonjudgmental, good listener. It was, like, 'Half an hour a week, I can talk to this person and they understand what I'm going through and they're not judging.' And that was what a lot of people liked about it."

The very anonymity of the platform-based work takes away some of the constraints that inhibit some clients, which may make some more honest about their struggles, but it also means that workers can get overwhelmed. "People will tell you a lot when it's, like, basically anonymous, you know?" Veronica said. "Like, you're just calling—you don't know who I am, I don't really know who you are. I've had people who were dealing with really serious trauma."

Nonetheless, the organizations do not seem to value their human connective labor very much; they do not reward it as a skill, and there is not a lot of training. "Training was very minimal in the beginning," Veronica recalled. "Um, like, there was no support really, except for from the other coaches. Um, and I was always just like, 'This is so hard.' Like, I'm starting a new job where I've never done this before, like, I've never been a peer counselor before. Um, I don't know, like, how to do this." She laughed. "All I can do is, like, my best guess. And then these people are telling me, like, really heavy things all day."

Finally, like other gig work, what looks like freedom to make your own schedule often transforms into the technology serving up an endless series of clients. "No, you did not have control over your workflow. Not at all," Veronica said. "That was another reason why I left, was because toward the end it was like—They were trying to, like, model everything using actual, like, algorithms and models, and it's just like, you can't account for the actual emotional burden of the job in those moments." She laughed again. "You just can't. And in order to make it work financially, like, you will need to give the people, you know, different coaches, as big a workload as possible. And not everybody can handle that."

When I talked to people in other contexts, whose connective labor felt supported, they mentioned leaders who prioritized their relationships, mentors who guided them, and, perhaps most important, a group of peers who could act as a sounding board, a safe place for them to hash out problems and try out new ideas. In other words, the relational design of their work was crucial, as we shall see in chapter 8. As a "company of one," however, workers individualized by the gig economy are missing the apparatus of people who can sustain and support their connective labor.[16]

Untrained, unsupported, overwhelmed, and overworked, connective labor in the new economy features many of the same problems we hear from ride-hail drivers and delivery workers across gigs and platforms. A busy, relatively advantaged population may be willing to get drive-by connective labor in return for its affordability and convenience, but this tier is costly for the practitioners.

A Different Kind of Alienation and Burnout

If mission-driven social architecture risked burnout due to the overwhelming demands it made of heroic practitioners in a context of resource scarcity, corporate social architecture risked burnout by asking too much of their professionals in service to profit, efficiency, or consumer pressures. The impulse to cultivate and respond to satisfaction scores, alongside the market pressures to shrink costs, often resulted in greater time demands on workers, who found themselves asked to be available on email or to extend their hours.

Jenna, the county hospital pediatrician, was also a patient at a large HMO. As a consumer there she was very satisfied, she said, even though she could see the toll it extracted from practitioners. "I'm so happy with GroupCare now," she said. "They have nighttime urgent care appointments now. I mean, you just call up and they're like, 'Sure, I can see you at 8:15.' I mean, the doctors are just burned to a crisp. The primary care docs there are ground into the ground. And the endless patient emails that they don't get compensated for. Endless. They can squeeze the physicians and just wring every last minute out of them."[17]

Ruthie found this to be her truth as well. After she had worked at the city practice for a while, managing the grueling schedule typical for primary care physicians everywhere, she too found herself getting burned out once again: depressed about work, crying with frustration, unable to talk about other things, and with nowhere to turn for help. Administrators seemed utterly uninterested in making any real change, she said, although they sent out invitations to "burnout conferences." "I actually went to one and it scared me," Ruthie said. "We lose four hundred docs a year to suicide. An entire medical school class to suicide, nationally."

One problem was the lack of solidarity in the corporate environment, she said. "Typically, our professionals become more and more isolated. As we bear down to survive, we literally ward each other off. There's no time. They have fifteen-minute slots for twenty patients." She went to her own physician, and the woman gave her ten minutes of her time, handed her a prescription, and never even sat down. "And I'm a doctor!" Ruthie was astounded. "Are you kidding me? This is happening to our patients on a daily basis. I realized later, I was talking to a colleague, and said, 'Thank God I have other resources.' If that were truly a patient in desperation . . . It's sad. It's just, that is the state of medicine."

She gained some solace in her own patients, she said.

> When I get really miserable, I go and spend time with a patient. And so that is, I think, what I've done to kind of stay in medicine and survive it, is to go back [to the patient]. In residency, when you lose that patient and you've done everything you can and you're just worn out, I couldn't do anything more, I just would go in, even if it was late at night or whatever, and I'd just spend time with my patients, sit at the bedside, and that's the only way I found I could get through it. Because, in a way, they give you the strength because they're going on.

Patients are inspiring because they endure, even in what seems like the worst of times. But they also get Ruthie to step up and out of her own frustrations because she thinks she might still be useful for them or for other people. "You hear their lives, and I may have something to offer.

Even if this other person is struggling, I gain the strength back and go in with another patient, if that makes sense," she said. "I still have a role, even if we failed everything."

Ultimately, Ruthie decided to leave the city practice. "I think the biggest piece to this, in an employed model, I'm just a cog in a wheel," she said. "My value and what I'm trying to do is not to be appreciated or valued by any of the folks." Primary care doctors today are treated as interchangeable, Ruthie said, workhorses who just need to put their heads down and labor harder and harder to pay for an ever-expanding layer of administrators. "We have less and less time with patients and work harder and harder because we're paying for three or four levels of management and the patients get this much time," she said, holding up two fingers a hair's breadth apart. "We don't like that. The patients don't like that. That's so wrong. So what I'm hearing more docs say is, 'Let's look at this differently.'"

Retainer for Hire: Personal-Service Social Architecture

Faced with the return of her crippling burnout, and at the recommendation of a friend, Ruthie went to a conference about direct primary care. There, among a group of primary care physicians who prioritized relationship, who were also dissatisfied with the conventional organization of medicine, it felt as if the speakers—physicians, and some patients—were talking right to her, about her. "I [saw] doctors happy for the first time in years and heard them echoing things that reached me. They talked about real things that matter," she said. "The docs got up there and they would give their little stories, and I identified with a lot of the stories," which were largely about cold-hearted administrators impeding the capacity of primary care physicians to invest in the relationships that were meaningful to them.

One patient told the crowd about her husband's death from liver cancer in an anonymous and unfeeling medical system. Ruthie recalled the patient saying, "'Now my direct primary care doctor works with me, knows me, knows my kids, he's available.' It was very heartfelt." She left the conference intrigued, and set about a career change.

Ruthie's third act—into direct primary care—represents her attempt to wrest her dreams of connective labor back from the institutions that she says ruined it for her. In the end, neither the strapped mission-driven community practice nor the efficient corporate model met her need for relationships with her patients that were both sustaining and sustainable. In the former, relationships were deeply rewarding but ultimately overwhelming, in a model that enlisted the efforts of dedicated professionals but allowed them to drown in unmet demand. In contrast, the latter model was more sustainable, asking for a bit less from the practitioners, but not nearly as sustaining, being quite limited in its meaningful rewards. "There have got to be better ways to get to what the dream was," she said. "I want the time and the space to hear what's underlying the ailment," she said. "To me the art of medicine is being fully present. If you just treat—give them a pill for the insomnia or give them a pill for this or a pill for that, you're not getting to what the problem is. If you don't hear about why they've got into that situation, you're not healing—it's basically creating more illness."

So Ruthie was hanging out her own shingle. "This is an opportunity to take back what should be for everybody," she said. "And I'm not an entrepreneur at all. I just want to take care of patients." When I protested that she was certainly acting like an entrepreneur, she answered, "Only because I have to. I believe in the concept of direct family care and I think it's an opportunity to get to what my dream is. [But] I'm not doing this because it's the right thing for health care and it's the right thing for America and whatever. I'm doing it because it's a goal, it's a means of achieving my dream of what medicine is. It's that I can't do this any other way."

Ruthie's latest turn involves another common organizational regime for connective labor practitioners: the personal-services social architecture of the retainer for hire. These workers serve far fewer people—sometimes even just one family or client, as do personal chefs or personal trainers. They are often well compensated and their work is granted ample time and space. They sometimes have peer groups of like professionals for support. Under these circumstances, connective labor feels artisanal rather than industrial or bureaucratic. Yet providing outstanding care to people who can afford you is satisfying in one dimension (the quality of the work on offer) and not in another (the sense of broader impact, the service to the public, the larger mandate that some consider their legacy). These underlying meanings shape a particular kind of connective culture, a fairly individualistic one that rewards advantage and separates out the value of connective labor from the social relations of who is giving it to whom.

While Ruthie's practice strived to be more affordable, most versions of this kind of connective labor are limited to affluent clients who can pay hefty fees, and the risk that these workers face is more existential: they feel like servants catering to the wealthy, rather than professionals serving those who need them. Gabriel Abelman, a California physician, told me patients in Palo Alto were paying $30,000 a year to join these kinds of practices. "My joke is doctors will pick up your dry cleaning," he said. "And it gives them what health care is not giving them, which is very personalized attention."

"There's this massively expanding sector of medicine, which is corrupting," said Nathan Weiss, the ex-physician and therapist. "The idea is that there are these places, usually for wealthier people, with better plans, you know. You can succeed your way into a place that people are willing to pay for [close attention], pay to have a little bit of hand-holding. Pay to have clinical intuition, and the more vaunted the place," he stopped and mentioned someone he knew in concierge medicine. While someone at Kaiser might have two thousand patients, he said, she had a hundred fifty. "That could be, you could see a person for a whole day, like twice in a year. That's like sixteen hours." You would have more than enough time to do your work well, he said.

The rise of concierge medicine reflects trends in connective labor across the economy, where the fastest-growing occupations were in what sociologist Rachel Sherman has called "lifestyle work," personal services in a variety of fields from personal investment advisors to personal coaches, all geared to helping the most affluent. Low-income people encounter scarcity of access to connective labor in overcommitted public settings, where AI apps are making significant inroads. And in the middle are platform workers giving connective labor as a gig—sometimes leading to long-term relationships, but without the training, regulations, or labor protections that make for a good job.[18]

In the emerging landscape of this work, then, we are seeing the rise of a servant economy, and a thinning out of connective labor by scripting, by increasing precarity, and by automation. There is inequality in who is on the receiving end of these different kinds of connective labor, as well as in who provides this labor and whose labor is standardized or automated. As Geoffrey Janney, an AI engineer, noted, the status of having a human being to serve you shapes the market for automated connective labor. "We might imagine there'd be a certain prestige in having a human personal assistant," he said, "and the masses will just have to deal with a digital one."

But serving as a status symbol raised profound identity questions for some workers. Bella Albrecht was a therapist in private practice who also had a full-time job managing a start-up incubator in Silicon Valley. She did not take insurance, and she fretted about serving only the most privileged. Yet when she looked at friends who worked in mission-driven organizations, "I see them as being in pretty broken agencies that don't give them the support that they need. I just kind of don't think that you can—I don't think that's a sustainable place." Without a life partner to share expenses, she "didn't want to be stressed about money for my entire life," she said. "I'm frustrated because I now have to choose an identity: am I a therapist in private practice who charges $250 an hour and serves only the most wealthy, but gets paid for my work? Or am I one who serves the people who really need therapy most, I think, the people who can't afford maybe $50 a week, maybe, and who need the most help? Am I that person? I don't know. I think I'd like to be, but I don't actually know that I am."

Charles Jiang was a primary care physician who worked for a major tech company in Silicon Valley that offered his services as a perk to its staff, who did not have to pay anything for their clinical visits or for many of the medications they stocked on-site. While he now actually worked for a corporation, his situation resembled that of the retainer-for-hire workers: he could serve only the engineers and product managers of his Big Tech employer, but he could offer them longer appointments.

"Here it's thirty minutes. Which is a luxury—in my previous practice it was fifteen, or even less because sometimes we overbooked," he recalled. "A lot of organizations where you're seeing patients every fifteen minutes, you really don't even have time to develop relationships. You just are kind of getting through. I used to be two hours behind and that was OK. I felt really bad because some of my patients waited two hours. And then I felt really bad because they waited two hours and I would spend more time, which would put me further behind. I was just like, 'This is not sustainable.'"

Charles also loves being free of paperwork, since the in-house clinic does not bill insurance for either medications or in-house procedures. While "efficiency" was not necessarily a priority for his work, nonetheless he told me, "I would describe this kind of model as probably the most efficient model that you'll see in our healthcare system. When they come through the clinic, we don't actually bill the insurance, so that insurance is already removed out of the equation. And we have our own little pharmacy. I know that the patient gets the medication because I give it to them, on-site, so that saves a lot of time and administrative work."

Now he is free to establish the relationships he values, Charles said. The only hitch? Those relationships are a part of the modern company town. If they want to stay on as his patient, they can't leave his employer. "The only caveat is, if they ever leave [the company], then that relationship ends. I say, 'Yes, as long as you understand that if you ever leave this company, I don't practice elsewhere, so you can't find me anywhere else.' So, that's the only caveat."

In addition to the existential risk of asking oneself "Am I a servant?," the retainer-for-hire social architecture comes with other costs: in particular, the personal tie between employer and employee means that boundaries are few and far between. Indeed, the onus is on workers to establish boundaries, because their clients are inclined to believe there aren't any. That means these workers are also susceptible to their own unique forms of burnout. If burnout is what happens when something gets in the way of relationships serving as sources of purpose or dignity, the burnout of retainers for hire results not from time constraints or overload but from the erosion of meaning stemming from their existential doubt.

Ruthie was a little worried about her own recent burnout—not a good starting point for embarking on a new business, she observed—but she thought she would be all right because she has always found relationships with patients rejuvenating. "I'm doing it because I want the time with the patient," she said. "My real passion is my elderly patients. I love them. They want to talk, they want the stories, the connections, that are so desperately needed. That's what I'm really doing in medicine." Ruthie was exiting conventional medicine to practice doctoring the way she thought was necessary, via connective labor.

Her plan for direct primary care differed from the concierge doctors; in the first place, it was much cheaper, on a model like a gym membership, where people would pay $80 a month. She would see six to eight patients per day instead of twenty, and, with her spouse serving as an office manager, she could still make ends meet. "You are just back to you and the patient, not all of the billers, coders, and insurance people in between." She was locating her new practice in a small town about a half hour away, she said. "I picked it because that's where my patient population primarily is and I want my rural patient relations. The problem is," she said slowly, "they seem to be the least open to the idea. They're skeptical. They don't like the idea of paying a membership fee for something [like this]." Ironically, the wealthy, consumerist patients she was fleeing were more likely to be comfortable with the idea, being more used to paying for their gyms, she observed wryly. "So this may or may not work," she conceded. "It may not fly."

I contacted Ruthie periodically after she started her new practice. About a year later, she wrote me to say it was going very well. The tension between caring for and catering to her patients appeared not to apply to her direct service model, perhaps mitigated by its reduced fees and more rural environs. Most recently, the clinic website posted a sign that read: "Thank you for your interest. Unfortunately, our practice is now full. If you would like to join the wait list, please click on the 'Get Started' button to the right." Direct primary care was working for Ruthie. "I have time to provide quality care to my patients, and I am actually enjoying my work again!" she wrote me. "I should have made a change long ago."

The Hidden Stakes

Social architecture shapes what it is like to deliver connective labor, beyond simply the skills or the relationship of the worker and client. Three types—mission-driven, corporate, and personal service—dominate the landscape of how we organize connective labor, and yet there's a crucial mismatch here in what might be good for the practitioner, good for the client, and good for the public. The artisanal retinue-for-hire model might be best for workers, if they can get past the sense that they might be essentially a well-paid personal servant. The corporate model might offer the best service for the already advantaged patient or student, although it promotes a consumerist approach to relationship. The mission-driven model might be best for "the public," if only because they do not get much connective labor in other settings. Yet all the caveats start to outweigh everything else; these may be the choices that shape how we see the other and how we are seen, but none of these choices seem very good. And for workers, burnout is a problem in all three—albeit for different reasons, either due to the overwhelming demands placed on their practitioners that preclude the relationships that might sustain them, or due to their disaffection from the terms of their work, their service to the overserved. In each setting, the model distorts the connections we are able to forge.

I listened for hours to doctors, teachers, and others as they told me about the costs of working in these environments, but the stakes were higher than even they knew. Practitioners in mission-driven contexts may have been talking about hands on doorknobs, the impaired diagnostics, their constant sense of not being enough, but their list was incomplete. Mission-driven social architecture, as it is currently experienced in so many schools and clinics, means that many low-income people have little to no access to humane connective labor, and it is that incontrovertible fact that underlies the "better than nothing" ideology so prevalent among tech enthusiasts. Those who draw and redraw the automation frontier are looking to where the social architecture is failing us to justify the mechanization of connective labor.

Yet social architectures do not appear on their own. They are constructed of the choices of organizations and their stewards, bending to the twin pressures of market and state, prioritizing efficiency over relationship, and hoping it won't matter in the end. The anguish of the practitioners, the perfunctory or dismissive connective labor on offer—these costs are produced by the way we are organizing this work, and by the way we are allowing ourselves to be organized in it. There are other choices, other ways to respond to its degradation, besides its mechanization; chapter 8 delves into some of these alternatives.

Before we get there, however, we need to consider a core challenge facing connective labor in diverse contexts. While social architecture may differ in fundamental ways, standardization is common to many, spreading far and wide, dictated by employers but also called upon by practitioners under conditions of considerable overload generated by an industrial logic. While it can offer some relief, it also transforms this work, laying the groundwork for its automation and challenging the search for meaning. The next chapter will examine how people negotiate the ascendance of standardization as they accomplish the personal, tailored, particular work of seeing the other.

6

SYSTEMS COME FOR CONNECTIVE LABOR

Pamela Moore was an energetic African American woman who had left a demeaning job in retail customer service to become a teacher, after years of watching others teach her own kids. Three years later she was in a public middle school classroom in a large California city, teaching seventh graders during the day and taking classes at night to finish her certification. Students were drawn to her warm charisma from the start. "I have an open personality, so kids talk to me," she said, which meant she heard all about their troubled lives. "There were times that I cried upon getting to know them . . . just saying, 'Hey, what's going on?' and kids just wringing themselves out as if they were a sponge."

The school had many children who were "going through a lot," she said. Almost all the students at the middle school were non-white, with 86 percent socioeconomically disadvantaged; a quarter of the school were English-language learners. She told a story of one pair of brothers "who just totally broke me down," who showed up in smelly clothes because someone had actually been murdered in their home and no one had cleaned up the crime scene. "So, there's blood, and gore, in this child's home and he and his brother had to try to clean it up," she said: "How many eleven-year-old boys do you know [who] can clean up anything? Let alone . . ." her voice trailed off.

> The reason he told me was that I was like, "What is going on with your clothes? Man, I gotta be honest with you, you kind of smell, man. What's going on?" And he was like, "Well, I haven't been able to take a shower and stuff, we've been cleaning my house." I said, "Cleaning what?" I look down and his pant legs were slightly red, and the white parts of his shoes were pink because he had been walking in blood. So, I'm like, "Hey, let me get you some sweats, I'm going to wash your clothes." Luckily, we have a washer and a dryer on-site. So, I'm like "Throw these PE clothes on. We're going to wash these clothes." I held it together, had to call CPS [Child Protective Services], got our principal, and was like, "Hey, this is what's going on with this family. They need some support from the counseling department. This is all that's going on," and then absolutely bawled when I got home.

She experienced a lot of what she called "vicarious trauma" while teaching there, she said. "I can understand why my instructional coach probably thought I was fragile," she said. "I needed to cry."

The relationships she forged with the kids were certainly useful for Pamela, letting her know what interested her students, helping her see when their lives were overwhelming their capacity to learn, even just enabling her to keep them motivated so they would return to her classroom. But they were also powerfully meaningful for her, an opportunity for her to be the kind of teacher who was once there for her.

When she was in middle school, Pamela told me, a teacher took the time to find out that her selective mutism was not an indicator of her actual capabilities, but instead a response to her family moving incessantly. "Kids get us talking to them all the time, [but not] listening and that's what she did, just listened," Pamela remembered. "I could have been tested and put into Special Ed. Instead I was tested and put into Gifted and Talented." That relationship inspired her own exhaustive efforts to connect with her students. "Relationships are so important," she said. "I thought, 'I want to be that teacher for my kids in [this city].

I want to be the teacher that I wanted, and that I needed, and that I finally got.'"

She started in on that plan the day she was hired, with a huge stack of papers from kids who had been asked to write down what they wanted in a science teacher. "These were little messages from the students who would be my students. They were sixth graders at the time, [and] they were going to be my seventh graders," she said. "I read every single one of them, I went and took notes, and I was highlighting and I was like, 'OK, this kid likes this and wants this.' I knew that relationships were really important. [In] my own education experience, relationships meant everything."

Relationships with her students may have been her goal, but many of her job requirements made it actually harder for her to connect to them. "You've got a curriculum, you've got a hundred eighty days, you have a certain number of academic instructional hours, you have testing, testing, testing, testing. It's hard because we need to do the social and emotional stuff, because for some kids? The test doesn't matter," Pamela declared. "I have a little boy right now, it's not working. And I'm like, there's attention issues, his diet, he's got a girlfriend, he doesn't care about school anymore. So, we have to think about these things, [but] we're so focused on curriculum and testing sometimes that we forget about those things for kids."

If the previous chapter was about how we organize connective labor—the broader landscape of the institutions that offer it, and the hard choices workers often face between a social architecture that offers unsustainable but meaningful work or one that generates work that feels soulless or subservient—this chapter zeroes in on the everyday systems that shape it: the templates and technologies that affect how people connect, and how workers cope with the demand. And while some of the systems were vital to Pamela being able to do her work well, many got in the way of the relationships that made her job both possible and meaningful.

By "systems," I mean phenomena that we might gather under two main umbrellas: scripting and counting. Scripting includes standards and protocols, which attempt to outline best practices and abstract principles, as well as manuals, checklists, templates, and scripts. Counting, on the other hand, includes quantification and data analytics, often relying on technologies that interrupt the connective labor encounter to insist upon information gathering for other people, such as insurers, administrators, or engineers. Both of these trends often accompany campaigns for efficiency and productivity gains, which might involve restructuring workflows, squeezing appointments, or increasing rosters, in service to metrics external to the relationships that these providers forge. Both are often tied to funding, such as insurance reimbursements, Medicare dictates, or state funding for schools that are linked to approved protocols and data metrics. While these phenomena have different histories and impact, they are also broadly related, in that each involves a sort of standardization, an abstracting out from the close attention to the particularities of another person that the work of connective labor entails. They are also each pervasive, overlapping, and powerful, even as connective labor workers often (but not always) see them as detrimental to their work.

The tragic paradox is that the crushing overload facing these workers, and the systems they put in place to address it, stem from the same cause: an industrial approach to work that relies on relationship. Employers and policymakers have taken the central insights of Fordist management and applied them to connective labor jobs, pushing for measurement, efficiency, and productivity, even for work that is personal and emotional. Faced with such overload, practitioners turn for relief to systems that originate from the same industrial model, finding them sometimes helpful and effective, while often time-consuming and alienating. Yet above all they are not optional.

Indeed, it is not just that employers force practitioners to adopt scripting and counting, or that such practices are tied to private funding or public subsidy. It is also that the values of efficient productivity and marketized virtues have taken hold and spread indiscriminately, so that even practitioners who might disagree with the relevance of such values for their work feel they have to respond to their framing. The industrial model has become a cultural juggernaut, finding its way even into the talk of people who forge idiosyncratic, particularistic relationships for a living.

Overload and the Intensification of Connective Labor

The biggest impediment to warm, empathic connective labor is sheer overload. Sometimes—often—workers have too many clients with too much need, while they face too many other tasks crowding in on their relationships, and they end up with too little time to listen deeply, to meet people where they are and reflect their truths. As we saw in the previous chapter, overload is an outgrowth of inadequate staffing and scant funding that is part and parcel of the mission-driven social architecture, such as at the public school where Pamela worked. "They work teachers to death, absolutely to death," Pamela told me. "It's the hardest thing I've ever done, and I've been a parent! A lot of fun too, but there's a hundred fifty little people, a lot of personalities, getting to know me, getting to know them." Yet overload was also customary in corporate social architecture, where the drive for efficiency and productivity ratcheted up what workers were expected to do. In both contexts, systems were both problem and solution.

Nathan Weiss was a middle-aged white ex-physician who had left that career to become a therapist; medicine had changed in his lifetime, he said, becoming much more demanding for practitioners, particularly in primary care.

> It used to be fifteen, twenty years ago, before managed care, before then, you had time. People didn't show up [missing their appointments]. You were expected to see less patients. You didn't have metrics, you didn't have t's to cross, you just had time. So I think the reason there's so much dissatisfaction inside of medicine is the change. It was relational. I mean, these are sort of stereotypes, but they didn't have to work [all the] time, they had—people didn't show up, they had easy patients and tough patients. Now easy patients see the nurses, right? So there's this type of partition: if it's simple medically, you get the nurse to see it. So there's concentration of complex.[1]

Like the automobile factories of the early twentieth century, medical care had been transformed by drives for efficiency, with the component parts of each job analyzed, deconstructed, and redistributed to maximize productivity. The upshot has been an intensification of the physician's work.[2]

Overload stems in part from accountability pressures. In industries like education and health care, funding is increasingly tied to whether organizations meet particular measurable metrics, from the number of clients served—students, procedures, billable hours—to outcomes like "adequate yearly progress" for schools or whether healthcare providers meet certain goals. Yet while it may be the money that makes metrics and rankings matter, the spread of counting and scripting also reflects a massive cultural change. These are cultural values that signify prized characteristics like objectivity and rationality, and that encourage management decisions to prioritize metrics even in the face of uncertainty about what is being measured, whether it is measurable, and the impact of such measurement.

This transformation is happening not only in medicine but in mental health care, in education, and other fields. Nathan told me he was witnessing the same revolution at his therapy practice. Before, he explained, therapists "[were] just getting by, doing the best you [could]. There [were] not a lot of policies or procedures or attempts to maximize profit. You [were] just doing it and there [was] not this . . ." he trailed off. "Everybody was doing their best but there wasn't an external pressure to show anybody results. [Now] it seems like there's just a different mentality."

All of a sudden, he reported, the emphasis at the therapeutic clinic was on efficiency, measurement, and progress. Nathan recalled: "The whole board started to have this feeling like 'Holy shit, we've got to start making money and we have to start . . .'—It feels like a light switch where people are like, 'Holy shit, somebody's going to . . .'—I think it's like this feeling that somebody is holding you to account, if it's like funders, or I don't know, insurance companies, there's somebody suddenly holding you to account." And yet the standard to which people were being held to account was not a clinical one, he noted, or rather, it was a clinical one that a nontherapist could understand, meaning a thin facsimile of their actual work. "I mean, it's the same forces in every other sector," he said. "More and more and more performance and production." If the "before" involved therapy aimed at a clinical priority, the "after" involved therapy that was constantly measured, evaluated, and rationed.

This kind of intensification is squeezing relational approaches to care, workers told me. Rebecca Rooney was a biracial woman in her forties who was a family medicine practitioner. Primary care physicians have had to add billing to the very long list of what they are charged with overseeing, she told me.

> So, we have been the typists, the social worker, the pharmacist, the counselor, the nutritionist, the physical therapist, for our patients for many, many years. That's how it's evolved and we somehow have to be mindful of billing and the business of medicine as well. It just was a lot of pressure for us, and so we became very burned out. In addition to that, our patients come to us with—I actually have this great screenshot of my schedule and it has all of my patients. The very first one says, "Reason for visit—Everything." It's like, yeah, that's what we do in primary care: everything.

She laughed and said she knew exactly who that patient was. (See figure 6.1 for a copy of Rebecca's screenshot.) "It is true, when she comes, she comes for everything. And those are the most—They end up populating our schedule the most. So, how do we deal with that?" Billing was just one care too many, Rebecca said, when the scope of responsibilities for the primary care physician could be summed up in one patient's word: everything.

The squeeze on worker time is very familiar to people who work in retail or other commercial settings. Valerie Clausen was a hairdresser, for example, who was normally self-employed, renting a booth in a high-end salon in southwest Virginia. But for a brief period when she moved to follow her husband, she had to take a job in a corporate salon with strict time rules. "I swear, every night when I left there, I wanted to just stick a fork in my eye," she laughed.

●	8:40 AM		Primary Care Office Visit || everything
●	9:00 AM		Primary Care Office Visit || Repeat Pap smear
●	9:20 AM		f/u
	10:00 AM		3yr well child
	10:40 AM		Primary Care Office Visit || Discuss about ob/gyn referral
	11:00 AM		12 mo WCC/bump list
●	1:00 PM		Primary Care Extended Office Visit || Annual Exam
●	1:40 PM		f/u on left arm pain
	2:20 PM		Cpe with Pap/ conf clinic at || pt conf appt md sparc
	3:00 PM		Pt's husband scheduled appt, spoke with Triage already, has been vomitting :
●	3:20 PM		1 mo f/u
●	4:00 PM		f/u ok per Dr.

FIGURE 6.1. A primary care physician's daily schedule, with reason given for each visit; for the first visit: "everything."

> It was like, "Oh, yeah, this reminds me of why this is not what I want to do." A lot of times in that atmosphere, you're limited to— They set a timer and if you don't achieve the results of whatever the service is in this time window, you're penalized for it. But they also tell you, "OK, you can only talk for this long and if your service time is slowing down, then stop the talking and get to doing what you need to do." You know, sometimes you need to stop and make real live contact with somebody, not just through the mirror if you're into that deep of a conversation. And they really don't want that to happen and they don't give you the opportunity to build a buffer into your own time schedule. Everything put together is twenty-two minutes. That's all you got. It doesn't matter what. Twenty-two minutes.

The twenty-two-minute standard chased Valerie away from the corporate salon.

Time compression came for many who were ensnared by corporate bottom lines and bureaucratic exigencies of efficiency and productivity. We will see in chapter 8 how some clinics and schools are able to forge another path, finding ways to expand the time people can make for connective labor with ample staffing, long appointments, and continuous relationships. Without those possibilities in mind, however, employers and workers turned to systems to "scale up" connective labor, resorting to templates and technology to solve the problems of supply, but sometimes, paradoxically exacerbating worker overload.

The Benefits of Scripts and Standards

Proponents of standardization itemize an extensive list of potential benefits. Scripts, templates, or checklists help workers coordinate with others and also get through their tasks quickly, particularly helpful as work balloons, they maintain. In medicine, Atul Gawande may have made the checklist famous for its impact on improved surgical and ER outcomes, but checklists have also been increasingly applied to interaction with patients. Some doctors have instituted what they call a BATHE script, for example, which begins with "So, what's going on in your life?" (Background) and ends with "That must be very difficult for you" (Empathy). Billed as an efficient way of achieving patient rapport and compliance, the script also elicits important context for their symptoms, with advocates promising that doctors can get some of the primary benefits of relationship without—and this is key for its allure—requiring any more interaction time.[3]

Standardization can promote fairness and access, overcoming the old-boys network or other forms of traditional practice that sometimes enshrine inequities. Scripting can also protect practitioners from demanding or chaotic situations or clients—an insight that comes from researchers studying McDonald's—and it requires less of a "cognitive load" than continually responding in unique ways to the specificities of the other. In medicine, for example, "standing orders and protocols represent tools to ease cognitive burden for physicians and their staff," wrote one research report, which found that high-value medical practices (those with high-quality care and low spending) relied on such protocols to direct nonphysicians to handle uncomplicated acute and chronic diseases.[4]

Scripts also promised to improve performance, and not just for novices. I spoke to Zelda Loring, a therapist in the VA system, who told me how much she appreciated that her employers had adopted Interpersonal Therapy (IPT), an evidence-based approach that she considered more relational than other options that the VA allowed. "Although it is still an evidence-based treatment, it was the first one that I did that felt like, 'OK, this is actually a fit for me.' It's much more, it's kind of a hybrid.

It's much more flexible and there's a lot more room, I think, for creativity and using your clinical judgment."

Nonetheless, she said, it was less the creativity and more the script that she found helpful when the hospital system first adopted the IPT approach. One of the first tasks it requires is an "interpersonal inventory," where you spend a few sessions talking about the most important people in the patient's life, she explained. Even though she had more than a decade of experience, Zelda said the script enabled her to get vital information about a patient's father that she had actually missed in two years of therapy with the woman. "I, you know, over time kind of put together, like, we had this critical father that she has kind of distanced herself from, you know, [and] wants his approval," Zelda said. "[But] I had never gotten a full history of her relationship with her father, or really asked the targeted questions. I remember thinking like, 'Oh,' like, 'This is—It would have been really helpful if I had known about this two years ago.'"

Jenna Marchand, a pediatrician, compared the procedures at GroupCare, a large health maintenance organization where she had done her residency, with her own employer, a county hospital.

> GroupCare takes a lot of thinking out of it. You know in medicine we have this stuff about checklists and pathways, and so GroupCare is just religious about implementing pathways and checklists because there's always human error. Somebody missed the training. The old-timey doctors give a liquid form of albuterol and no one does that anymore. But if you've been practicing for fifty years, you wouldn't know. And so GroupCare kind of takes that away by forcing you into this. . . . In my hospital I'll still see kids that came from such and such a clinic where they're still getting liquid albuterol, and I'm like, "Oh my God, who is this person?"

By taking the thinking out of it, the standardization protected against human error, even by seasoned professionals, Jenna noted.

The capacity of templates to bolster performance was especially important for new practitioners, of course. I observed an adolescent therapy class for a semester with Russell Gray, the master therapist. His trainees—who were all seeing patients as part of the school's free clinic—told him they had been glad to take a class in cognitive behavioral therapy (CBT), a "manualized" approach that featured checklists and time limits, taught by an instructor named Tabitha, early on in the program. "Tabitha's class was helpful as a first therapy class," said one student. "There were a lot of manualized how-tos. It made me more comfortable being able to see clients."

Scripts helped people practice new ways of talking. Hank Rhodes was a military-vet-turned-hospital-chaplain who liked to learn new "modalities," or interventions, to try out with different clients. I asked him if he ever felt like he was using a script in his work. "When I'm trying something new, I do that all the time," he said. Lately, he had been learning Nonviolent Communication, a template that involved four steps (Observation, Feelings, Needs, Requests). "That's new for me," he said, reminding me about his background. "Military, martial arts, church, I mean, it's all about demanding, telling you what to do, no feelings," he said.

> I was feeling illiterate till six months ago. For a Marine to say "sad" or "scared"? I mean, I had to really do some soul-searching there, and prayer, and releasing stuff. So I stuck to that script. I listen for that script. And I do that in the [patient] rooms now, because I'm practicing. It's OK. I'm learning here. And I'll even tell the patient, "What I just said there didn't sound right, did it? It was a little awkward." And they said "yeah." And I said, "Do you mind if I try again?"

The scripts let him try it out before he knew what he was doing, Hank said.

Standardizing systems did not just involve scripts or techniques, however, but often also decreed how much time a given practice would take. For therapists, that meant limits on how many sessions as dictated by the CBT manual or paid for by insurance. While Russell Gray denounced those limits as not borne out by research ("We're mostly not curing people in thirteen weeks. We're doing thirteen weeks of therapy but [people are not cured]"), others found some utility in therapy that had a finite end. Zelda, of the VA hospital, noted that the time limits helped clear out her schedule. "Earlier on, I would take these patients and then just start seeing them forever, and we have all these people on wait lists and it was like I was full and sometimes it wasn't like there was the best use being made of the therapeutic hour. But they were my patients and they would become attached to me and I would become attached to them, and then if you don't have a frame for an ending, like, how do you end things?" Time limits introduced by a new system enabled her to end care that needed to end, Zelda thought.

The limits also helped focus patient efforts. "It's like 'We have twelve sessions to work, what do you want to focus on?'" said Sarah Merced, another VA therapist. "There's something about that structure that can be really containing for people, can help people show up in ways that they might not if it was just, like, this endless treatment." Zelda agreed, saying, "In IPT, time is both your friend and your enemy. It helps, you know, [when] 'We're halfway through and you still haven't been able to talk about your deceased mother, you know, that's something we're really going to need to talk about if your depression is going to have a chance of getting better.' It can be a way of pushing a little bit, but it's also your enemy in that: What if you're getting to this breakthrough and you don't have any sessions left?" The time limits pressured both parties in the clinical encounter, sometimes for good and sometimes for ill. Systems designed to improve therapist productivity sometimes improved client productivity as well.

Debating Standards in Therapy

But scripting has its detractors, plentiful among those who pursue a caring witnessing of other people for a living. A debate about standardization and relationships has been at the heart of a roiling conflict in the field of psychology for years. On the one end are those who aspire for their work to be less art and more science, subscribing to therapeutic approaches that are "manualized," often more scripted and time-delimited, and that lay claim to "evidence-based" results backed by randomized controlled trials and adopted by government systems and insurance companies; CBT is among the most influential of these methods. On the other end are those who prefer to be less structured in how they approach the patient, relying on approaches that emphasize relationship, emotion, and the connection between patient and practitioner; so-called attachment-based methods are typical.[5]

This debate came to life among the therapists-in-training I observed in the course on adolescent therapy. The students were learning both manualized and less manualized approaches on their way to becoming therapists, and Russell Gray, the course instructor, downplayed the differences among their faculty. "Ethan [another professor] came from CBT, I came from psychoanalytic. You can't tell we're different now," Russell told them. Still, he said, "I have an ambivalent relationship with manuals. All adolescent therapy is relationship therapy. If it's manualized, it's not going to be a good relationship. Relationship requires sensitivity to the other person, and that's very hard to script. If you're acting really scripted, no matter how good a script, that's a problem." Scripts got in the way of the relationships that were vital for good therapy with teenagers.

At the same time, he said, he understood the impetus to systematize the field. "Therapists have spent a lot of time doing things that don't work," he told the students. And for certain problems, manualized treatments were absolutely warranted. "If someone has a panic disorder, it would be malpractice not to do manualized therapy," he declared, referring to a series of CBT sessions that involve exposure exercises (such as going to crowded places or driving in traffic) and covering particular themes in talk therapy. "You can largely cure those in a short period of time. Thirteen sessions largely work for those," Russell said. Of course, he added, even in those situations therapists should always tweak the dictates to the specifics of the situation. "Session Three doesn't have to be Session Three."

Beyond a specific list of disorders, however, he thought that the emphasis on reproducible results had led the field, in tandem with major institutions and insurers, to turn away from other approaches that were more promising. He said: "It's as if we're looking for our keys in the parking lot, and we only look under the streetlight because that's where we see best." Therapists should be prepared to use different techniques for different patients, he maintained. But he cautioned against a certain dilettantism, saying, "None of this is the therapist who says, 'Well, I do a little of this and a little of that,'" urging the students to be "more thoughtful than that. This is about getting beyond looking under the streetlight, and just know there's other stuff out there."

Russell's warning about those therapists who "do a little of this and a little of that" echoed exactly what "science"-oriented therapists criticized about more unstructured styles. Gerard Juliano was an avid CBT advocate at the VA hospital who was scornful of more intuitive approaches. "To me, we have too much evidence. We know too much about the human mind in terms of the biases that we hold and how our own experiences color our interactions with other people to just fully trust our intuition. And if I have someone in my care, that is a huge responsibility, and I don't trust that completely to my own intuition and really personal judgment."

But Wanda Coombs, the African American therapist in private practice whose story opened chapter 4, was deeply suspicious of "evidence-based practice" because of what it erased about her own work. "As far as standardization, I think evidence-based practice is great. I also think it's like a way for like systems and bureaucracies to be able to determine what they will and will not pay for, and pigeonhole therapists away from like being their full and whole self with clients into doing a certain thing." Notwithstanding her prefatory comment ("I think evidence-based practice is great"), Wanda clearly did not actually think the standardization that often comes out of it was great, and gave full-throated defense of her artisanal approach to therapy. "We can't define what we do. We're just in it and we know what works and we develop what works for us, and then that's a way to work with clients."

For Wanda, as we saw in chapter 4, "what she did" included relying on intuition to know when she was getting it right. "I think in my training they were very clear that there is no standard way," she said. "Like, they said, 'We teach you this as if it's cookie-cutter and it's not.' Now I pretty much have, like, what I do, but then it's also . . . I just kind of see where things take me, and I'm very transparent with clients about that.

I just have to trust that I'm going in the right space, and if I stop and question [myself] like, 'Why am I asking [the client] these questions?,' I have no idea." While her intuitive approach could make it seem like she was stumbling, Wanda was a far cry from a novice; recall from chapter 4 how she knew that she had to establish trust early on with a client, and did so by helping them have some kind of epiphany in the first session. The distinctions between Wanda and Gerard—both therapists, but with diametrically opposed views—rested on their stances toward the templates, or scripts, dictated by manualized therapy.

Despite the ring of scientific validity that reverberates in the words "evidence-based," the conclusion to this debate is actually not obvious. While the efficacy of CBT and other methods are supported by research, particularly for specific disorders, there are also multiple meta-reviews that find that overall the particular kind of approach that the therapist uses does not matter so much as the strength of the therapeutic alliance—in other words, the connective labor on offer. Psychologists call this paradox—in which all therapeutic styles seem to achieve similar outcomes—the Dodo Bird Verdict, after the scene in *Alice in Wonderland* in which the dodo bird announces that "everybody has won and all must have prizes."[6]

Part of the problem is that the medical gold standard of the "randomized controlled trial" is simply not that workable for evaluating therapy, because it is ultimately an interaction, not a hypodermic needle subject to strict controls. As one review put it, clients influence therapists' behaviors "in subtle ways despite rigorous attempts to [standardize] interventions." The authors said that the goal of maintaining a pure version of a given style of therapy, regardless of client input, was "by its very nature flawed."[7]

Furthermore, this debate is relevant to more than just therapy. Many other fields are engaged in similar conversation, with advocates of systemic approaches facing off against those who approach their work as an artisanal craft. Indeed, these very same oppositions characterized the advent of scientific methods in bricklaying more than a century ago. The reckoning of these trends for interpersonal human work, however, is happening now.

The Problem with Scripts and Systems

As Russell Gray noted, some research has found that templates and scripts actually worsen connective labor work. One large study of more than fifteen thousand text conversations hosted by the Crisis Text Hotline service discovered that successful counselors respond to texters more creatively, and the degree to which they avoided generic or "templated" responses predicted whether the texter felt better after their exchange. Other studies suggest the scripting of interactive service work threatens innovation and autonomy, alienates workers from their own feeling, and demoralizes workers, draining the work of its meaning. Templates mandate a certain sameness, and to the degree that connective labor is about honoring another person's authentic feeling by recognizing it, scripts make that more difficult.[8]

Katya Moudry was a brand-new therapist, hired out of graduate school as an intake counselor by a busy county hospital in California. Part of her current job was to screen patients for mental health problems. The hospital assumed the task would take her fifteen minutes for each patient, and gave her a questionnaire to help the process along. But she grew to see both the clock and the survey as a form of violence.[9] "I'm the first person they're talking to about mental health, and we have to do some stupid questionnaire, and we have to ask about suicidal ideation. I've had somebody totally shut down during that part," Katya recalled. "I was doing a suicide risk assessment, and when it came to the gun part, he said, 'I'm not answering anymore.' I thought, 'Oh shit, I've lost him.' Whatever connection we had was just severed in that moment."

In a standard suicide risk assessment tool, the "gun part" is Question 14, which asks whether the person has taken any specific steps toward acquiring a gun, pills, or other means of harm, after a host of questions about thoughts of harming or killing oneself or others. Katya told me that some 80 percent of the people she saw in the triage area recorded a "yes" on Question 2 ("Have you ever had suicidal thoughts?"). "It's extremely high," she said. But instead of the fifteen minutes allotted, it often took her forty-five minutes or more to get their story. "It's not just 'So, what do you need? Let me connect you with that person,'" Katya said. "I'm the type of person who will be right there with them, who will be 100 percent with someone and feel their emotions. It is what I have to give to somebody."[10]

The survey that the hospital had tasked Katya with administering was designed to be a convenient, speedy means of tackling one of the clinician's more laborious tasks: translating someone's story into relevant categories for assessing potential treatments. It sought to help the listener identify what services the speaker would need, and to trigger a series of responses—an appointment or referral, potential medication. It also enabled the hospital to keep track of what kind of needs they were facing. But Katya felt that the questionnaire, and the pressure to complete it quickly, inhibited her capacity to respond to that patient's despair like a human being. Indeed its benefits depended on actions—going quickly through a series of prewritten questions—that posed a direct threat to the relationship she was trying to establish with the man by "seeing" him and his story, to reach him, to provide the solace of talking it out with someone on his side.

"How can you—I will just not follow typical protocol, when it's their first time they've approached mental health, and they're a male and they're crying, and just say to them—" she mimicked a perky singsong—"'OK, what services would you like? Come back in three weeks.' It's such a disservice to the vulnerability that they've expressed. If that means somebody has to wait a lot longer, I'd rather offer that full part of me rather than pump people through."

Katya objected to the survey as a form of scripting, a standardization of her work that was supposed to make it easier for her to finish her job faster, and she disliked how the survey reduced the complexities of the people she saw so as to better fit them in preexisting slots for service and treatment. The survey was also a form of data collection designed to generate measurable, comparable metrics about the man and his suicidal ideation ("How often has that occurred?") as well as about the stream of individuals—the 80 percent—of which he was a part. But for Katya, the survey exerted a corrosive impact on her capacity to see him well.

Even for new practitioners who were enthusiastic about them, templates sometimes got in the way of their learning how to listen. Carrie

Koppel was a therapist at a VA hospital, and she recalled hearing a trainee assiduously invoking the various steps of CBT, which is premised on the notion that patients are hindered by negative "core beliefs" about the world. "I had this one trainee and he was going to be like a neurobiological researcher, and I would see his tapes and then I would hear him down the hall because I would be here finishing up notes. I could hear him [exclaim], 'That's a core belief!'" she recalled. "He was just trying to pound the 'core belief' concepts into the patient. And I just knew that he was not going to . . . ," she trailed off. The zeal this novice had for the script impeded his capacity to see the other, she thought.

Standardization also reduced the capacity of seasoned professionals to use their judgment and expertise. When Sarah Merced, the therapist, talked about the mandated treatment plans dictated from above at the VA hospital, I was struck by her wistfulness. "I guess my wish as a clinician would be to be able to use some actual clinical thinking," she said. "Like, I went to school for many years to kind of figure out what best treatments work for most people, and so it would be lovely to, like, be able to choose when to use these tools and when not, versus, like, having them mandated."

Sometimes the combined impacts of standardization—with its reductive dictates removed from context, its priorities of time and efficiency, and its limits on professional expertise, autonomy, and discretion—led to a profound alienation among practitioners. Owen Erbudak was a hospitalist—a generalist physician assigned to a hospital rather than a clinic—at a large, prestigious academic institution. When he first started his job, he was living his dream, Owen said. "The first year I started attending, it was great. The job was amazing. I won awards for teaching and stuff. The first couple of years—I remember I'm walking to my car and I'm thinking, 'I'm getting paid for this. I hope they don't take this job away from me.'" But when the hospital gained new management, his job altered radically. "The hospital changed the focus," Owen said. "We used to be about patients and teaching. Now it's about money and discharge."

The new CEO—"who's never cared for a human being," Owen added darkly—put in systems that had a direct effect on Owen's daily work. "They decided that we need to come up with a more efficient way of getting patients out of the hospital," he said with bitterness. Efficiency experts observed the physicians and told them how they could improve. The hospital even distributed a corporate bible to the staff, Owen said disgustedly.

> They actually bought us books, the actual *Toyota Way*, and gave us all the books. I swear to God. I'm going to go burn it one of these days. I never read it. I gave it to my dad as a joke.
>
> Here's the beauty of it: they use the analogy of *trains*, where we come at a certain time and we leave at a certain time. The goal isn't about care. The goal is about discharge date and how do we meet that discharge date, so the goal is discharge, and to discharge them before noon. Discharge them as soon as possible. That's what it's about. It's not about better care, better outcomes, or any of that stuff.

But the *Toyota Way* was never going to work, Owen said. "The thing about this was they never could realize the reason why it didn't work. Because people are not cars. They were trying to standardize care for human beings. It was the most absolutely ridiculous thing I've ever seen."

Finally, scripts could make the work feel inhuman, not just by hindering relationships but also by making workers feel like automatons. Recall from chapter 3 the diabetes app that Helena Edwards was working on, aiming to relieve the telehealth service staffed by nurses. "What we hear: there's a ton of burnout. In large part, it's because they're just repeating the same thing over and over and over again and it's boring. They're nurses. They want to work on the more interesting cases."

While Helena's app is an example of "better than nothing" arguments advocating for socioemotional AI and apps, it is also evidence of how scripts act as a two-step trigger for automating connective labor. First, as soon as connective labor gets broken down into itemized steps—say, the scripts for diabetes care repeated again and again by the burned-out nurses, or the steps of a CBT protocol—those scripts can then be more easily entered into an algorithm to be delivered by an app or virtual agent. Scripts make it easier to mechanize the work.

Second, and perhaps more important, scripts serve to degrade the connective labor on offer. The more scripted the labor, the less meaningful the connection, especially if it comes across as canned or standardized. "People are trying to formalize this. You'll see this in a sad way. A lot of human interactions over help or chat, there's a veneer of relational work and [the lines are] being chosen by a human, and they're forced into some kind of 'Oh, I'm very sorry you're having this problem,'" said Geoffrey Janney, an engineer. "A lot of tasks are a person's job is to portray they care about you, where they don't. So the question becomes: To what extent does authenticity really matter?"

When the nurse's job is boring and repetitive, and when any connective labor they offer is scripted and performative, automation surely looks like an appealing alternative, especially if we somehow address concerns about job insecurity for the laid off. But it is a false choice, one generated by decisions made even earlier—to overload practitioners, to shrink the time they have for their tasks, to script their work so they can take on more—decisions that degrade the connective labor that the nurses are charged with giving. There is another path, one that involves respecting the power of connective labor to help motivate people to take care of themselves, as well as its capacity to create other social goods like shared dignity, purpose, and understanding. But if we opt to script this work, and to respond to workers' inevitable alienation with apps and AI, we cannot then be surprised by a depersonalization crisis.[11]

Scripts—the standardization of language, bodies, and time—makes robots of humans, which shortens the path to robotic connective labor. From a relational perspective, then, their use is a bit ironic. If overload is already a symptom of prioritizing efficiency over connecting, opting for templates and scripts to save time is acting like the seventeenth-century doctors who hoped to cure King Charles II's chronic gout with bloodletting: the cure resembles the disease.[12]

The Primacy of Data

Scripting may be one way that systems shape connective labor, but another important one is counting, or the data collection and analysis that is infiltrating even these interpersonal jobs. From the introduction of K-12 assessments by No Child Left Behind to the spread of the electronic health record (EHR) in medicine, practitioners felt like their work was dominated by the dominance of data as authority.

Like scripting, counting has its uses. Data collection allows for planning and evaluating, ideally enabling us to discern between good and better teachers, doctors, or other practitioners—although that discernment depends on not just whether data is collected but who controls the information and its transparency. Counting also lends greater legitimacy to many kinds of service work. Ultimately, counting is a human activity, just like relationship, argues Karen Levy, a Cornell University sociologist. "People constitute and enact their relations with one another through the use and exchange of data," Levy wrote. Counting is a human language of power, used by some people to make other people move.[13]

Yet some scholars argue that such data collection makes little sense when it comes to interpersonal work, because the connections that form the basis of teaching someone poetry or giving someone a massage are not easily measurable. Instead, inspired by the industrial model, employers and others assess outcomes, proxies that can themselves be difficult to measure or even identify: the degree to which someone can parse a poem or doesn't like their haircut or feels better after a massage does not easily translate into numbers either.

Furthermore, measurement—what David Beer called "metric power"—does not just calculate a standard; it also establishes one. Not being counted can render something invisible and unvalued, while sometimes being counted can be hazardous. Data analytics both reduce and exacerbate race and class inequalities, as sociologist Sarah Brayne has argued, noting that data-intensive police surveillance "widens the criminal justice dragnet unequally."[14]

Indeed, the simple beauty of measurement—and the trust people put into its reductions—can "change the very thing one is trying to measure," something known by anyone who has watched college administrators try to game the *US News & World Report* rankings. Deployed for reassurance, data streams paradoxically contribute to uncertainty, and "comfortably crowd out alternative ways of thinking and hustle conceptions of what is worthwhile." As a reading specialist told one research team, "I don't always know them [students] by face; I know them by data."[15]

Practitioners repeatedly told me that data collection and measurement, far from helping their work, quickly seemed to become the very point of it. "Boards or certifying bodies, or auditing bodies, just look at more and more and more different things," said Nathan Weiss, the ex-physician and therapist. "Everybody is just feverishly checking it all off and they have no time to actually look at the person."

Abraham Verghese, the physician and novelist, has written powerfully about how the reliance on technology can blind physicians to the body's messages. "Disease is easier to recognize than the individual with the disease, but recognition of the individual whose care is entrusted to us is vital to both parties. There are some simple rules: First, we must go to the bedside, for that is where the patient is. . . . It simply isn't possible for the patient to feel recognized and cared for when they feel unattended; the fact that their data is getting a lot of attention in a room full of computer monitors where doctors sit does not satisfy." The clinical exam was a lost art, Verghese lamented, while physicians were enchanted by the promise of data. "We are chained to the medical record, and every added keystroke adds another link in the chain. We must be unchained."[16]

For many doctors, the moment that the electronic health record (EHR) was adopted was a pivotal one. Introduced in 1972, the EHR has become standard procedure for most practicing physicians only in the last ten years; 86 percent of physicians reported using an EHR in 2017. Meanwhile, a 2016 study found that doctors spend two hours on the computer for every hour they spend with a patient, with one additional hour after work. In addition, when they are in the exam room, physicians spend about half the time on direct "clinical face time," and more than a third of the time on the computer. Research suggests the EHR is now a primary contributor to the burnout attested to by more than half of the primary care practitioners.[17]

Sadie Aronson's father, a small-town doctor, was one of those. Sadie, now a family medicine physician, remembered listening to tapes about foot fungus in the car on family road trips, bearing witness to her father's devotion to his practice. Yet even he eventually left the field, chased out by its corporatization; the proximate cause was the EHR, she said. "It destroyed his concept of what it meant to be a doctor. He was no longer able to pay attention to his patients, and he felt this very clear separation between what he had to do, which was documentation in this medical record that didn't make sense and didn't work, and then also continuing his relationship with his patients and what that meant and how that affected his joy of practice, and it just destroyed his joy."

What the record did was take a patient's story and reduce it to individual parts, said Ruthie Carlson, whose journey from mountain clinic to direct primary care was chronicled in the previous chapter. "It's all rote and there's nothing personal about that document anymore. There was a story and there was a point to making a story. It's totally gone now," she said. "Instead of being a tool to facilitate patient care, it's become an absolute forest for us to get lost in."

Collecting the data—in such tiny, unrelated component parts—not only crowded out the time to connect but also limited the clinician's capacity to think holistically, said Nathan Weiss, the ex-physician and therapist.

> It's like the more information there is in medicine, more drugs, more tests, more algorithms, and the more measurement there is essentially . . . physicians, their consciousness goes towards that. So, the agenda, because they're being measured, just gets broader and broader. There's more—If you went to an inpatient nursing document, it's like: if you printed it out it'd be like this long [*he stretched his arms far apart*]. Right? All the things they have to check off. Like, "don't have ulcers," you know, "isn't holding a gun," "isn't—" It's just ridiculous, it's just, like, every possible thing that could go wrong.

The outcome was a morass of unintegrated data. As a result, the work of the primary care physician went from approaching the patient as a whole, in which the individual components of a checkup were an indivisible ritual, to an itemizing of increments as small as rabbit pellets. Along the way, the patient's particularities, and what the physician did to assess them, were standardized. "[Even] if there wasn't anything abnormal,

I still have to document all these normal [tasks completed], like, 'What did I do?'" said Simon, the primary care physician who in the previous chapter decried the "hamster wheel" of fifteen-minute appointments. "'Did I check their ears?' 'Ears, normal,' you know, 'mouth, normal,' you know, all that has to be documented, so that you get credit for 'I did that.'"

The impact of this system was devastating for someone who values relationship, Simon said. "It'd be one thing if it was just the joy of interacting with patients and then there wasn't also this 'Oh, my gosh, I've got to do a hundred different tasks as a result of this visit and get it documented and entered into the electronic health system.' I mean, all of that is really, I would just say, is tangential to the actual role of the physician, as healer and diagnostician." Simon was very concerned with how the primacy of data crowded out what was more essential for his job. But the problem is not just that the EHR adds to the physicians' workday, he said. Rather, it is the combination of the exhaustive effort required for effective connective labor with patients, intermittently interrupted by mandatory work processes they don't believe in.

> It's that the relationship and the role that you're in is really exhausting. It takes sort of a hundred percent of your concentration, and then to throw on this—So, it's one thing if you control that and you're like, 'OK, I want to be doing this,' and you're realizing it's exhausting but you're getting immediate joy, and you can be present.
>
> But then imagine it's like you had, you know, psychological dog experiments where, like, you're getting shocks throughout that. Like, you're having all these things that you don't actually value that someone else's controlling and you have to do, like, oh, you need to bill this or the computer system is really a distraction and unwanted and something outside of your control.

In this recounting, there are actually several different themes, including the exhausting effort it takes to provide connective labor well, the reward of "immediate joy" that makes it sustaining, the sense of time scarcity, and finally the frustration for this elite professional of not being in control of his own activities. The result, he says, is that he feels like he's in some sort of cruel experiment, as an animal getting periodically and unpredictably shocked.

Simon was bitter when he remembered the EHR's early promise. "It's like we have this new technology, this electronic health record that will make everything more, you know, standardized, streamlined, and the reality is it, from our perspective, from the cynical perspective, is really just driven by billing. You want it billed," he said. "So, yes, it's probably helped the system procure more money . . . and I mean, some of it's medically necessary billing, but it's not in the main scope of what physicians would embrace and why they're doing the job, which is, you know, the relationship with the patient."

The irony was that later cohorts of physicians did not know what was missing, Ruthie said. "The new docs coming in don't recognize the change. That's what they've learned on. They don't see what's missing. The newer docs that we've hired are probably seeing twenty to twenty-four patients. They're billing higher than any of the rest of us. They're getting out early and going home to family. They're using templates and quick phrases to meet the demands of what's asked of them." The system was producing a particular way of being a primary care physician—one that her description perhaps inadvertently made sound more viable for two-income couples and physicians with "second-shift" duties at home—while those who did not fit exited.[18]

The data revolution extended well beyond medicine. Teachers' antipathy to the new priorities of data gathering, which surged after legislation linked school monies to standardized testing, is well known; most research finds that "increased testing and teacher accountability have altered teacher–student relationships and reduced the possibilities for teachers to develop and nurture caring relationships with their students."[19]

Even some therapists were having to reckon with the data imperative. I sat in on periodic meetings of a group of therapists in a VA hospital in a West Coast city, and one day the discussion centered on people's data collection practices. "We have a lot of boxes, no, literally, that we have to click on," said Carrie Koppel, "questions that we have to ask them, so the computer is telling me to ask about smoking and alcohol and suicide." "I know some people do their work on the computer while the patient is right there," another therapist offered. "I do my documentation on Sundays," Carrie added. "I'll usually stay late, but now I'm trying to carve out time in the day," someone else said. "It's a work in progress." "Same for me," chimed in another. "I'm not able to take on too many clients because of the limited time I have to write the notes." The unit manager, George, noted that "some of the guidelines were decided based on medicine, not mental health, and they affect our ability to be as patient-centered as we want." Sandy, a social worker, made the distinction. "There's the best-practices box, and then there's the bureaucratic box [of stuff we have to do]."

Some therapists were up in arms about another form of data gathering that was creeping through the industry. While doctors complained about data collection in the EHR, they were often resigned to another kind of data: the customer satisfaction scores that were becoming commonplace. But a small movement to incorporate satisfaction data had begun in some therapeutic practices, and clinicians were revolting against it. Many refused, reflecting their professional power. Brian Jacobsen was a therapist who had trained with a clinical practice whose management attempted to collect customer satisfaction surveys.

> Because a funder wanted more measurable results, they instituted a very CBT kind of thing, where the patient was supposed to fill out this, you know, "circle from 1 to 5" on a form on these four axes before the session, and a different form after. And it was like, "Do you feel like you addressed the thing you wanted to? Do you feel like your therapist listened to you? Do you feel—" I can't remember. And that you would score blah, blah, blah.

In response, "there was like a revolt," he said, "and some of the supervisors said they wouldn't supervise there anymore." The clinic backed down; in some settings, clearly, therapists still had the professional autonomy to refuse to participate.[20]

I asked Brian to articulate what the problem was with the form. It was not necessarily the questions, Brian said, it was how they were being asked. He labored to explain. "Well, how honest are—You know, you're going to grade me after the—like—or you're going to be—You know, it's like, I want it and I don't want it on the form. I want it in the relationship." His convoluted speech reflects some ambivalence here—does he actually want to be graded?—but he ends with a definitive answer: the problem is that the answers are being treated like assessment data, produced for an external observer (the funder) as opposed to feedback within the confines of the therapeutic alliance.

"I think there's so much stuff that you can't quantify," said Wanda Coombs. "I know that I do good work with my clients. I can't give you a way to measure that." Some practitioners viewed quantification askance, suspicious of the priority it made of managing data—collecting it, cleaning it, analyzing it—that they feared would interfere with the reflective resonance that connective labor provided.

At best, scripting and counting reduced messy individual facts to more abstract principles that could help the practitioners—or their supervisors or funders—figure out what needed to be done and to do it. While sometimes improving access or training, however, they risked making workers worse at their job (e.g., by impeding sensitivity and innovation, or intruding on their time or concentration) or making the job worse for them (e.g., by increasing alienation, adding to their workday, or introducing a new ethos of client-as-consumer). Most important, standardizing practices had direct impact on the relationships practitioners could forge, directing their attention in particular ways and sometimes getting in the way of their connective labor. At its worst, standardization was a tool for managing overload that itself contributed to it, frequently taking over the encounter altogether to radically reshape and curtail the way people could bear witness to each other.

How Workers Cope with Overload

Every morning, Pamela, the charismatic middle school teacher, relied on a trick to ascertain her students' moods quickly and efficiently. "I check in with my students, we do something called an SEL [social-emotional learning] check-in and it's our 'Do Now' in the morning," she explained. "I have them write about how their day, their week was, their weekend. And I say, 'Hey, does anyone want to share out?' A couple of kids share out and then I say, 'OK, on a fist to five, fist being the worst day ever—like, apocalypse—and five being 'woohoo, I'm going to Disneyland,' just quietly let me know: Are you a two, a one? Where are you today?' And I let them know where I am."[21]

Pamela said sometimes it's been a bad day for her, and she shares that with the kids also. In those cases, she said,

> I'm like, "To be honest, today I'm a two. I didn't get a lot of sleep, the kids were really cranky," whatever is going on. "I'm having a rough day; can you be a little kind to me?" And then I look at the kids and I say, "Look around our room, look at the hands that are up. If you see a classmate who is a two or lower, can you be kind to them? Be gentle, give them a smile if they don't have one, give them one of yours. So, you know, be kind to each other, OK? We're a community."

Practitioners struggling with overload, their work stretching way beyond the end of the day, often criticized the systems they thought contributed to it. Yet those same workers sometimes implemented their own scripting and counting to manage the increasing need they faced. Coping often meant using some of the very tools that they railed against, an adoption of templates, techniques, and technologies as their commitment to relationship collided with the burden their employers assigned to them.

For Pamela, the fist-to-five check-in sometimes got surprising results. One day, a student raised her hand and said, "'I'm not feeling it right now, I'm feeling kind of depressed,'" Pamela recalled. "So, I said, 'OK, let us get started on our lab and I'll come check in with you.' And then I got the lab started and checked in with her [later], and she had been cutting herself." At this point in the narrative, Pamela stopped and looked at me, reliving the shock. "And she told you?" I asked. "She told me," she said, and paused. "That was really hard, right?" She remembered the moment:

> I asked [the other students], "Like hey, do you mind if we take a moment?" I didn't—I tell all the kids, "I'm a mandated reporter, I have to for my job, and then as a human being I must get you help. I must

> make sure that you're OK." So, I said that I was going to have to tell somebody, to tell a counselor to get [her] some help. And I said, "We can go outside, I can have another teacher come, and we can go to the nurse's office." She'd been cutting herself places that she could cover up. On her arms, on her hips because no one could really see that.

Pamela was shocked to discover the cutting, yet another example of her students' extreme challenges. Yet she may never have known about it at all had she not established the fist-to-five system and the language it gave her students to reach out.

Quantifying information in this manner—"Are you a two, a one? Where are you today?"—is a fairly popular teacher shortcut, an "SEL check-in" that brings a certain efficiency to detecting mood among some number of students quickly, even though it likely misses any who refuse to share in such a public way. Proponents of the SEL check-in say that it builds student awareness of their own feelings, as the first step toward being able to regulate them, that it normalizes all kinds of feelings, and that it builds community. While Pamela wants to know how students are feeling, classroom time is compressed, with the school assessed in part by how the students perform on statewide tests of a mandated curriculum that is not about recognizing or regulating feelings. So Pamela needs a fast way to elicit emotional information—and to communicate that she sees her students—without diverting the class from their primary learning tasks. Pamela also uses the tool to create some emotional meaning, to convey that she sees her students as whole people, and to establish the terms of her relationship with them, which she views as extending beyond the classroom.

Yet Pamela's check-in is also an instrument of standardization. While the fist-to-five check-in "normalizes" feelings, it does so by ensuring that they are made legible to others, both qualitatively—"I'm feeling kind of depressed"—and quantitatively—"To be honest, today I'm a two." The messy inarticulate particularity of feeling is rendered smooth and modular, communicable through a common language, and "measurable" by common metrics. Critics consider SEL curriculum as a way for schools to control student emotions, excluding violent or intense ones as inappropriate and those who feel them as emotionally illiterate, even when they might stem from violent or intense circumstances. Ultimately, these critics argue, SEL programs assign the children individual responsibility for feelings that arise out of socially structured inequalities.[22]

Furthermore, Pamela would not be doing it at all if she had more time or fewer students. The fist-to-five check-in reflects her competing mandates of doing connective labor within a mission-driven social architecture: she simultaneously tries to elicit feelings quickly while subordinating them to meet the dictates of her employer. Many contexts feature similar approaches. I observed Karen Ellsworth, a young education professor, teaching graduate students to become school counselors; she showed a "Feelings Wheel" to her trainees so that they could better reflect client moods, with a hundred thirty labeled feelings to choose from (see figure 6.2). Medical professionals commonly direct patients to identify their pain levels on a pain scale chart; one chronic pain patient told me that far from using the same number to mean the same degree of pain, she modified the number depending on the practitioner, their demeanor, and what they understood about her complex medical history. "I don't even know what that number means," she said. Abstract measurements of feeling happen in all kinds of contexts, masking similar uncertainties.

Providers also adopted other systemic approaches for coping. Pamela had only been teaching three years when I talked to her, but she had already amassed a host of techniques for connecting with "a hundred fifty little people, a lot of personalities." Something she called "the 10:30" was yet another tool in her capacious kit of tips and tricks. For ten days in a row, the teacher and individual child have a thirty-second conversation that is not about the official curriculum, nor is it a reprimand, "not about your behavior, not about you forgetting your homework, cussing at someone," she said. "So, the next day, if Trevon says that he likes football, 'Hey man, did you see the game, did you see the Raiders yesterday? Awesome game, right?' As he's coming in the classroom, or as he's getting something out," she explained. She was enthusiastic about the trick, but as I listened, I found myself stuck on how all

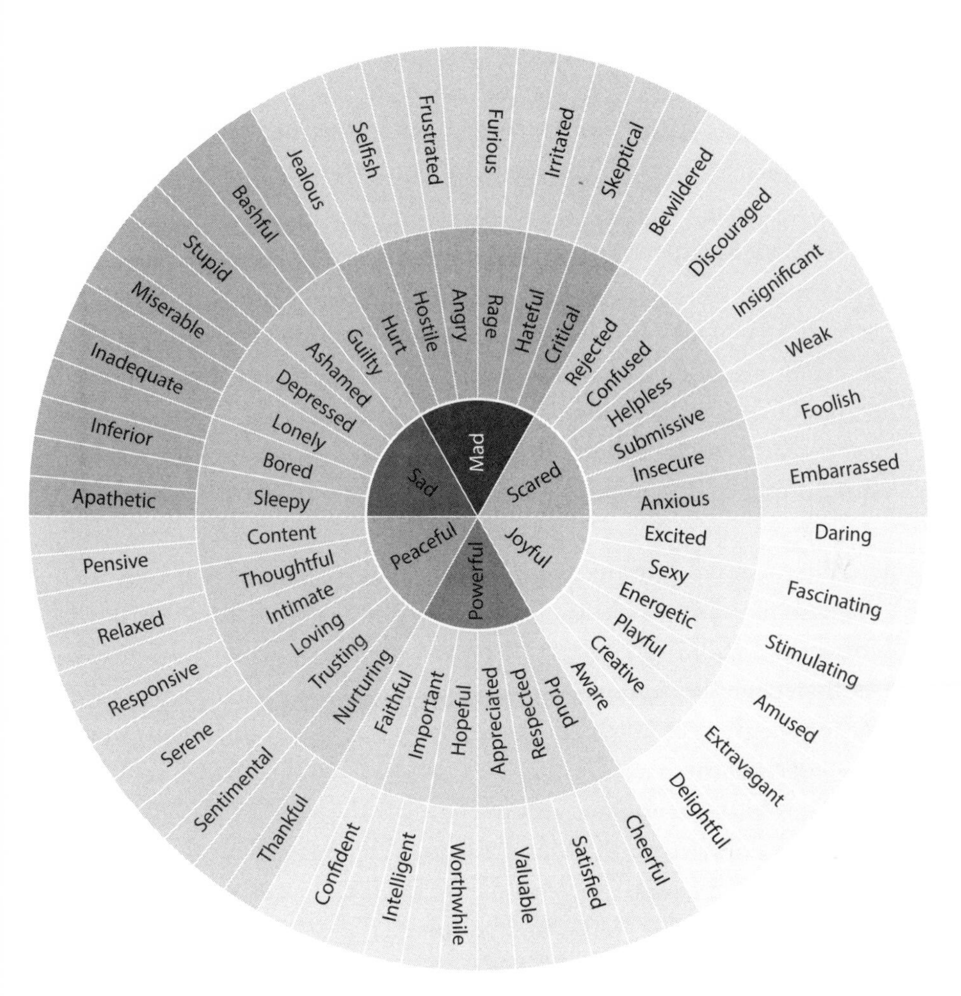

FIGURE 6.2. The Feeling Wheel.
Source: https://allthefeelz.app/cc/feeling-wheel/. This image is licensed under the Creative Commons Attribution-Share Alike 4.0 International license.

those thirty seconds could add up, for teachers who had a hundred fifty students to attend to.

It was not for all hundred fifty students, Pamela assured me. "Typically, it's for those kids that have had the behavior that's been like completely off the track or off the rails," she said. It's a way of "getting them back. If you are reprimanding them, [the student will feel,] 'You don't like me, you don't care, you've reprimanded me.' But if you come back to the classroom and there's something you need at school and there's something you need from me, and I need to reinforce that, I need to rebuild the relationship. And [convey that 'I] like you, and you're valuable.'" She may have had a large number of students, but that did not mitigate the importance she put on relationship.

Yet a system that relies on individual heroism is one that does not value connective labor, because at some point the heroes will run out. Most practitioners I spoke to engaged in some combination of scripting and counting, often dictated by their employer and sometimes deployed on their own. Despite the tricks, however, the overload was always hovering nearby, because for many, ultimately, connecting relied on their sheer presence, or the combination of time and attention to the other, and the practitioners were themselves a finite resource.

When I asked Mutulu Acoli, who worked with high-need kids in a different middle school, how he managed to connect with students, his answer was about extending that presence. "Extracurriculars, going to see their counselors, going to see their games, talking to them about, like, having them in for lunch and going beyond the teaching realm to get more of the individual," he said. He coached basketball after school. "If teachers didn't—if teachers only interacted with [kids] in the classroom, even with the systems, those probably wouldn't have relationships with half the kids," he said. "So you do have to go that extra step to kind of figure out what other things they're doing."

"You have to see them in other environments to get a fuller picture of them," said Ken Margolis, who taught sixth grade at another public school. "I love seeing them in class too, but it's just nice to see them outside of class because you get more of their personality." I asked him for an example.

> There's like a boy who is very, like, super flamboyant. You would never know that in class. He's just, like, kind of quiet and to himself, and then he gets into the dance studio, and his movements are big, and he's kick—you know, jumping up and touching his toes. It's just, like, I would never see that. It's also true with some of the students who, like, there's multiple Asian girls who are, like, very quiet

> in class, but they are so loud with their friends, and it's like, if you don't go outside and if you don't see them outside of class, you wouldn't see that.

What does he do with that fuller picture? "Then it just allows for—It's just further connection," he said.

His school was extraordinarily diverse, he said, with kids speaking different languages, kids of different ethnicities, "kids on free and reduced lunch, kids who, you know, like, their parents own wineries." When he said he felt very lucky to work there, I noted that many teachers might comment on how hard that combination was. "Oh, I am struck with how hard it is," he said. "My job is impossible. But at least it's impossible in a beautiful way, I think, or in this ideal way."

Ken had a lot of leadership roles in the school, including heading up the History Department and the sixth-grade "family," and serving on the school's instructional leadership team—all responsibilities that were part of his regular job. But he also taught dance to kids at the end of the school day. At first, "it was like an idea of equity. Like I could give these kids something that the other kids get for free," he said. It took six years to raise money to outfit the space with mirrors and a stereo, which was "horrible." Ken's devotion to his students, and his all-hours commitment, was eventually too much. He ended up getting burnt out and clinically depressed. Now on medication, he was careful to preserve boundaries with the students and designate time for his own exercise. Nonetheless, he had returned to teaching dance, he said, because he loved seeing them outside of class. Even at the risk of his mental health, Ken took on extra to forge those connections.

Like Ken's dance classes, Pamela was inspired by equity to plan the two-day camping trip for her students. "It's an experience the kids haven't had," she said. But the cost of all that connecting, even with her bag of tricks like the 10:30 and the fist-to-five check-in, might have been too high for Pamela. Toward the end of our interview, Pamela told me that she had been recently diagnosed with multiple sclerosis. "I was dropping things, I started to trip, and I'd joke about it but I was getting really concerned," she said. "Sometimes especially when I'm tired, I forget words. And luckily, I can kind of play it off with students, I'm like 'So, that vocabulary word, that word . . .' and I can point to the word wall and the kids say 'Membrane!' and I'm like 'there you go!'" But the doctor had told her that MS was aggravated by stress, she said, confessing that she had "put out feelers" for a transfer. She had been running to stave off overload, but she might have to hop off the treadmill.

Other workers withdrew before taking on the extreme demands, by checking out emotionally or actually exiting the job. "Well, it's, yeah, it's just using people up," said Simon, the primary care physician. "So people—every person finds their way to survive. I have colleagues who, like, do very little charting and they leave as soon as they can and, you know, pay is good and they have a secure job and that's it." While he did not love the charting, he saw those colleagues as retreating from the core demands of the job, by declining to go the extra mile for people. "For me, that's not the joy of seeing patients."

Even before the pandemic, Birdie Mueller, a nurse practitioner, said overload was getting worse in her community clinic.

> So, we just had a very tense conversation that we have, you know, budgetary constraints and we need to see more—It's the same everywhere, we need to see more patients. We need to see more patients and we get a set number of hours as admin hours because not only are we seeing patients every fifteen minutes, but then I have to review all their labs, you know, all their specialty visits, all their X-rays. I'm getting tasks of, like, "They need this—this person has this issue, can they manage this, et cetera." So, there's just a never-ending supply and list and so, no, there's not enough time to do that. And then you're charting like twenty minutes of writing all these things and clicking all that, and so it's—there's a lot of burnout for some people, to kind of do that.

The organization of connective labor can make overload acute, and burnout a common outcome.

Widespread burnout and exits by highly trained and competent professionals is a major cost to the way we organize connective labor. In medicine, for example, a 2017 federal study found that "more than half of primary care physicians report feeling stressed because of time pressures and other work conditions." Nearly a third said they needed 50 percent more time for physical exams and follow-up care. The report concluded: "Work conditions, such as time pressure, chaotic environments, low control over work pace, and unfavorable organizational culture, were strongly associated with physicians' feelings of dissatisfaction, stress, burnout, and intent to leave the practice."[23]

COVID increased these burdens; an Urban Institute study found that primary care physicians noted "rising levels of burnout and exhaustion for . . . themselves and their staff." In 2022, an international review reported that compared to other healthcare workers, primary care doctors reported higher levels of stress, depression, and burnout, and lower job satisfaction than other specialty groups. Meanwhile at least half of physicians aged fifty-five years or older said they would stop seeing patients in the next three years, according to a Commonwealth Fund survey. The story for teachers is not better; a 2022 Gallup poll stated that 44 percent of K-12 teachers reported burnout. The crush of patients and the tasks that accompany their care threatened to overwhelm providers, and for some, pushed them out of the field.[24]

The Cultural Juggernaut of the Industrial Model

Accompanying the spread of scripting and counting, the industrial model has become a cultural juggernaut, dominating the ways that people think and talk about value. Even connective labor practitioners sometimes find themselves trapped within its interpretive framework, forced to frame their contributions and constraints, however reluctantly, with its alien grammar.

This sort of influence was evident when I talked to Alan Krupner, the head of the chaplain group. As Erin demonstrated in chapter 1, the chaplains were told they had to enter the data from their visits into three separate platforms, and I wasn't sure why they had to enter into any at all, let alone three. What did spiritual life have to do with the EHR? I wondered. Alan had some answers.

The most important reason, Alan said, was that someday the hospital was going to decide the future of the chaplain service, and data supporting what they did would help chaplains have a hand in steering their own future. He said:

> I felt like chaplains needed to learn how to count something, because we live in a world increasingly driven by evidence. And one day somebody's going to tell us what the evidence points to, and "This is what you're going to do." The question is, Are we going to be the ones who say these things or is someone else going to tell us that "this is what you're going to do, and we drive the bus, so you're going to do what we tell you to"? Because the work is being done.
>
> Now there's no malice in it, but if we're not interested in working with health care in ways that health care can appreciate, interact with, and understand, that's OK, they'll go on. They'll go on without. But eventually they're going to come up with something that if it includes us, they're going to tell us what to do. If it doesn't include us, guess what? We're either going to be gone or we're going to be this scarcer resource.

Data, and its quantification, was the only way that chaplains were going to be able to be a part of an inevitable evidence-based future, Alan said, which required that chaplains learn "to count something."

While he was enthusiastic about promoting chaplain note-taking in the EHR and other platforms, Alan was not naïve about whether it would make a difference in that future. "Now, to the healthcare system, there is no such thing as a spiritual care goal affecting a discharge outcome," he said. "So I don't know that us coming up with outcomes the way that health care can understand it will have much effect." But it was possible that the system could someday recognize the impact of "psycho-social-spiritual influences," he said, and that the chaplains might be seen as important for that. "But to do that," he said, "we need to get over ourselves."

By "get over ourselves," Alan meant that chaplains had to stop believing in "the uniqueness that every patient encounter is special." He continued: "Every nurse will tell you the same thing, every doctor will tell you the same thing, that every patient encounter is special and unique. But at the same time they've figured out some way to recognize the similarities in the parameters of their care, and be able to identify and articulate those parameters of care, and talk about these unique aspects inside of the parameters. I think we need to do that." The standardization required for the EHR applied just as well to a patient's spiritual distress as to their physical disease and treatment, Alan thought.

Alan's argument is one thing—a statement about the inevitability of quantitative data deciding the place of a spiritual team in the hospital that attests to medicine's institutional muscle, as well as the intellectual hegemony of evidence-based thinking in his world. But his language—chaplains need "to learn how to count something" and "we need to get over ourselves," with its casual mockery about chaplains as number-averse feelers—is another: evidence of a certain cultural supremacy of standardization.

This cultural dominance was also evident, I thought, in the acquiescence of many professionals to their own overload. Despite the steady drumbeat of desperation about time from the physicians and allied professionals with whom I spoke, I was also struck by how many seemed resigned to it. I often observed the Connect group, a weekly meeting of about ten physicians and psychologists who shared the goal of making medicine more humanistic; I joined the group for eight months.

In one session, an oncologist named Layla Shoshani gave an extended presentation about the time her own mother had spent in the hospital. Due to her condition, her mother lay prone on the bed, on which she had to be rotated periodically. "It became my family's duty for people to know who my mom was, that she was not a rotisserie chicken," Layla said. "My family and I gave my mom a face—via pictures, via our presence. I knew my mom needed to be a person in everyone's eyes. Otherwise she would be a woman with an unusual name while this [string] of people came in and out. We made that waiting room our living room. We brought a box of pictures, and later, people kept remembering the sheep of her childhood." Layla's presentation took most of the session, and she highlighted some dehumanizing interactions her mother had experienced. Most of all, she said, the goal should be to figure out how to make a person of the patient, as well as out of the caregiver, and she decried the hospital's practice of putting patients in tents and hallways when they were oversubscribed. She concluded with the concept of dignity therapy, a brainchild of Canadian psychiatrist Harvey Chochinov, who urged physicians to ask their patients: "What should I know about you as a person to help me take the best care of you that I can?"[25]

After she was done, the Connect group focused first on the tents and hallways problem. Owen, the hospitalist, acknowledged that "the hospital is over capacity and it does affect the patient-physician relationship." But when it came to thinking about how to make their care more personal, time was a limiting condition they all could agree on. The group wondered whether there was a way to put the patients' hobbies in the EPIC electronic health record system. "From a mindfulness perspective, we can't be mindful eighteen hours a day," Layla conceded, while the group nodded. "We need to think of things that we can do that don't take a lot of time, because time is our main issue. The 'what should I know about you' question, that one can really help." The cultural dominance of efficiency and productivity led even these physicians, so dedicated to humane and patient-centered practice, to limit their vision of what was possible.

Finally, even Russell Gray, the master therapist so dedicated to teaching his trainees how to have a relational practice, gave some evidence of the cultural juggernaut. At one point he offered guidance to his students about how to handle teenagers in therapy. Speaking from more than thirty years of experience, he said: "What you do with adolescents, you don't want to talk about what's making them depressed." Instead you talk in metaphors, he said. He described his sessions with a teenager who was being abused at home; they spent much of their time talking about how to handle it when her friends would not pay for gas when she gave them rides, he said. Talking about it was analogous to her situation at home, in which she was "dealing with how you are used by people who don't see you. She's looking for a way of talking about the feelings that's generated by the abuse that's safer." He relied on her affect to tell him whether his oblique references were right, he said. "If you assume she's talking about the abuse, you can actually talk them through the abuse quite effectively. I could say things that were emotionally about abuse, and her face lit up. My statements would have seemed odd if we were just talking about cars. She's got these feelings about a topic that is too threatening to talk about."

Most important, he told the students, you have to be careful not to push the teenagers. "You've got fifty minutes. Don't feel like every minute you have Medicaid looking over your shoulder saying, 'What are you doing? That's not therapy, that's basketball.' The bulb's gotta want to change, you gotta have the adolescent on board with you."

Then he added: "This is not manualized, it's not empirically validated, but it works," he said, and concluded: "Take it for what it is worth."

It was the "Take it for what it is worth" and the defensive cast to "This is not manualized, it's not empirically validated, but it works," that I found more than a little heartbreaking in how Russell framed his advice. The language of evidence-based methods even served to attenuate the claim of an expert therapist, in the very moment that he offered unique and valuable guidance, based on his extensive clinical experience, to his own trainees, who were expressly there to learn from his wisdom. "Take it for what it is worth," he cautioned, as though it were not worth much, even though it was clear that it was worth a lot.

The degree to which systems—checklists, memory prompts, protocols—are welcomed is evidence that they alleviate some of the worker's terrible burden. For Pamela Moore, for example, within severe time constraints that she did not perceive as changeable, the fist-to-five check-in served as a way to siphon her connective labor, to direct her efforts to see and bear witness to her students' emotional struggles. The tool came at some cost: it opened her up to additional "vicarious trauma," and introduced greater state surveillance into the lives of troubled students and their families. But as an instrument for systematizing connective labor, the fist-to-five check-in established her relationship to the students as a group of people to whom she bore some emotional responsibility, and did not stop at asking how they feel, but took some action.

But if the overload these workers face is itself an outgrowth of industrial logic—the pressures of productivity and efficiency, applied to relationship—so too are the systems that practitioners use to solve it. Modern medicine squeezes the physician's connective labor into ever smaller units of time, for example, while the EHR and other forms of data analytics expand their domain. Yet even the BATHE checklist ("That must be very difficult for you" [Empathy])—as much as we want doctors to practice more empathy—is a way to acquiesce to that squeezing without admitting to its costs.

The deep meaning that practitioners glean from connective labor is not reducible, however, a fact that shapes their perception of those systems as an alien grammar. Indeed, many experience standardization as an impediment to the work they value, and in fact sometimes the combination of overload and system generate a connective labor that fails. Such failures are more often borne by disadvantaged others, for whom connective labor is already a fraught blend of risk and reward. In the next chapter, we shall explore what happens when connective labor goes awry.

7

CONNECTING ACROSS DIFFERENCE

The Power and Peril of Inequality

Rebecca Rooney was a warm primary care physician, a woman in her forties who identified as biracial, with such a caring, positive energy in our interview that I could tell why she was assigned to play "the good cop" at her old clinic. She told me she sometimes paired up with a clinical pharmacist there, Dana, who was particularly stern. "I would be the warm, because I'm good at warm-fuzzy: (*in a gentle voice*) 'Are you OK? Don't work too hard, you're fine, I'm sorry you have to go through this.' And then [Dana] was like (*in a charged, dramatic voice*), 'You know what, you're going to die if you don't take your medications.'" "Warm-fuzzy" was Rebecca's modus operandi.

Rebecca was especially proud of being a family care doctor, a primary care specialty that saw patients throughout the life course and actually trained its practitioners in relationship building. Her trainers quickly recognized her gifts; she remembered a supervisor observing her at work and saying, "Oh, yeah, you've got this." Nonetheless, during her entire first year, she was videotaped and monitored through a one-way mirror, with an observer asking her, "Did you notice how your body gestures or their body position changed?" and weighing in on when she could have done better. One lesson in particular stood out, she said. "This was a gentleman, an older African American gentleman, very tall, and he started talking about the baseball game, but I was like, 'And so how long have you had the blood pressure and la, la, la,' and at the very end I said something like 'Who's playing tonight?' Or something like that, and he just lit up." Rebecca watched the session recording later with her supervisor. "She said, 'Did you notice how he lit up at the end when you asked about baseball?' I'll never forget that," Rebecca said. "I was like 'Ohhhh, that's the magic.'"

Her solicitous, persistent, caring style made the "magic" happen for many, cultivating a reflective resonance. "I've been told I have a knack for this," she smiled, as we talked in her office. Her practice surveyed patients on whether they would be "likely to recommend" their physician to others, and Rebecca kept close track of her scores. "I have a very big practice and my likelihood of recommend had been 100 percent for most of this year," she said, noting that recently it dipped, although just to 97.5 percent.

As we spoke, it seemed to me that this seasoned and intuitive provider would not have a lot to say in response to my query about any mistakes she may have made—not medical errors but those involving the relationship. And yet one of her biggest failures—"particularly traumatic for me," she said—involved a young patient whom she had cultivated slowly and over a number of years, but who ultimately left her practice, full of shame. "The child was very emotionally disturbed, his mother has HIV, and she's African American. He looked like he was mixed. I don't think he knows his dad. My colleague used to take care of them, and eventually the mom transferred the care to me because of issues of alienation and judgment," Rebecca said. "These things are—you know, it's really important." In other words, the family had chosen her, indeed had sought her out for their care, because they perceived her as a safer option, less judgmental than her colleague, better able to help them feel connected.

Rebecca saw the family for years, slowly trying to win their trust. The boy had some issues—he was very emotionally disturbed, a previous provider had told her he frequently smelled of urine, and "he was quite tall and large, like a big guy for his age, you know, he was this child, yet he looked like a man," Rebecca said. "You could just keep listing the things that he had going against him."

When he was ten or eleven, the mother consulted her about his worrisome weight gain. "Now, the first thing—*Now* I know, [but] I didn't know that then—but with childhood obesity, the first thing you do is target the parents," Rebecca told me, her missteps in the case clearly still troubling her. "You know, [if] you're going to talk about the *child* losing weight, you've got to talk about the *family* losing weight. That was definitely relevant in this situation. She [the mother] was overweight. So, that means that there was junk food in the house, and sodas in the house, all that stuff. It's a family affair." She may not have known to take that approach then, but she did know to get him to a specialist. There was a highly recommended clinic at a nearby academic hospital that was not part of the same health care system, so Rebecca spent hours trying to get the boy into the program. After getting nowhere, she then found out through the mother that the clinic had just flatly denied him admittance, with no further explanation; she remembered being very frustrated by the news.

But the mother continued pressuring her to do something about it, she recalled. "And I'm also like, 'It's the right thing to do, childhood obesity is such a problem.' So, then I started counseling him myself. I don't know if I [used the] motivational interview, or if I did a little bit more of mom-telling-me-to-tell-him-to-stop-drinking-sodas. That was pretty much what I was doing at the end of the visit and he ended up in tears."[1]

And that was it. "Then, that just—it was actually after [years] of building up this trusting relationship with this boy who had been so emotionally charged, I lost all trust with him," Rebecca said sadly. "And the mom ended up very politely—She really liked me and she understood I was trying—but she did transition his care over to [a nearby HMO]. I mean it was just, it was really sad to see that happen."

Frustrated by the specialty clinic's refusal, pressured to do something, and without adequate training or the time to treat the patient with the delicate touch she understood now as necessary, Rebecca's counseling that day conveyed a negative judgment that appeared to lay the fault for the boy's weight gain at his feet. Distressing as it was for

Rebecca, for whom such ruptures were exceedingly rare, we can imagine the visit may have been even more so for the boy, so much so he cried in his own doctor's office. After years of treating them, it was the last time she saw the family.

Frequently, people-to-people jobs involve connecting across difference, and at times the differences can be great. Providers with many years of education, high incomes, and other advantages bear witness to others without, while workers with less education and pay—for example, those engaged in personal body care, such as hair stylists or personal trainers—are charged with seeing people of great privilege. Across these chasms, some manage to connect anyway. But in other cases, as for Rebecca and her young patient, connective labor goes awry. And while standards and systems such as we saw in the previous chapter are often introduced to reduce the impact of such inequalities—for example, K-12 testing to identify low-performing schools, or checklists to mitigate bias in hiring—they sometimes have the opposite effect. This chapter explores the mishaps and misrecognitions that arise when people fail to see each other well, how they vary depending on whether the provider or client has more advantage, and how industrial logics and the pressures of overload can contribute to their incidence.

Risk, Vulnerability, and the Ambivalence of Power

There are myriad ways that connective labor can go wrong. Some of them—shaming, negative judgment—involve practitioners seeing the other and finding them lacking, conveying their disappointment or disapproval in a way that lands like a slap. Other times people misunderstand clients or students completely or in part, so that they fail to feel seen. Shame and misrecognition were at the heart of the rupture between Rebecca and her overweight patient. As Rebecca recounts it, rather than a safe space where the boy could articulate his goals and challenges, the clinical exam became a site of his disgrace, with the judgment coming from someone he may have thought he could trust.

In reality, however, every connective labor interaction, even those quite small or routine, involves some risk of negative judgment or misrecognition; every such encounter incurs the possibility of shame. When people open themselves up to these practitioners, they take the chance that the workers might not approve or like what they see, that they might communicate that verdict, and, furthermore, that their opinion might matter. These risks are more acute with therapists or doctors—experts and professionals who see us at our most vulnerable—but even encounters that are more mundane can pose a hazard of judgment. The hairdresser can deliver a haircut that somehow—helmet-like, with lots of hairspray—conveys she sees you as an old woman, the car salesman can proffer a minivan when you were trying to signal you lead a life of bold adventure, the realtor can steer you away from a neighborhood you liked because, as he says impishly, "you're not cool enough." To be sure, these not-quite-slights are not nearly as damning as the shame endured by Rebecca's patient. Nonetheless, while we might shrug off judgments from people who don't know us well, they can also feel like momentary spotlights shining pitilessly on our fantasies of who we are.

It is a bit harder to shrug off the misrecognition when it comes from someone with more social power, or when the encounter involves true intimacy. As Rebecca's patient illustrates, inequalities of race, class, gender, age, and other dimensions make those judgments cut more deeply. The crux of the problem here, both the reason why connective labor is hard to automate and the reason that it is sometimes perilous, is that it only works, these encounters only manage to achieve ostensible goals while fostering some dignity or purpose along the way, because the worker's opinion matters. We would not be seeking someone's services if we did not think they had something to offer us, and in jobs that involve connective labor, part of what they offer is recognition. The degree to which their judgment matters—either because we consider it valuable or because it is enforced by real-world consequences like grades or court action—is also the degree to which we are vulnerable to it.

This vulnerability speaks to the considerable power wielded by these providers, particularly those at the top of the occupational ladder, like physicians or therapists. People in positions of authority or influence are often tempted to use it, which points to yet another risk that inequality exacerbates in connective labor: the paternalistic impulse to "save" the other. Workers wrestle with where to draw the line between reflecting and intervening, between inspiring and rescuing. We all know that with great power comes the potential for abuse by the malevolent, but it is more complicated, and perhaps more unsettling, to think about the problems of such power in the hands of the well intentioned.

Throughout my research, I was often struck with how similar people in quite disparate occupations sounded—the funeral director and the teacher, the hairdresser and the therapist—all echoing how they saw and reflected the other in their daily work. But when it came to how things went wrong, their stories often diverged, reflecting where they stood on the advantage divide. And while the inequalities of these encounters may be inevitable, how they are experienced and deployed is shaped in part by its social architecture, including the standardization and systems that organize this work today. Like Rebecca, professionals and experts with years of education and acknowledged skill dealt with shame and judgment and silences, navigating the dictates of scheduling and performance metrics while they searched for a way to behave honorably with their considerable influence. When providers had less privilege than the people they served, however, a different kind of struggle emerged, raising the question: what is connective labor on demand?

Shame, Social Inequality, and Shortcuts

The potential for shame hangs over connective labor encounters. Practitioners talked about things they couldn't say to people, about dancing carefully around particular subjects, about hiding their judgments. Even when it went unnamed by either practitioner or client, the prospect of dishonor shadowed them both.

One therapist in private practice, Brian, told me he had a client who made sure to pay him his fee in the waiting room before walking in to sit down. The adult child of alcoholics, the woman was not sure that she deserved to claim his attention, he thought, and he interpreted her habit as her way of making sure she could take up his time. "But it would be years before you could talk [to her] about that," he said. "Because you want to—you don't want to shame people and you don't want them to hear you say you're doing it wrong,' and you don't want to humiliate them."

Sarah, the VA therapist who talked in chapter 4 about the redemptive benefits of owning up to mistakes with clients, agreed that a long time had to pass before she could invite people to talk about what caused them shame. "I mean it can take—For example, if someone has—It can take several years. I mean, if somebody has something, say, a trauma, an experience that they had years ago they never shared with anyone because of psychological reasons like shame or guilt, right? It can take people many, many years to kind of figure out even just a level of trust to feel like they can share that."

Yet while the prospect of shame and judgment are everywhere, their risks are not evenly distributed, nor is the capacity to mitigate them; they loom over cases where disadvantaged clients meet with more-advantaged practitioners. At stake in these encounters is not just hurt feelings but even health and well-being. Reproductive justice scholar Patrice Wright tells the story of Courtney, a pregnant African American woman who was on a prenatal visit when the doctor made some comments about her keeping the weight off and about the government's WIC program, a food subsidy for poor mothers and children. The comments revealed to Courtney that the doctor assumed she was receiving WIC assistance, didn't know basic nutrition, and would tend to gain too much weight. But Courtney was a graduate student, who may have been low-income but was not receiving WIC, knew a lot about nutrition, and was not overweight. Courtney felt profoundly unseen and did not return to the doctor; the misrecognition on display there—what therapists would call an empathic failure—caused her stress, anger, and anxiety, according to Wright, whose work documents how these encounters plausibly contribute to the "weathering" phenomenon and the maternal health disparities documented among women of color. Misrecognition matters, particularly for marginalized populations.[2]

The example illustrates another crucial point: many times, misrecognition stems from assumptions that can get in the way of clear perceptions of the other. What Patricia Hill Collins called "controlling images" can become what people see, for example, the poor Black mother who doesn't know basic nutrition. Racism, sexism, classism—forms of organized hierarchy with cultural messages attached—serve up the images that cloud one's vision.[3] Even if well intentioned, recognition is often a brutally imperfect affair. If it is a "vital human need," as the philosopher Charles Taylor maintains, it is also one that forces the other to be legible—to fit in the procrustean boxes of a priori cultural meanings—in order to be seen. But sometimes that legibility involves generalizations about the other that gather their force from interlocking systems of inequality and oppression.[4]

The Magical Listening Ears

If shame and judgment are part of the risks posed by connective labor, one can imagine that people—particularly disadvantaged people—would do what they can to avoid these practitioners. And to be sure, some do. When sociologist Freeden Blume Oeur studied a school serving mostly Black low-income boys, for example, he found that while some sought respect or dignity, others actually wanted to "be unknown," an urge strongest among those boys with prior formal contact with the criminal justice system. To them, relative anonymity felt like a privilege, the privacy of being free from others' presumptions, a way of belonging to their communities without the mark of a criminal. In another study, teachers who used individual, personalized relationships to deliver support were viewed by some high school students as "trying to know my business." Even while they appreciated the teachers' willingness to extend themselves and their interest in students as individuals, many of the students were wary of teachers managing their sensitive personal information. "[I'm] here for science, for math, and you're trying to know me," said Lupe, a seventeen-year-old. Alongside those who express skepticism, however, there are also many others who look to be witnessed, hunger to be seen, yearn to tell their stories to those who will listen.[5]

Most practitioners I talked to reported what felt like a vast need for reflecting work among the people they see. They felt that the people they met were desperate for their witnessing gaze and further, given their own finite resources, that they could never effectively meet that need well enough. Students, patients, and clients might not like judgment, but they were nonetheless picky about who was doing the seeing, and they seemed to want these providers in particular.

Jenna, a county hospital pediatrician, said she faced enormous need on the part of her patients, who appeared to want only her to bear witness to their struggles. "So many other people could do that," she told me. "Anybody could do that. But the patients don't want just anybody. They want me." I made some noises about how perhaps patients were responding to her uniquely empathic abilities, but she demurred. "It's just that one of the values society has placed on doctors is that somehow our ears are magical listening ears and that when we hear the story we're going to magically extract whatever needs to be extracted and give it back to them in a way that's going to make everything better." The very thing that they were afraid of—her judgment—is what gave her "magical listening ears."

Greta was another pediatrician working for a large HMO in a West Coast city. A white woman in her early forties, she had a real charisma, and her warm attention felt like the sun shining onto you when she turned in your direction. Yet so much of what she did could be done by other people, she said. "Everybody has real problems, and what people need is an hour with you. I mean, seriously." She paused and gestured at me. "With a good listener, it doesn't matter what their training [is]."

Like many primary care doctors, Greta said she was not "practicing at the top of her license." As it turns out, what she spent most of her time doing was the work of witnessing. "Eighty percent of what I do is not what I learned in medical school," Greta said. "In some ways I'm more of a therapist or a human connector or something. A relation person." She was a pediatrician, she reminded me, "not a mom doctor"; in other words, she was not there simply to console and soothe the parents of her actual patients, who were the kids. "But in many ways," she added wryly, "I'm a mom doctor." She lamented that she was spending her time answering questions about car seats and solid foods, or comforting inordinately anxious mothers. "I mean, this is not medicine. This is like a buddy. Any monkey can do it, a relational monkey," she added, chuckling. "Anyway, it doesn't require an MD to do that, but they feel value in that. A huge value." Ultimately, even though others could do the witnessing, patients wanted her, she said. "We know that if my patients see me, they are more likely to get better. If they don't see me, they utilize more services. They're more likely to come back and want to see me."

It is not coincidental that people sought out witnessing from practitioners who could render judgment, I would argue. Instead, there is both unease and appeal here, a potent combination that hearkens back to a concept of "pastoral power" developed by the French philosopher Michel Foucault. In that notion, the pastor gains an intimate understanding of another, the "knowledge of people's secrets," and harnesses it to the goal of saving their souls. In modern secular contexts, we might say these workers take that same intimate understanding and use it to take responsibility for others' mental and physical well-being. The concept suggests a mutual affair, the people proffering up their stories in return for the "pastor" taking charge of their welfare, but Foucault diagnoses the paternalism in that responsibility, when practitioners take it upon themselves to save others. Foucault also highlights the emotional dimension to this power—how the pastor can make the other feel better about their vulnerability, and how pastors are also made to feel accountable. "The point is not that everything is bad," he said, "but that everything is dangerous."[6]

To the professional practitioners I spoke to—physicians, teachers, and the like—this vulnerability and power came together with particular force when they were working with less-advantaged people. Of course, being seen is part of a meaningful life for people up and down the class ladder. Recall in chapter 2 how Paula, the chaplain, acknowledged the profound effect of hearing her fellow chaplains seeing "that wild, silky part" of her, which she said was easier to perceive when it was being reflected back. Yet from the vantage point of practitioners I spoke to, the need for recognition seemed particularly intense when people could not ensure a good supply. Jenna described her patient population as very disadvantaged. "Tons of refugees and immigrants. People who haven't had access to health care in a long time," she said. "Even though they just live in the far-flung suburbs, they're just really poor and they don't have a car. So if their kid's not deathly ill, they don't go to the doctor. So, yeah, it's really intense."

"With working-class children, who come from working-class backgrounds, relationships are incredibly important," said Majeeda Nur, a middle school resource teacher who taught in a West Coast city for more than two decades. "You have a very difficult time getting kids from that type of a background to learn from you if you don't build the relationships. I mean, other kids it's important too, but [it's] really important with working-class kids."

Peter Almond, a Stanford researcher, agreed. "At [a local community college], they get a lot of people who are a little more like from the vocational side and [from lower-income] backgrounds and so on, and I heard multiple times that it very much makes a difference when the instructor notices the absence of one of the students. And like even just sends an email or something. That that is really important. That somebody cares. You know, even just that fact."

Research bears out their impression. Studies have found that while all students benefit from better relationships with teachers, less-advantaged students who have a good relationship with their teachers benefit even more than their advantaged peers. Likewise, medical research documents that although disadvantaged patients are less likely to have strong relationships with their physicians, such relationships can have a bigger impact on those patients than upon more-advantaged patients, attenuating health disparities. Moments of being seen, and being supported while being seen, seem to have particularly dramatic impact for those who often go without, a terrible irony.[7]

We can look to the "magical listening ears" to begin to understand this paradox. People hope to be seen by others whose opinion they value, and the inequality that makes someone more vulnerable to another's judgment is the same force that makes that other a valuable judge. Practitioners advantaged by training and the accumulation of expertise, if not an amalgam of other dimensions such as class, race, and gender, often face people who are not. When Jenna's patients said they wanted her and only her to hear what they were going through, they sought not just her medical advice but her witnessing, and the reflective resonance it elicited. To Jenna, they acted as though their other options—perhaps a medical assistant, the office administrator, a neighbor—would not be as powerful an audience, both in the judgment they might render and the effect of their empathic reflection.

The practitioners I spoke to took these disparities seriously. In a classroom session for future school psychologists that I observed in Virginia, the professor counseled the trainees to be extra careful when esteem was in doubt. "The less privilege or advantage your client has had, the more respectful you need to be," the professor said. "Pull out the southern 'yes ma'am, no ma'am.' What you are doing is having the client actually see you as someone who might actually care, as opposed to not even seeing them."

We might consider the southern "yes ma'am, no ma'am" a form of what I came to consider "dignity practices," esteem-building efforts by practitioners facing the challenges inequality posed to their capacity to deliver meaningful connective labor. Another of these practices included inviting aspects of the "whole person" into the classroom or clinic. Ken Margolis, a middle school teacher in Oakland, had students handle attendance so he could walk around the classroom. "You just notice aberrations, [like] if some kid is always bubbly and they're not bubbly that day," he said. "And you have to ask very specific questions. So you say something like, 'Is it a friend? Is it an enemy? Is it your stomach? Is it your head?' Because they need to feel—everybody needs to feel cared for. And respected, right?" Dignity practices were also one way practitioners fought the growing systematization of care, by disrupting the boundaries around what counted as "relevant" information. Aiming to convince themselves and others of autonomy, safety, and equity, even in its relative absence, practitioners described it as if they were finding the humane in themselves while insisting upon it for others.[8]

Stigma, Judgment, and Misrecognition

The power of the expert audience reverberates, but not just when they get it right: it does so also when they get it wrong. Advantaged practitioners were haunted by encounters that they considered mistakes, and their stories seemed to hinge on an unevenness between themselves and their clients. Connective labor made a social space for the powerful to impute their own (pejorative) notions upon others, with potentially devastating psychological effect. When such judgments came from connective labor workers, it landed like a betrayal of trust.[9]

"Sometimes I just miss it. I just miss who they are," Grace Bailey, a white therapist, told me. Recently she was treating a woman, a "serial monogamist" whose philosophy was to leave "when a relationship stops being good." When the woman's partner became ill, Grace was interested in what that meant for her plans. "I asked her how that would affect the state of their relationship, and she was really, really hurt by that."

Why did the question hurt her? I asked. "Because I don't know who she is," Grace said. Still unsure, I asked her whether the problem was that Grace had intimated that her client's "philosophy" signified that the woman was actually not a committed person. Grace nodded: "Exactly." Grace's initial question suggested that she was still unclear about whether the woman's intentional approach to relationships—that she left when they were not "good"—would mean she might abandon someone when they were ill. While she and I agreed that the question may have been a reasonable one, I noted it was nonetheless one the client likely had to defend against all the time. "Absolutely," Grace said. "And I fell right in that camp. Then the next time I saw her, I said, 'I think I hurt your feelings.' And she was like, 'You really did.'" Grace's misrecognition of the woman, whose identity clearly rested on an ethical redefinition of commitment, involved conflating her unconventional approach to relationships with unreliability, made all the more hurtful because it was a charge the patient continually had to deflect.

Memories of their misrecognition linger for these workers. A white man in his late thirties, Conrad Auerbach was a middle school teacher in a large Virginia city, still troubled by the memory of how he treated a student when he was just starting out more than a decade ago. The teen was an African American boy in his class who "was having a really rough time in his life, in general," he said. "I wasn't too good at connecting with middle school students yet. So, I used to be very confrontational, so if students weren't doing their work right away, I would get up in their face, and that was the wrong thing to do with him," he said. "He would get very defensive and upset back at me, right away. Just, kind of, shouting and I would have to kick him out."

Conrad approached the boy harshly, as someone who was not measuring up to his standards, a less-than-sympathetic take whose message was likely compounded by the inequality that suffused the context. Black and brown children are more likely to be suspended, expelled, and referred to the criminal justice system, for example, while research documents how low-income children of color are subject to punitive treatment that spans their schools and neighborhoods. We cannot know for sure, but Conrad's student may have seen his teacher's harsh treatment as of a piece with the rest of his experience in a white space whose denizens did not understand what he was going through, did not see him beyond his racialized gender.[10]

Now, Conrad said, he would approach the boy differently. "Well, if I had gotten to understand him, and talked to him more, and connect with him more, I wouldn't have been so confrontational with him for not really, you know, getting to work right away, or being in a great mood first thing in the morning in class. I would have been able to work with him more and he would've, most definitely, been more successful." Conrad regrets how he treated this student, how he confronted him with his negative judgment rather than connecting with him in order to ease his way into the experience of learning. Recognition harms lingered in the memory, as practitioners grieved judgments rendered and repented.

Judgment Withheld

Inequality's challenge, from the perspective of the less-advantaged patient or client, may have been shame and its corrosive effects. But for their providers, it posed a different problem: silence. If clients or patients felt judged, they shut down about whatever problems they might be facing; withholding their story was sometimes the only recourse in the face of the provider's social power. To the practitioner, their silence made it much harder to "see" them. The connective labor exchange relied on client storytelling, the raw material for the seer, and the task of the advantaged practitioner was to keep the stories flowing.

Gunika was a primary care physician who said she had changed the way she talked to patients to avoid shaming them. She tried to make sure her exam room was "a safe environment to talk about things like 'I can't afford' or 'I don't like this' or 'I don't know how to cook it' or 'I don't have time to do it,' or whatever the reason is that it's not happening even though they know it would be something beneficial to them. If we don't get to the root cause, if we don't get to what's the barrier, we can't remove the barrier." Ultimately, her goal was to have her patients know "it's OK to tell me that, because I'm not going to judge you and I'm not going to make you feel bad for saying that," she said.

Rebecca, the primary care physician who regretted the loss of trust with her overweight patient, said medicine had moved away from shaming patients for what they were unable to accomplish. That move involved shifting from provider goals to patient goals, all part of an uneven but industry-wide turn to "patient-centered medicine." The approach represented a big transition for existing physicians, she said, recalling one presentation when a doctor in the audience raised her hand. "[The doctor] asked—and I really commend her for asking—she's like, 'What do you mean "the patient's stated goals?" Like, what if my patient's goal is to drink a six-pack every night, and my goal is for him to bring down his A1C [a blood test to determine blood sugar levels, usually for diabetics]?' Yeah, so she was like, 'What do you mean I'm going to talk about his goals? I'm going to talk about the fact that he needs to bring his A1C down.' But Rebecca remembered how the presenter responded to the doctor's query. "Without missing a beat," Rebecca recalled, "he said: 'And how is that working out for you?'" She laughed at the memory, at the simplicity of the question, and at how we all knew the answer: not well. "And that's it," Rebecca said. "It's not effective." Judgment did not get the right results, she said.

> In that situation, proselytizing and telling him how much he has to stop drinking, that really alienates him, if I were to do that. [But] if I were to just sit with him and, "OK, yeah, you're drinking and you're

> not doing what you want. All of us are doing that. We have something that we do that [isn't recommended]." So, just keep talking about it and walking that journey with him. Then, when he's ready to make that change, he's going to feel a lot more comfortable, and a lot more willing to take that risk with me. If he fails, no big deal. If he doesn't, I don't have any judgment or expectations of him. I'm just there to try to help him.

This story, told as a parable to support the new patient-centered regime, nonetheless points to a continuing tension in contemporary medicine. Rebecca may have had no expectations of the patient who was drinking too much, but her own judgment was certainly embedded in the idea that he ought to make changes, that she could see already the changes he had to be comfortable enough to make. Like Foucault's pastor, she folded intimate knowledge and judgment together, and her own professional expertise shaped her view of what would count as his salvation: in this case, drinking less. While she did judge, then, for the most part she was willing to try to hide it from the client for his own good.

The "magical listening ears" of doctors, teachers, and the like come as a result of their training and experience, and thus carry with them a risk of judgment. It is not simply that practitioners offer unconditional positive regard; it is that they *could* be judging. That possibility—the underlying risk of shame—lends a frisson of peril to the encounter, and the emotional intensity of the experience is related to the intensity of the perceived risk. Ultimately, what makes for successful connective labor, then, appears to be judgment withheld.

That said, not all advantaged practitioners shunned the prospect of shame. While those in many different occupations talked about their strategies to sidestep or suppress it, therapists stood apart. Unlike the physicians, who talked about avoiding shame because it might shut down good information flow, for therapists, shame was a symptom, a signal of where the work needed to take place, much like a masseuse might view a big knot in a muscle. Shame was the knot the therapist had to work on without doing violence, an indicator of where to go—carefully—rather than where to avoid. Therapists used to use the language of "resistance" to talk about ruptures between them and patients, Brian Jacobsen explained, but what he liked better was the notion of empathic failure, and the implications it held for shame. "Like I didn't say [something] in the way you could hear without triggering the shame—but the thing is, it's—We're not going to avoid that. We're not going to avoid that and we—But 'can we, like, live through it together?' is the idea." In this view, shame is certain—"we're not going to avoid that"—but it is not the end of the story, rather the beginning; even after shame is triggered, therapist and client maintain their connection.

Yet for many other providers, the risk of judgment loomed over every disclosure, the more sensitive the more threatening. Disadvantage of many kinds makes such judgments more painful, and disclosures more hazardous. The palpable risk suggests that what witnessing might actually come down to, then, is managing other people's defenselessness.

Systems, Inequality, and Misrecognition

Connective labor involves inherent risks, but campaigns to make it more systematic can increase them. Of course, on some level, the risk of judgment is a constant. Because humans meet humans in these moments, their backgrounds intrude, prior understandings shaped by inequalities that determine who is on which side of the desk, by who is the practitioner and who the client. All these erstwhile meanings drift into the witnessing moment like clouds of steam, making it harder for people to see each other in ways that might diverge from preexisting definitions. Yet some conditions—particularly those involving time pressure, stemming from the push for productivity, and exacerbated by systematization—make those clouds more likely. These conditions encourage people to cut corners on relationship building, resort to shortcuts and controlling images, and otherwise reduce the individual to generalities. The industrial model can exacerbate inequality.[11]

It is not that introducing systems necessarily amplifies disparities. We know that women and people of color can benefit from bureaucracies, which tend to elevate rules over traditions that sometimes involved backroom deals or rewarded old hierarchies. The efficiencies of big-box optometry or chain hairdressers bring service to far more people who would not be able to afford it otherwise. Accountability pressures can sometimes have a democratizing effect; the personal relationships of yore—the small-town GP, the community police officer—reflected and enforced an old social order of race, class, and gender inequities.

Yet we have also seen that practitioners are often rushed or overloaded in settings that serve low-income people, who do not get the artisanal connective labor that some affluent people make sure to secure for themselves. Furthermore, conditions of overload can aggravate the impact of inequality, as busy practitioners rely on controlling images to navigate complex situations quickly. One study demonstrated that time constraints led doctors to be more aggressive in their diagnoses of white patients, and less so when it came to Black patients, for example. Time pressure, patient load, and overcrowding exacerbated "implicit racial bias," making it more consequential in physician decision-making.[12]

In fact, even though standards are celebrated for fairness, standardization itself can sometimes magnify inequalities; the same controlling images invoked by time pressures and overcrowding are embedded in policies and procedures that govern connective labor. One day, for example, a guest speaker from Child Protective Services (CPS) visited the adolescent therapy course I was observing to advise the therapist trainees on when to call in a report of child abuse or maltreatment. Most people working with children are mandated reporters, she reminded her listeners. "I know you want to keep that therapeutic relationship with your folks [the clients]. A lot of people don't want to report; you want to keep your therapeutic relationship with your family," she said. "It happens here, I know you're saying, 'I would never do that,' but it happens here, because they'd rather salvage that relationship." She urged the students to act differently. "Whenever you have a question, call it in," she said, and went over a long list of possible situations, from "injuries in weird parts of the body, where it's not going to happen accidentally," to underage drinking, to patterns of emotional maltreatment. "You're going to be hearing people's deepest darkest secrets," she said, recommending that they "call and consult" with CPS about them.

After she left, however, Russell Gray, the course supervisor, was critical. "That's the most aggressive view of CPS I've ever heard," he said. "She's basically saying, 'Anything that's bad, tell us.' What was lacking was any sense of how aversive this process is experienced by families. And how well do services work when they are coerced, or experienced as coerced? What does it do to the therapeutic relationship? She didn't have that balance at all."

A call to CPS should be made judiciously, Russell said. It "sends the message that that behavior is not OK. You can do it well," by which he meant defray the impact on the relationship with the patient, "but it's aversive. Half the time it's OK, half the time it ruins the therapy." The guest speaker was urging them to apply a decontextualized standard of when to call in the agency; as Russell said, the visitor was "a true believer in CPS." "My line is way far from hers," he said. "My line is if there's clear evidence and suspicion of harm. But not like 'this might be harmful to kids.'" He urged them to consult their supervisors or ask a colleague. "You don't have to make that call on your own." When they encountered suspected neglect or maltreatment, he recommended a different approach, one that relied on professional discretion.

CPS is not a race-neutral entity, and as legal scholar Dorothy Roberts has written persuasively, its apparatus serves as part of a state surveillance system for Black families in tandem with the carceral state. Neither Russell (who emphasized how a referral would threaten the relationship with families) nor the visitor (who argued that a referral would protect children) mentioned race, although the visitor did urge the students to "try to separate poverty from neglect." The example serves to remind us, however, that standardization is not just a vector of efficiency, threatening real connection. Instead it can act as a chute for ideologies that are weaponized by the state to define and control marginalized others, especially low-income people of color. To be sure, standards are a bulwark against the discretionary power of parochial interests and entrenched elites answerable to none. As the CPS example suggests, however, bureaucratic action can render unequal violence, by shaping what practitioners see, the meaning they are asked to make of it, and the consequences of that seeing. Sometimes the wrong that connective labor does is generated by the systems in which it takes place.[13]

From Savior to Witness

Experts and professionals who see others for a living wield considerable power, both that which is ceded to them by the intense need of others seeking recognition, and that which they accrue through the standards and systems that institutionalize their roles. Many of them struggled at some point with the question of what it meant to use that influence well or badly. When Hank, a hospital chaplain, told me his story, I knew what I was hearing: a sometimes-painful quest for integrity when it feels like people are handing you the reins.

Shaped like a bullet, Hank had a tightly held energy that made some people think he was a police officer. It hung around him, a whiff of the military, a holdover from the five years he spent as a Marine before he left to work first in the trades, then retail, then as a minister in a megachurch outside Washington, DC. Hank was resigned to giving off that impression. "It's just like, 'Yeah, you have a look of a cop.' I'm like, 'Great, that's another detriment—white male that looks like a cop, great,'" he said.

He liked being a chaplain, Hank said, but he was nonetheless pretty sure that he was meant for different work. Speaking with me in a small room in the bowels of the medical library, he sought to be precise: "I want to make sure I'm saying this is very meaningful work," he said. "But if I'm honest, I'm called to get out there."

"Out there" for Hank was epitomized by a mission house in a small city about forty minutes away, where he went to worship whenever he could on Sundays. "It's horrible. I love that," he said, grinning.

> There's a little laptop where they play music. We sing really bad. The congregation is homeless, recovering drug addicts or active drug addicts, veterans who are homeless and getting out of prison, jail, so it's rough and tough and nasty and smelly. I miss the smell of Section 8 housing. There's a smell. I can't describe it to you. But it's not a smell

> you would have in your home and would be repugnant to most people. But having worked out there, it smells very powerful. It's like "OK, I'm where I'm supposed to be." It's like a cigarette-stale kind of smell and stuff.

He was drawn to serving the most marginalized of people in the hidden corners of society, he said. His passion for being "out there" started when he was a minister, charged with outreach to low-income communities. "I just started stumbling out into Section 8 housing, projects and stuff, and working with at-risk kids" who were intimidated by his big, monied, white church, he said. "After I would get done helping with our congregation on Sundays, I would take church out to them."

Hank doesn't talk much about his origins, but his father was a wealthy lawyer—he called him a "limousine liberal"—who once asked him, "What the hell is a conservative guy like you doing hanging around with a bunch of poor Black people?" Hank laughed mirthlessly as he recalled the question. "That was his . . . he just didn't get it. It's like, 'Why are you doing that?' And I said, 'It's just where Jesus would be.'" At first, he headed to those neighborhoods without any particular training, or even much of a plan. "They just wanted to be loved," he said. "I just knew I had this supernatural ability to love people I wouldn't have loved prior to my calling. And so I'm learning on the way. I mean, I have no recipe. I'm just going, 'OK, seat-of-the-pants stuff.'"

Hank's personal charisma was strong—he exuded confidence, leavened by an energetic bemused curiosity, a bit like a friendly drill sergeant with a sense of humor. He founded some sports camps, then four tutoring centers, and after some initial skepticism, youth started to come regularly, responding to his magnetism. "We had a hundred kids," he said. "It was busting at the seams." He described a typical afternoon: "Fifteen teenagers sitting on these little rails with me, just talking, telling me about this girl drama at the school, and 'Oh, my football practice is going eh.' Girls and boys just talking and joking with each other, trash talking to each other, then coming back and asking me stuff."

One memory captured how cherished and valuable he felt there. "I said, 'Let me see some interim report [cards] because I know you got them, see how you're all doing,'" he said. "So I sat there. And it was like I was Santa Claus. They were bringing their interim reports to me and they were horrible. Ds and Fs and Cs and incompletes. But they brought them and they were like . . . Kids went to the dumpster and got theirs out that they threw away to bring them to me and got in line. And none of them were good, but I was like Santa Claus, right, in the mall." Like Ruthie treasuring how valued she felt by the community at her mountain practice, Hank marveled at the pull his attention exerted upon the kids.

In those early days as a church pastor, however, Frank enacted a "savior model" that clearly relied on his own advantageous position as someone who could offer any sort of "help" to "at-risk kids," coming from where it was "rough and tough and nasty and smelly." He was particularly proud of forging those connections despite being a white guy who looked like a cop, nonetheless embraced by the Black and brown people in a poor urban community. "I was all about fixing and rescuing people back then," he said. While their connection clearly met some need—in him as well as them—when he looked back, he could see the problem of this early approach. "I was a Marine, right? So go out there, throw yourself on a bunch of grenades."

While Hank started out with the conviction that he could save others, the person who ultimately forced him to rethink that surety was a woman who had been a high school teacher, with "the most infectious smile you would ever see," Hank recalled. But she was battling heroin, and it ended her career. "She ended up having babies with several different men. You know, that's the life. Drug dealers living in the home. And her kids were being abused, and CPS trouble, and all of this stuff. And I was just sitting there," he said. "She was like, 'Can you help me, Pastor?' And it just felt like it just . . . whatever I was saying just fell to the ground, that it had no life to it." His words had no impact upon the woman, who had hit rock bottom. "And that disturbed me. I was actually having a little theological crisis, like why isn't the word of God just overwhelming?" he asked, and answered himself in the next breath: "Because I wasn't meeting her where she was at."

The crisis got him doing his own research. He started looking for books that would help, and discovering the names and diagnoses for some of the phenomena he was seeing, such as "attachment disorders." "So that kind of started leading me down a path. I'm a voracious reader, and so if I get stuck somewhere, I'll go self-teach," he said. He learned about different approaches to counseling, such as cognitive behavioral therapy, and brought them to bear in his pastoral work.

One could imagine that the people in these communities would view with great suspicion someone with such privilege, a "white male that looks like a cop," coming in with newly self-taught "tools" to "help them." For some reason, however, they allowed him into their lives to try out the methods; Hank thinks it was because he was already a trusted insider. "I think the love that we had in that way solved a lot of issues, because [the former high school teacher] trusted me. And this is not just her now, it's a lot of families, because of my reputation of just being a nice guy who cared, who can help them with different things. There's grace because they knew I was doing it out of love." Given that we only have Hank's account, we don't know how they actually felt, but it sounds like their emotional connection allowed Hank to stride into their lives with "tools" he had picked up as "a voracious reader."[14]

Talking with me years later in the hospital, Hank had come a long way from sitting on the banister with the teenagers. By the time I caught up with him, he had amassed many degrees and years of training; "I do have a lot of interventions," he told me, half-proud and half-sheepish. Perhaps surprisingly, however, after all his training, after collecting all of his "interventions," he now practiced an entirely different approach. He characterized the change: "I used to get in the water with them and swim for them, right?" he said. "And let them say, 'I'm in the water, I'm drowning.' Instead of getting them to a place, using that metaphor, to 'Hey, yeah, I'm in deep water, but look at me, I'm swimming, and Hank's swimming alongside of me.'"

Hank was still as ebullient as in his megachurch days, still energetic, even cocky, but part of his ability to change so dramatically likely stemmed from experiencing what he saw as a major failure—a recent job loss—that steeped him in humility. Now a chaplain, he was reexamining not just his own trajectory but also how he approached his work, from a savior model with him at the center, rescuing those in need, to one in which he simply saw and reflected their own resistance and survival. His story—from the ex-Marine hell-bent on fixing other people's problems to a chaplain whose best practice was listening well—was ultimately one of his discovery about the importance of taking power seriously in connective labor.

I asked Hank what he would do differently today, after so many years of skills acquisition, if he met the woman with the heroin addiction and the sweet smile. "A lot," was his answer. "My approach now is way more witness-based and contemplative than interventionist. I could have walked with her differently, meaning just to witness to her and not try to solve her or fix her." His reflective approach not only reflected new knowledge about what "worked," but also enacted a new understanding of the other, however disadvantaged they were. As he described it, his task now was "to put them in charge, and honor their strengths, and honor their ability," he said. "And if they are in pain or they are in despair, honor them in that instead of trying to change it." The savior had become a witness.

When Workers Have Less Advantage: Doing Deference

Despite their apparent gifts, Hank and Rebecca represent some of the risks of connective labor when workers have more advantage than the people whom they see: judgment, misrecognition, and paternalistic rescue. Yet many people who do this work do not have the years of education, the high incomes, or the social status of doctors or teachers. Unlike physicians or therapists, some connective labor workers have less privilege than most of the people they see. Massage therapists, hairstylists, home healthcare aides—these people told me about other risks, other hazards stemming from these encounters.

If the account of Rebecca and her overweight patient was one of the entanglement of power and shame, the story Betty Sinclair told me was more about what it feels like to provide connective labor on demand. A forty-seven-year-old home health aide who immigrated to New York City from Guyana in 2011, Betty talked to me in a café in Manhattan's tony

Upper East Side after she finished with a client. She told me she had done the same work part-time in Guyana, but in New York she cobbled together a full-time schedule working for two different agencies, three days with one client and three days with another, a schedule that was relatively common among aides since employers frequently did not give them enough hours to qualify for health insurance. Her job involved taking care of elderly and infirm clients in their homes, preparing food, bathing, dressing, and accompanying them outside. Having a good relationship with clients was important for being able to do her job well, she told me. "When I put myself in their position, it motivates me more," she said. "Yes, it motivates me more to try to help. And sometimes you tell yourself that they will pull through."

Betty was so trusted as an employee that the agencies gave her the "hard" cases, those who were openly rude or disrespectful—about 10–15 percent of the clients, she estimated. She found that the "hard" ones were often simply unhappy, frequently because they had just lost their mobility. "So you better understand what really happening to them, or why they behaving like that," she said. "I try to meet them halfway, to make them feel comfortable, to make them feel they're at home, or they're living at their place." Her empathy for them meant that Betty extended them inordinate grace, telling me, "I never met a client that I don't like. Never."

One such case was a couple she had seen recently, where the man was officially her patient while his wife was healthy and well. The woman was so disrespectful that aides quit every day, until finally the agency sent Betty. "The supervisor tell me the wife is a lot of problem," Betty said. "But she said, 'Betty, you go. You got a lot of patience with clients, so if the client [misbehaves], then tell me and let me assess it.'" Betty was prepared for the worst, and so when the wife was discourteous on their first phone call, she was not surprised. "So normally, if they give me a case, I like to call, and I would call and say, 'Good afternoon, my name is Betty. I'm the new home health aide who'll be working with you,' just to get a feel, you know, like who you're going to work with," she explained. "And I said, 'I would like to speak to Mark, Mr. Mark.' So the wife was like, 'Everybody who calling always calling and asking for

Mark, Mark.'" A bit taken aback, Betty did not blink. "And I was like, 'She rude,' but I didn't take it in."

When she showed up the next morning, she helped the man get ready for the day, and then wheeled him into the dining room. "So I said, 'What he will like to have for breakfast.' And he told me, and then I asked the wife, I said, 'Willa, you need the same thing?' And she turned and she looked at me, you know." It turned out to be the question that pierced right to the heart of the matter. Willa was angry that the other aides had never included her in any of the food they were preparing, Betty said. "And after none of the aides never offer her any breakfast, she's got to get up and go and make the breakfast," Betty said. "She said, 'They're coming into my home and sitting down in here, and they kind of give me [just] a cup of water, you know, and things like that get to me.'"

Betty could see where Willa was coming from. "It's hard for me going to a home and see a husband and a wife and don't give the [other person] their breakfast," she told me. "I can't do that. They're eating the same thing, they live in the same apartment, they're sleeping in the same bed." She ended up making twice the amount of whatever food she made for the husband, which violated the agency's policy—the supervisor told her, "Betty, you're training the client bad"—but it worked: after Betty had sorted out the problem, she trained another aide, and that person ended up staying with the couple for years.

Betty's account has some similarities to those of the physicians, therapists, and teachers we have heard from in this book so far; at their core, they all deployed connective labor to do their jobs well. Like them, Betty tried to "understand what really happening to them, or why they behaving like that," using her own emotional antennae to bear witness to the client's perspective, and then responding with what she gleaned from them to help them feel comfortable. Like the others we have heard from, Betty also identified personally with her work; she brought herself—her awareness of the other and her expertise in managing difficult personalities—to her job and was gratified when she did her job well.

But there are some important differences as well. In some occupations, practitioners are the advantaged ones, and face questions of how to make sure they do not exploit or harm others, but in others, workers like Betty find themselves subordinated to the people they serve and vulnerable to exploitation and harm. While Betty was a citizen regularly employed by an agency, she was an immigrant woman of color with a high school education, taking care of elderly people on New York's Upper East Side who were white, educated, and unusually affluent. Betty was actually more privileged than many of her undocumented peers who endured irregular or off-the-books employment, but unlike many teachers, therapists, and physicians, she was uniformly disadvantaged compared to the people whom she served every day.

She also was clearly an expert in managing complicated relationships, although these capacities were not enough to get home health care designated by the US Department of Labor as more than low-skilled or unskilled work. Handling the demands of multiple people at once, for example, formed part of the complexity of Betty's job, which involved satisfying not just the client but also the person who contracted her services—often the client's adult child. She actually considered that the most difficult part of her job: not handling the elderly clients, even the "hard" ones, she told me, but dealing with their children. Betty used emotional labor to negotiate these relationships: managing her own feelings to find empathy for and suppress more critical reactions. If Betty—a "hard cases" specialist—never met a client she didn't like, as she told me, that testimony likely said less about the clients than it did about Betty and her capacity to handle them in a way that still allowed her to connect.

But the relational complexity that Betty faced was also saturated with imbalances of class and race and immigrant status. Betty was able to solve the case of the disrespectful wife, for example, but only by cooking for her as well, in essence doing more work than she was contractually obligated to do; in this case, Betty's service of cooking for her was also a version of seeing her, and furthermore it was Betty's emotional attunement that helped to recognize the woman's need to be seen as a person to be served. While we can perhaps, like Betty, muster up some sympathy for Willa, whose house was invaded by a string of aides serving her husband, it is true that her expectations and her disrespect reflect the anticipated subordination of the aides, low-income women of color who were not serving a white woman in her own home. There may be any number of factors explaining Willa's position—maybe she was an exhausted caregiver to her sick husband, Mark; maybe she was herself in need of care; maybe she was tired of being unseen by all the aides—but in truth, if a physician or therapist came daily to attend to Mark, it was unlikely that Willa would expect to be treated also, and implausible that she would feel affronted enough to stage a daily conflict about it. Preexisting inequalities created the problem and made Betty's exploitation its easiest solution (for the wife and the agency, if not for Betty).

The mystery we are left with is why Willa wanted to be "seen" by these low-income women of color, who had neither the privilege nor occupational status that helped create Jenna's "magical listening ears." It is not, of course, that Betty was utterly without power. When she told me that "you don't force them to eat, you don't force them to do nothing that they don't want to do," implicit in her statement was the possibility that she could actually, as a younger, more able person, "force" clients to do things. Furthermore, she told me, she could always refuse a client. "If you go to a client and you don't like their behavior, you could decline, just like the patient [is] allowed to decline you too," she said.

As the "hard cases" specialist, however, Betty had mastered an easygoing personal style that perhaps made it easier for her to slough off such encounters. We can see this intentionality in how Betty described her first phone call with Willa: "And I was like, 'She rude,'" Betty said, "but I didn't take it in." Faced with apparent discourtesy, she decides, at that moment, not to challenge it, but instead just to let it go unacknowledged.[15]

Betty and other disadvantaged workers faced the risk of being treated like a servant, of deference demanded and rudeness unchecked. Even as Betty's tolerance is her active choice, it reflects something of the terms of her employment; if she did a lot of confronting, Betty would undoubtedly be less useful to the agency. While patients or clients afraid of judgment silenced themselves when facing an advantaged practitioner, the one who is perhaps silenced here, who chooses not to confront the insolent other, is Betty. The wife was mollified only when she proved able to command her own acknowledgment. Betty may be the "hard cases" specialist whom the agency relied on for her relationship expertise, but to Willa her recognition was something to be extracted.[16]

Relationship Work Denied

While workers without much education often faced demands for recognition from advantaged or affluent clients, the warmth of connective labor relationships sometimes made those interactions meaningful for them anyway, similar to the "love rhetoric" draped over domestic workers' relations with their employers, as researchers Raka Ray and Seemin Qayum argue. Yet all that meaning led to more wretchedness when and if clients violated the norms of friendly relations. Valerie Clausen was a white hairdresser in a small town in Virginia who had worked for twenty years, "through several marriages—theirs [her clients] and mine," she laughed. She could handle the demanding clients, but stopped wanting to when they treated her more like a servant than a friend.[17]

Valerie knew she wanted to be a hairdresser since she was a child "butchering Barbie dolls." "My mother would get so mad because in the bath at night, I would take all the bottles and mix them together to make pretty colors," she said. "I was put on rations for shampoo." She worked in a day care center to pay her way through cosmetology school, but as soon as she graduated, she quit to work in a "mall salon" for six weeks, hoping to pick up a few tricks while learning on less choosy clientele. She spent Tuesday nights at a higher-end salon, however, where the hairdressers let her watch them while they plied their trade late into the night, until an opening came up and she joined their staff. After a couple of years, she left to apprentice at an elite New York City salon that did magazine photo shoots and had contracts with a modeling agency. "It was very humbling to go from being this hairdresser here where you're starting to be the big fish in the little pond, and you go there and you aren't even a little fish in a big pond. You're this minuscule tadpole just swimming around trying . . ." New York was an adjustment, but the heady experience meant that when she returned to Virginia, she was a very big fish indeed. She ended up working for herself, renting a "chair" in a high-end salon where she worked alongside the owner.

When Valerie described what she did with her clients, it sounded just like the connective labor that we've heard about from so many others in this book. "Part of our relationship, besides giving them the service of hair, we're almost like a psychotherapist," Valerie said. "A lot of them come to us and unload. 'Oh, my God. My husband's lost a job. I don't know how we're going to do this.' 'Oh, my son came out as gay. How are we going to handle this?' You know, that type of thing." The hairdressing was important, but the conversation was just as vital, she said. "The hair, at that point, you're doing what you do, that you know how to do: cutting or coloring, and you're talking, engaging, and giving them almost a shoulder to cry on or somebody to bounce ideas off of." Clients just wanted to be heard, she said, so she just tried to reflect what they were saying.

"I have a client right now, her daughter has been cutting herself," Valerie said. "She's been in residential therapy for over a year now. I forget what they—something-personality. I don't remember. I'm just a listening ear and go, 'You did a good job with her. This is something—and you're there for her.' I don't offer solutions. You're just trying to reaffirm to them and get them to come off a little, you know?" The conversation was intimate, deeply personal, and it was clear that what Valerie gave her clients went well beyond their hair. "I'll feed you a listening ear and sometimes a lot of people, they just need somebody to get it out of their chest and their mind and verbalize it and say, 'It's all right.' Especially when it's issues with their kids."

These relationships surely had a financial dimension—Valerie mentioned how they built customer loyalty and how she was able to charge more—but they were also what gave her work meaning. "You get to know a person," she said. "You know their preferences and what they ultimately want to achieve, where their insecurities are that you want to help build away from and achieve that look when they look in the mirror. 'Oh, I am fabulous.'" She laughed. "That's the big draw for me is helping people look in the mirror and go, 'Oh my God, I'm fabulous.'"

Valerie's adept blending of business and friendship was readily apparent, even when she took a phone call in her salon, her side of the exchange illustrating her emotional attunement. In a cheerful tone, her voice showing hints of a Southern twang, she answered the call:

> Can I help you? . . . It is . . . Hey, Denise! How are you?! . . . Good . . . I am—I'm doing very well . . . What can I do you for? . . . Sure! I've

> got—let me see—I've got a three thirty or a four forty-five . . . Which would you prefer? . . . Oh, OK . . . umm . . . [mumbles] We can do a twelve thirty? Is that OK? . . . Cool! [cheerful excitement] This is for you? OK. [laughs mirthfully] I know you call for multiple people, you know? [more laughter] How's your momma doin'? . . . [switches to a lower concerned tone] Oh, no . . . is she all right? . . . Oh . . . oh my goodness! . . . Oh my goodness! [sympathetic tone] . . . Well, that's good . . . Well, I will add her to my thoughts and prayers—and you too . . . Yeah . . . [becomes louder . . . and more businesslike] See you this Monday the 23rd at twelve thirty! Thank you! Have a good weekend. Thanks . . . Buh-bye.

For Valerie, the relational component of her job—the exchange of confidences, the warm conversation—transformed it from a transaction to something else, more like a valuable favor provided to a friend. People sought her recognition, but because they did so in relationship, most of the time it did not feel extracted. "I love doing hair, but I love entering into relationships with people, and that's really a plus of this job," Valerie said. "You can teach a monkey to do hair, but a real plus of it is the humans and interacting together and learning each other. Mostly me learning my client, but my clients learn me too, and that's a real plus of this job."

Although she had only a high school education, Valerie's elite salon training and affluent clientele meant that she was more privileged than many, insulated from many of the peremptory interactions that other service workers endured. Nonetheless, she talked about imperious clients—mostly people demanding looks that their hair could not pull off—and how she handled them, coaxing some, persuading others. "Many when I tell them, 'OK, you can't do that,' they respect me and have faith in my abilities there," Valerie said, but others rejected her know-how. "They want this picture and they want somebody that's going to do this picture. They don't understand that my expertise tells them, 'Your hair's not going to do this,'" she said. "I've had a couple of people want me to do something big and drastic like that on their hair. When I tell them the cons of it and refuse to do the service, they kind of get mad and never come back." Valerie's job was to remake the client in the image they sought; when clients wanted an image she judged was too distant from their reality, she paid a certain price for holding fast.

Valerie also had bad memories from the "mall salon" when she was just starting out, where relationship was discouraged and her time was measured. "Both [kinds of salons], you're dealing with people's vanities, but in a lot of the corporate salons, you basically walk in, you check in, you sit down, you wait for your number to be called," she said. When she worked in the mall, she felt like an automaton, and both the employer and the clients contributed to it. "A big part of it is I want to talk to get to know somebody. Don't tell me, 'I want a haircut. I want to do this, this, this. And don't invest yourself any further than that.' I don't like that." Connective labor is what made her more than a pair of scissors to be deployed on demand, and when clients refused to play their part, she felt demeaned.

Now she worked in a more artisanal environment, making her own schedule and taking her time. Where connective labor went most awry for Valerie, however, was not demanding clients but those who violated the implicit bargain of their friend-like connection, making her feel more like hired help than confidant. She described writing off two clients who were "heavy chemical services" who would repeatedly book three-hour appointments and then cancel at the last minute. "I mean, I'll gladly give you a free service," Valerie said, in a nod to the amorphous friend-not-friend terrain she was traversing, "but when you're doing it by strong-arming, 'Hey, I'm not going to show up, and screw you,'" she started feeling like "they weren't respecting me professionally, [and instead were] totally taking advantage of my generosity." After they each canceled five times—"a day and a half of pay out of my pocket," she decided to charge nonrefundable deposits and they both left. "When they're letting their toxic behavior affect my book—because I hate to say it—at the bottom end there, I work because I love it, but I also have to pay the bills," Valerie said. Each of the difficult clients "was going through some stuff, both of them, because they had shared with me," which made their conduct both harder to sanction and harder to stomach. Connective labor forged bonds that made bad behavior—and transactional consequences—feel like betrayal.

When Connection Is Performance

Betty and Valerie derived satisfaction from their ability to connect with clients; while their disadvantages made them sometimes vulnerable to peremptory or rude behavior, their sense of themselves as good people was closely aligned with the work they did "helping out the elderly" or making people feel "fabulous." But some connective labor jobs are not necessarily for the other's well-being; some jobs do not necessarily offer workers social honor from the practice of seeing the other. These workers talked more about "performing" a certain stance toward the other; inauthenticity was clearly more salient for them.

People who worked in sales, for example, sometimes had to convince clients that they were more connected than they actually were. Keith Joffe was an outgoing white man who had worked in sales for nine years; he described relationships with his clients on a scale of one to ten. "A ten would be . . . like they would be in my life maybe forever, I would say would be a ten," he explained. "And then a one would be completely transactional, like I don't need to see them at all." Given this span, he said, "they would have to feel like we were a six at least, I think, to get a large deal done." But he would not have to feel the same. "I would be OK with it feeling like a three, but I'd be putting on a show that I cared more than I really did. I don't feel like I need much emotionally invested in order for me to sell."

What does "putting on a show" look like? "I give it more time," Keith said. "Like I'm maybe calling them more, I'm asking more questions about their life, I'm sending them gifts. I'm understanding what are little things that they care about outside of work and making those happen. From my perspective, it's less that I'm genuinely trying to connect with them and it's more of I know what that will do for me getting work done with them." The connective labor here relied on a simulation of intimacy, so that Keith is seeing the other, but it is not clear that the other is seeing him. On the one hand, we might wonder—as did some early readers—whether this kind of witnessing "counts" as connective labor. Certainly it does not have quite the reciprocal attunement explored in chapter 2, such as when Wendy, the public school teacher, felt proud when her student told her she was "not the white person I expected you to be." Yet as Keith's description suggests, the client may "see" some version of him, just one that he was projecting, "to feel like we were a six at least." Performing is particularly salient when the work involves not necessarily serving the other's well-being, but instead luring or enticing them within a consumer culture environment.

Yet even more "altruistic" forms of connective labor need not require utter authenticity. Recall, for example, when Russell Gray mentioned pretending to be enthralled in *Grand Theft Auto* with teenagers in therapy; in that case the emotions of the overarching relationship might have been authentic, while that momentary interest was fabricated. Critical care-work scholars in particular reject the notion that skilled practitioners have to gin up authentic love for the other to be able to do their job well. Ultimately, I have come to consider authenticity in connective labor—and the degree to which the seeing is shared—on a continuum. Some encounters on one end become mutual reflections of the authentic selves of each party, and some on the other end involve performing closeness in service to seeing the other, but some form of witnessing is taking place in both cases and in those in between.[18]

When performance is extracted, however, it reflects some of the core inequalities centered in this chapter. I was brought up short by the notion of inauthenticity one afternoon when I talked with a very wealthy white man about my project. We were in a group having lunch as I explained what connective labor was and described the interviews and observations in the research. "Aren't they all performing?" he asked me. At the time, I viewed his remark as revealing much about his own dilemma—how do you know people are being authentic with you when you have so much money it distorts the social environment? In fact, however, the rich man's comment stayed with me as I contemplated the vagaries of connective labor under duress: the degree to which it is forcibly obtained—by the demands of either the job or the client—contributes to the degree to which it is more likely to involve performance.

Inequalities of Seeing and Being Seen

Many mistakes of misrecognition or shame that advantaged practitioners recounted involved interactions with low-income people of color, suggesting that racialized, class-based assumptions may have been clouding their mirror. Many struggle with the dictates of the power they are given, debating the duty to rescue and the call to witness. Meanwhile, low-income workers faced domineering others, the potential for abuse, perhaps an alienation from self. In both cases, inequality serves to suppress or silence the other in the interactional moment. For less-advantaged workers, even those like Betty, whose expertise usefully diagnosed the issue of Willa's recalcitrance, this silencing looks like her accommodation to "hard" clients by doing twice as much work. In the case of less-advantaged others, it looks like the overweight child crying in the clinic when his doctor lectured him about not having soda.

The prospect of this silencing was also what underlies the core uncertainty that advantaged clients might have about whether their personal-services workers were performing, and that advantaged workers had about whether or not they were hearing "the whole truth" from the people they were seeing. It was behind efforts such as Gunika's (and many other practitioners) to communicate: "It's OK to tell me that, because I'm not going to judge you and I'm not going to make you feel bad for saying that." The power disparity muzzled the honest exchange of difficult information coming from either side of the desk, alienating one set of workers as they suppressed their very selves, while making the more-advantaged interlocutor unsure about the veracity of what they are hearing. Connective labor is more likely to go awry when one party is silenced.

Here we see the answer to why people seek connective labor from people whom they outrank in social status. In the case of more-advantaged practitioners, the possibility that they might judge, but did not, lent power to their connective labor, rendering it a welter of risk and reward. For more-advantaged clients, on the other hand, their capacity to command the deference of another—to silence their judgment for a wage—appeared to make these encounters valuable for them. Inequality was not just reflected in these moments; it was also made there, its meanings created and contested. When Willa, who was Betty's "hard case," made it impossible for aides to stay until Betty recognized her need to be included, she was extracting recognition from low-income women of color, via accommodations she likely would not have demanded from more-advantaged practitioners such as physicians or therapists. Through their personalized service, the aides would recognize her as someone worth serving, bolstering a claim to honor that rested on a hierarchy of race, class, and gender. They could refuse that conferral—and until Betty, they did—which made Willa stage an extended protest by withholding her civility until she secured it. When practitioners had less advantage than clients, then, it seemed their recognition otherwise suppressed was part of what was bought and sold.

As we saw at the start of this chapter, the experience of connective labor feeds on these power imbalances. These moments of seeing each other only matter for those who are seen because the workers' opinions have significance, although they do so in different ways. When people go to doctors or teachers, the encounters are rife with the potential for shame, a risk that makes it all the more powerful when practitioners convey empathic reflection. When home healthcare workers or personal assistants do the reflecting, it is their withheld judgment that is for sale, as advantaged clients derive some meaning from the capacity to compel their acquiescence.

This tangle of human judgment and human value also bears implications for automated connective labor, the AI therapist or nurse allegedly free of stigma or judgment. By selling shame-free services, technologists are taking aim at one of the core dilemmas of interpersonal work: the risk of judgment by another human being. Yet by offering recognition without human judgment, the cases in this chapter suggest, engineers are also ignoring something about human value. It is not at all clear that you can have one without the other, because even as the risk of judgment is daunting, it also appears to be part of what makes connective labor powerful.

We saw in earlier chapters how connective labor can be an oft-profound experience. Somehow, sometimes, people transcend the vast gulfs that divide them to offer and experience the power of seeing. While connective labor is valuable, however, it can also pose hazards to practitioners and the people they see. Some of these hazards are what happens when human beings meet other human beings, and two sets of cultural meanings collide. Some are inherent in the inequalities that structure the encounter, as where this research took place, in the contemporary United States. Some, however, stem from the effects of scripting and counting on connective labor, the industrial model shaping a social architecture that shrinks time, emphasizes metrics of assessment and evaluation, and enforces a standardization codifying an unequal order. But there is another choice. The next chapter reveals what it looks like when an alternative social architecture creates the space for connective labor to thrive.

8

DOING IT RIGHT

Building a Social Architecture That Works

For years, Catalina Maldonado had worked as a medical assistant in a harried community clinic much like the ones we saw in chapter 5, with the "hamster wheel" of appointments one after another. "In and out, in and out," recalled Catalina, a compact woman with long black hair who immigrated to the United States from Mexico as a child. "Physicals, hearing test, eye test, urine test, immunizations, the doctor's visit, all in fifteen minutes they had to be done." She loved serving the largely Latino clientele there in California's Central Valley, but was also troubled by the medical care they were getting. "It was like, 'Medication and you'll follow up in a month,' and it was just like this circle over and over again. And nobody—I felt like nobody really ever got better. It was so fast, you hardly ever had time to do anything extra." But the circle came to a stop at her new employer, who had figured out how to halt the hamster wheel.

Catalina now worked for a new clinic—University Transformational Care—reserved for patients with complex medical histories, a facility that made relationships the center of their practice. The founders who designed the unit were guided by a focus group of thirty patients, and how they answered four questions: What's the worst moment you've ever had with regard to your illness, how did you survive, what did health care do that helped, and what did health care do that hurt?

Their answers led to a radical experiment in primary care. Instead of fifteen-minute appointments, the clinic offered first-time patients two hours (one hour with the physician, bookended by a half hour with the medical assistant); subsequent visits were a full hour. At the Central Valley clinic, Catalina saw thirty-two patients a day. At University Transformational Care, she saw six. And when patients were referred to specialists, Catalina was even available to go with them, helping them ask questions, clarifying answers, and relaying what was said back to the primary care physician.

The clinic served a select group of patients, to be sure, but not the easy ones—instead only those people who had three or more chronic conditions and who used five or more medications could visit it. These were the patients who tended to use the hospital more, according to a federal government report, with risk factors about two and a half times higher than the average primary care patients in the system, and health care spending three to four times the average. All were affiliated with the local university—another condition of entry—which acted as their insurer and thus had a strong incentive to save money on their care. For these particularly difficult cases, clinic founders argued, the key catalyst for change was their relationship to their doctors. "If you look at the 5 percent of the patients who cause 50 percent of the money [healthcare spending] in any given year, almost always relational stuff matters and health care has blown it," said Gabriel Abelman, a physician who co-founded the clinic with his wife, another physician, in 2011. The way to help them get healthier, and along the way reduce their spending, was to start with connection, he told me.

I spoke to Gabriel one day at a sunny California picnic table, listening as he narrated the clinic's tale. Short, bald, and intense, he had the driving energy of the charismatic visionary, and his stories were replete with pride and humor and scorn for those who would stand in their way. I asked Gabriel how much that initial panel of thirty patients talked about relationship, when the clinic was in its design phase. It was "almost universal," he said. How did they phrase it? "'Someone who knows me and cares about me,'" he recounted. "'Someone who has my back.'"

Catalina was one of University Transformational Care's secret weapons—the reason why they were able to offer so much time—in the battle for their patients' hearts. In most practices, the medical assistant is the person who puts you in a room, takes a few measures of weight or blood pressure, and then gets out of the way. Their training is fairly minimal compared to other healthcare professionals, and they need just a high school education. But at the new clinic, they were asked to do much more. "So, what we did was, [we asked,] 'What are all the functions that people need, based on our interviews and our knowledge, and then how do we put most of the functions that are not requiring a physician in one person?'" Gabriel explained. "And so that became the medical assistants, and they're the person that—rather than a series of tasks that they do, they have responsibility for people."

That responsibility meant that the medical assistants, called "care transformers" at the clinic, stayed in the room while the physician was there, served as a scribe, managed phone calls and appointments from patients, presented their details in all-clinic meetings, and accompanied them to specialist visits. "Most of all," Gabriel said, "because they were part of what we call 'crucial conversations,' the patients trusted them, so when the patient would call in and get Delilah on the phone, they weren't saying, 'Just have Dr. Abelman call me back,' it was 'Delilah, can you help me?'"

The expansion of the medical assistant job was at the core of the University Transformational Care model, what allowed them to offer so much time to the individual patients without breaking the bank. To get there required convincing not just patients about the potential and possibilities of medical assistants, but also doctors. Gabriel described an exercise where they put physicians and medical assistants in the same room, talking about how they ended up there. "The doctors' jaws fell open when they realized the depth and passion of the MAs, which was every bit as vital [as their own]," Gabriel recalled.

> The doctors would say, "When I was ten, you know, my parents are doctors, my uncle's a doctor, my aunt's a doctor, all my siblings—I was always going to do it." Versus these MAs who said, "Well I was undocumented, I came—You know, I was a DREAMer, I came in at age four, my grandmother was sick and I cared for my grandmother. Then in high school I realized I couldn't go to college because I'm not

> documented, so [being an] MA was the best I could do and here I am." And you just sat there and went—

His jaw dropped in simulated shock. "And so it changed the doctors' view of who they were working with."

The clinic hired medical assistants of a particular kind—"people who were getting in trouble for doing too much," Gabriel said—and trained them for about a month, including mentoring and shadowing. They embraced the work, he said. "[It was a] big job, they loved it, and the job satisfaction [was] through the roof. I mean they wanted, 'Can't we be on salary so I can work at night and weekend?'" In contrast to many primary care providers, there was almost no burnout reported. Was that because the clinic was careful to restrict their hours? I asked him. "No, it's because people have meaning and relationships in work," he said. The relationship model fed the workers as well as the patient.

Catalina told me she could not imagine loving her job more. "It feels very rewarding," she said. "Like I go home and I feel like I've accomplished something. Even when work is hectic, I can sit at my desk and say, 'Wow, I love what I do.'" We were talking in an exam room in the clinic, as she sat next to her computer. The contrast between the new clinic model and her old employer was stark, she said. "I feel that [the Central Valley patients] didn't understand really what medical and health care really is. And the importance of taking care of yourself and kind of taking charge. Here we kind of want to push our patients and we help them, right? But there, it just seemed like if you didn't help them, they just kind of got lost." Suddenly her voice broke, and she looked down. She wished her own parents could get this kind of care, she said, tearing up. "Everybody deserves this."

Connective labor has enormous power, in conveying meaning, dignity, and purpose for workers and their charges. Yet its power does not solely depend on a particular worker's skill or a client's receptiveness. Instead, as chapter 5 outlined, organizations have a social architecture that can make it easier or harder to see each other well. As we saw there, too often firms are getting it wrong, cutting corners in service to profit or efficiency campaigns that ultimately weaken the relationship and drain it of its power. The demands of scripting and counting crowd out the seeing and hearing, and the degradation of connective labor makes automation more appealing, even more likely. But is anyone getting it right?

This chapter will explore the contexts where connective labor thrives, and identify some of the factors that help it to do so. Where and how is this work supported? What does that entail? How do people build a social architecture that works?

A Social Architecture That Works

Doing the research for this book meant that I had the opportunity to meet and talk with scores of people who provide connective labor for a living. An unusually warm, reflective, engaging group, they talked about helping people grow, seeing their dignity, finding the powerful meaning and purpose of their work. It felt sometimes like bathing in human light.

Many of them came to seem like heroes to me, battling it out for connection to others in need, despite expanding client rolls and demands for metrics. Katya Moudry, whom we saw in chapter 6 lamenting the survey she had to give to a military veteran who was thinking about suicide, was a good example. A therapist who recently joined the VA hospital, she was chafing at their requirements that she see more people more efficiently. "It is a lot of mental effort to fully engage, to give that part of yourself to another person for a while, and that's the part I'm not willing to give up," she said. "I've been questioning lately, 'Is this sustainable? I can't just jump in and be a hundred percent with someone and feel their emotion. It is what I have to give to somebody, but if I have twenty people, do I have the capacity to give it to them and to my personal life?' If I had to choose between seeing more people and not doing that, I'd rather bow out of the profession than not give it all." In the face of productivity pressure, Katya faced two daunting choices: to either make her connective labor more efficient, betraying her commitment to "be a hundred percent with someone," or to take on an unsustainable load as an individual hero.

Yet an organization that impedes connective labor, and relies on its employees to forge relations on their own without support, will surely not be able to offer connections to its clients very well, or for long. Stories of people's suffering and redemption encourage us to think about workers as individual heroes, as people with the "magical listening ears" that we heard about in chapter 7. As psychologists tell us, however, empathy is not wholly innate but in part depends on conditions that produce it. Studies in schools, for example, have documented particular circumstances that help support teacher-student relationships, which generally involve increasing the time and opportunity that teachers have to connect with particular students, from smaller classes to block scheduling to protecting them from the impact of accountability measures. In short, the conditions of connective labor shape the kind of work people are able to do, and ultimately, no one works alone.[1]

Drawing from my own research as well as that of others, there appear to be three vital components to supporting connective labor: an organization's *relational design, connective culture,* and *resource distribution.* By relational design, I mean the ways that people are put in relation to one another, such as through leadership, mentors, and peer groups acting as sounding boards. By connective culture, I mean the shared values, practices, and rituals that serve to bind individuals to these meaning systems. Finally, the resources that matter start with the time an organization allows for a given interaction, but also includes the ratios of worker to client, the cognitive load workers face, the extent of technology and data use—any of the material conditions that underlie and shape people's connection.

"A Gigantic Dose of Relationship": Relational Design

Before a client even enters the room, the people of an organization exist in relationship to each other, and the relational design, or the way people are positioned vis-à-vis each other, can impede or support the connective labor that the client receives. Relational design matters in three particular ways: when leaders can convey a relationship-centered vision, when mentors can help shepherd connective labor of others, and when peers are able to build a community of practice for feedback and consultation. A relational design helps others—leaders, mentors, and peers—provide connective labor to the practitioners themselves.

Gabriel Abelman, the cofounder of Catalina's clinic, was a relationship evangelist. A friend and another physician, Mark Aaron, himself an influential figure in state and federal medical policymaking, noted how unusual Gabriel was. "Gabriel thinks that it's all about relationships and that if you don't at the very beginning signal that you are personally invested in this person and care about them, it's not going to work," Mark said. "He believes that particularly for people who are struggling—and many of whom have been wounded before by bad relationships in health care—what they need is a gigantic dose of relationship."

The reforms that Gabriel and his cofounders were proposing as part of setting up University Transformational Care were not uncontroversial. "They hated him on campus because he would insist on spending so much time with each patient at the beginning," Mark told me. "He actually would draw them out on depersonalized adverse experiences they'd had in health care. He said getting it out on the table and letting them emote helps them then re-trust somebody."

Gabriel's approach to relationship in medicine was forged in his own desperation. Sitting at the picnic table, he narrated for me how he had initially gained notice for his novel approach to diabetes patients, an approach he developed almost by accident. Diabetes was the disease that most frightened him in medical school, he said, because a close friend's father had had it—"I loved the guy because he was really funny, and I watched him go blind, lose his legs, lose his kidneys, and die"—and because the medical approach was so primitive—"'You're going to die about twenty years early. Minimize medical mistakes and do the best you can' was kind of the message."

Shortly thereafter, he found out that he himself had the disease. Frantic to do something different, he knew the technology to self-monitor glucose was just emerging, and became an avid experimenter; at the same time, almost instantaneously, his clinic roster was full of diabetes patients. "They all came to me saying, 'You understand us,' [while] I kept saying, 'How the hell do I live with this?' to them, so the usual doctor-patient separation broke down and we were all just a peer group," he said. "Over thirty years, they all did well, you know, pretty much." What he learned about patient self-management he was able to implement on a community-wide level, he said, where deaths from diabetes declined by 29 percent over eight years. But the path to that understanding—the patients helping him with their accumulated knowledge—was a partnership model that he since turned to again and again.

Leaders are vital for generating a connective culture—perhaps more than one might think, given that they are not actually party to any particular interaction between workers and clients. By leaders, of course, I mean not just managers—people in a position to organize other people's jobs—but those who have the "moral, intellectual, and social skills required" to influence others. It is not that people look to leaders for particular tips on how to conduct relationships. It is not even that leaders who value their workers necessarily produce workers who value other people. Instead, leaders are important for developing a shared vision in which connective labor matters, for establishing or reinforcing a culture that conveys that seeing each other is worth the effort, and for creating the time, space, and recognition for such emotional collaboration.[2]

It's a truism that inspiring leaders offer a vision that rallies people; it is when that vision centers relationship that they create a social architecture that works. Scholars have written about the importance of flexible, relationship-focused skills such as "trust-building" as a fundamental part of strong leaders, who know what motivates employees, what concerns them, and what conditions they require to do their best. These findings tell us more about how leaders themselves do connective labor, however, rather than how they support the connective labor of others.[3]

But when does leadership help other people make stronger connections themselves? Education scholars have tackled this very issue, for example, as they try to crack the mystery of the principal, who rarely teaches yet profoundly affects student learning. Research suggests that a good principal is responsible for 25 percent of a school's impact. Scholars have developed a model of "transformational leadership," through which school leaders are said to affect teacher motivation and commitment, which then in turn affects student learning. Wrote one influential review: "The principal's efforts become apparent in . . . school conditions that produce changes in people, rather than in promoting specific instructional practices." Principals affect teachers' emotions at the organizational level "because they promote a culture of care in the schools." While principals can influence individual teachers' emotional well-being, by affecting how an individual teacher feels about their work, they also create a particular emotional climate in the school overall, with implications for morale, engagement, and a shared culture focused on relationship.[4]

Bert Juster was one such principal and a founder of a new all-boys independent school in a West Coast city. He had worked at a vibrant all-girls private school in the same town for years, but noticed that "I was working to prepare these really confident and capable women, and then they would go to high school, and kind of be with schmucks," by which he meant boys who were sexist jerks. He found himself inspired by the book *Boys Will Be Men* (1999), which described how communities can raise boys in a collective committed to racial justice. "I actually think my calling is to work with men, because that's what our world needs more from me," he said, and with a woman colleague, he set out to found the new school. Ten years later, the school had a strong reputation and ringing testimonials on local parent websites.

Relationships were fundamental to Bert's vision of good teaching. "Teaching is kind of like the art of relationship to some degree," he said. "Like, it is this dance of kind of working magic with their brains, with them understanding themselves better, and you creating a platform for that." As a head of school, his task was to help make those relationships happen, and he talked about the intentional practices they had put in place to emphasize them, from daily greetings to "restorative justice circles." "Education is such a relational business," he said. "Middle school, loving kids is almost more important than content. Yeah, relationships are super important in this world."

Bert's focus on relationship made him more open to hiring people who did not have teaching experience or credentials, he said. "I'd be willing to take a risk on someone who's super exciting," he said, offering as one example someone he hired to teach welding, design, and fabrication, who had "never taught a day in his life," but "I could tell pretty early on that the boys were listening to him, responding to him. He came in and it was a little touch and go, there was a pretty steep learning curve, but he worked under somebody who was really good at guiding him, and now he's one of the best teachers we have." For Bert, a teacher's capacity to connect to the kids was such a priority that he was willing to risk bringing on someone who had no formal training in how to teach.

In contrast, some leaders seemed less committed to relationship, or perhaps incapable of building it in as a systematic priority. Conrad Auerbach taught at a public middle school in a large Virginia city. He talked about how he strove to see his students, to get to know them, and how his teaching had changed over time as he got better at it. Not everyone in his school was on board, however, he said. "One of the recent things my school did, it's kind of funny. I think they got wind of a few teachers who were just putting absolutely no effort into relationships, so I think our principal printed out a list of twenty things—they left it in the men's staff room in front of the toilet, I'm not sure if it was on purpose—but [it was a list of] about twenty or twenty-five ways to build relationships with students." Conrad's principal may have been just as dedicated to relationship as Bert Juster, but stymied by a few recalcitrant refusals. Still, hanging a list of twenty-five tips in the men's staff room in front of the toilet was not going to transform much, and I said so to Conrad.

"But when, when would we do that?" he asked. "It's just impossible for teachers to be trained on all the things. We have a few days of professional development and there's a million things that we're supposed to be talking about. Sure, we do some, but . . ." He trailed off. At Conrad's school, relationship was another on a long list of items for which teachers needed training; the "How to Connect to Students" list tacked up in the restroom was merely an attempt to use absolutely every minute for that training. Leaders communicated their priorities, and what they made space and time for conveyed that vision.

Mentoring and shadowing also contributed to people's capacity to see each other, when the relational design enabled a trainee to work closely with someone more experienced. As a clinical practice, connective labor was often trained via apprenticeship, like the teachers and therapists who worked under supervision for months and sometimes years. Apprenticeships combined experiential learning, considered vital for practical skills, with the connection to a mentor or supervisor who could nurture the development of connective labor. Indeed, mentoring—itself a form of connective labor, requiring understanding and reflecting the other so they learn more about themselves—can actually enhance people's capacity for self-awareness, empathy, and communication.[5]

People also valued mentoring for the opportunity it offered to talk about different approaches. Bella Albrecht was a therapist who—unusually—started her own private practice right after graduate school. Most of her classmates joined a clinic or school, with some sort of structure around them, guiding them as they started, but Bella was much more isolated, and thus very dependent on her supervisor—someone with his own practice who agreed to meet with her periodically—as confidant and guide. When Bella went to him about a recalcitrant client, he urged her to share more of her own feelings with the man in the sessions—feelings like "I'm getting frustrated"—but she was skeptical, thinking that kind of messaging would be threatening even for her close friends to hear. She ended up switching supervisors to someone who offered a different framing: the new one urged her instead to simply be more human with her client. "This was the difference that changed things," Bella told me. Her new supervisor said, "'You don't need to be his therapist, you just need to be a person that he feels is supporting him and that's it.' For me, all of the anxiety and the not knowing what to do and feeling like I'm not getting anywhere with him sort of dissolved, like, 'Oh yeah, I can just be a human, great. Easy.'"

And with the client, the new framing opened the spigots. "It's so funny how quickly these things can snap," Bella marveled. "He finally shared some things that he'd never spoken about before. He's never told me some of the things in this last month and a half under this new supervisor." Mentoring had gotten her unstuck, and as a result her client was also freer. "So even that [mentoring] relationship that I have external to my clients," Bella observed, "it has already profoundly impacted my work and my relationships with my clients." Mentoring provided the crucial guidance to allow people to frame and refine their connective practice, making space for trying out new ideas, acting as a sounding board to enable change.

Mentoring was also an opportunity to pursue relationship as a value, to find support for making connection a cultural priority. Andy Cooper was a physician dedicated to humanistic practice, with his own research investigating how to help physician empathy and mitigate burnout; his warm mildness made it hard to imagine he would ever need guidance on connecting to patients. Yet he credited his own path to early mentors.

A gentle giant of a man, Andy was about sixty, white, with an air of calm expectation, like you might be about to tell him a small joke or some good news. He had toyed with the idea of a research career, he said, but always came back to "that human connection I found actually more satisfying." An early trip to Nicaragua to help a study of diarrhea conducted by a physician there—a man "very, just so humanistic and so caring as a pediatrician"—cemented his decision. "That just continued to make me realize that another passion of mine was health care for those who need it the most. I started getting more involved with homeless healthcare and then realized that I really wanted to be a doctor." Along the way, he kept seeing the kind of caring practice he aspired to, he said. "I had the opportunity to, you know, work with a lot of doctors who are very humanistic as a role model," he said. "So I felt it was supported. I mean obviously there were parts of med school where you felt like in certain rotations it was a grind. There was, you know, no sleep and you were sort of stripped of your humanity sometimes. But that was pretty rare."

Andy had also served as a mentor himself with many medical students over the years; one day, I observed him with students in his own practice. I joined Andy and the student as we stepped into the clinical exam room to see Mr. Thompson, a seventy-five-year-old African American man with a host of chronic diseases who was sitting against the wall. Andy sat down next to him while the student positioned herself in front of the computer, ready to take notes (I sat down on the other side of the tiny room). "I can just tell you what's going on with him," Andy said to the student, and recited a lengthy list of facts about his patient. "He has heart failure, kidney failure, he had prostate cancer but that's cured, he has edema but it's stable."

Mr. Thompson broke in: "I've had that for sixteen months."

"That can cause him to gain weight, but he's lost weight, if anything," Andy continued.

"I'm 177 now, I used to weigh 220," Mr. Thompson added.

"They've got him on Ensure," Andy kept going, and Mr. Thompson chimed in, "But it's been six weeks since I've had no Ensure. They've ordered it."

"They've ordered it," Andy repeated, adding for the scribe, "He's followed by a dietitian." At that point he asked the student "what was going on metabolically," which she answered, mentioning "electrolytes." Andy responded, "Well, sure, but what else does that cause?" The student replied, "Muscle wasting."

Mr. Thompson appeared to be enjoying Andy's performance. "He knows all about me. For sixteen years he's been my doc. Before that, Dr. Eton was my doctor." Andy smiled, and noted to me as an aside, with gentle amusement, "Dr. Eton is now the chief of staff," an exalted position at the top of the medical center food chain.

The student sat facing the computer, entering the data gleaned from the visit, but she also asked some questions of her own—"So you are new to [the town]?" "Are you sleeping well?" "Do you take your blood pressure at home?" After the patient would answer, Andy would sometimes interpret it for the student, saying, "Social work can help with that," for example, in references to his complaints about car troubles getting in the way of his access to his appointments. The visit continued, with Mr. Thompson talking about some new worries—"I have trouble with my throat"—and Andy discussing the medications he was on. "You keep my medications up to date," Mr. Thompson said. "Thank God for you!"

When it came time to take his blood pressure, Andy and the student both leaned into Mr. Thompson, reading the number on the monitor. A stillness descended upon the room while the man's heart slowly pumped the blood out to his toes and back again. I had the sense of a sacred ritual being played out before me, and we all stayed quiet until Andy pulled back and announced, "Blood pressure of 162 over 82."

"It will be OK as soon as I leave here," Mr. Thompson said. He seemed to be trying to get Andy and the student not to worry or take any prolonged action. Andy said instead that they would take it again, but in the meantime they discussed whether he could see the social worker to get help with the Ensure. “The woman who brought me here, she’s got someplace to go,” Mr. Thompson said doubtfully, referring to his ride. When Andy asked whether she could wait, Mr. Thompson talked about how she had forgotten that he even had an appointment that day. She had told him, “Well, everybody forgets,” Mr. Thompson said, but then he observed primly: “She should have written it down if she was like that.” Still, he added, philosophically, “You can’t say that to someone, you have to be calm and collected, otherwise people will be through with you pretty quick.” Surprised, Andy chuckled with real enjoyment at this bit of folk wisdom.

They talked about some pills he was taking, and that he needed to get a colonoscopy. Andy leaned in and touched the man’s knee. “Can you try to reschedule the colonoscopy appointment, and get a ride with someone?” Mr. Thompson heard him, the touch getting his attention. “Yes,” he said. Later on, in an interview, Andy told me he used the touch on purpose, to accentuate Mr. Thompson’s role. “I think if there’s something really important and they actually need to do something and be proactive, you have to sort of really connect,” he said. “I wasn’t facing him but I tried to, I think, and touched him. Because there’s a certain responsibility. Like getting those tests like the colonoscopy that he needs to do that I can try to troubleshoot. But it actually takes a responsibility for him to get the ride. I can’t really organize that. It takes time for him to call the GI suite to schedule it for his time frame. So there’s some important things that are really his responsibility.”

After Mr. Thompson held forth for some time about his financial troubles—his car had broken down, and he couldn’t pay his utility bills, and Andy told him and the student that the social worker could help with housing and transportation—they rechecked the blood pressure, this time rolling up his shirt sleeve. The same stillness returned to the scene. “Perfect,” Andy announced. “138 over 63.”

Mr. Thompson moved to the examining table, and Andy had the student feel his chest through his shirt. “Feel this, he has a heave, feel the chest here, that’s from surgery, the generator is up here, but feel that right there, pressing against you, it is the right ventricular heave, the heart is very big," Andy said, to the student, and then to Mr. Thompson, joking, "Your heart is very big, Mr. T. We've known that for some time." "Is it worse?" Mr. Thompson asked, disregarding the joke. "You're a very nice man," Andy said, explaining the joke, but then, soothingly, "It's about the same." He and the student put on their stethoscopes to listen to his chest. "I didn't hear anything. He usually has these little crackles," Andy said. The student said she heard the crackles. "That's a good pickup," Andy said to her. "The fine crackles, it's not unusual for heart failure patients."

Andy spoke to Mr. Thompson: "We may have dried you up too much, we'll give you less of the Lasix. I know you don't like that Lasix." "I can't get enough water," Mr. Thompson said. "And your weight is down, and the kidney function is a little worse," Andy said. "But we don't want you to have too much water because you'll end up in the hospital like last time."[6]

That day he concluded the visit by asking Mr. Thompson, "Did we address everything?" "I think so, thoroughly," the man replied.

The visit was a study in connective labor by a master of the craft, and like us, the student was in a front row seat. In addition to the blood pressure rituals, we can see Andy's attunement to Mr. Thompson's worries about his finances, about being "too dry," about how he managed the uncertainties of his ride. We also can see his use of touch to elicit Mr. Thompson's attention. It was highly collaborative, with Mr. Thompson chiming in and attempting to shape the way they understood his first blood pressure metric, all the way to the end, when Andy checked in on whether Mr. Thompson thought they were finished. It also featured a gentle open-hearted sort of affection, a positive regard between both of them, built up over sixteen years.

Andy's mentoring was similarly deft. He watched—and stepped back a bit—as the student asked the patient questions, trying to both connect to him (Was his new apartment in a new town for him?) and to elicit medically relevant information (Was the blood pressure the same at home?). Andy gave low-key quizzes (about the electrolytes), some helpful asides about what to worry about (retaking the blood pressure) and what not to (tasks the social worker can take care of), and positive affirmation ("That's a good pickup"). Students slow you down at first, he told me later, but are more helpful as time goes on; this student even put some orders through on the computer while Andy was attending to Mr. Thompson.

"I've had some students where I thought: 'Well, God, this is going to be a struggle,'" he said, raising his eyebrows. One such student was planning on being a radiation oncologist, and "they actually need to have some hands-on skills too, because it's not just the radiation, but they are oncologists and they do need to examine the patients," he recalled. "I worried because he struggled with his physical exam skills and he failed one of our major examinations that we do here. I worked with him and worked with him and he actually got better, and it was just a matter of him slowing down and being meticulous about each component of the exam and me just guiding him through that again." Later Andy got an email from the student just as he had finished medical school, "you know, just thanking me and saying, 'Guess what? I got the intern award for the best clinician.' And I thought, 'Wow.'" Andy took mentoring seriously.

Mentoring seemed valuable less for conveying particular tips about how to connect to patients or clients, and instead for supporting the notion that relationship mattered. "I actually felt lucky," Andy said, recalling his own training. "I felt like I worked with a lot of physicians in my training that were very humanistic. I remember it was at the height of the AIDS epidemic. I worked in an HIV clinic at the time. Initially, they really didn't know what it was. But the doctors there were just so dedicated. And I could see them as role models." Mentoring helped light the path toward a relationship-centered practice.

While leaders and mentors are important, practitioners also seemed to hunger for groups of peers with whom they could share ideas and struggles. Of all the groups I observed, the chaplains had the most occasions for this kind of exchange, from formal presentations and feedback to more ad hoc processing of moments during and after their shifts, and I watched many instances where they shared their struggles with each other to derive guidance or comfort. They gathered regularly to listen to each other, sitting around a conference table in a tiny room they borrowed weekly from a sleep science research group, which had posted a few signs in the hallway about being quiet.

One day, they were discussing how to serve new parents when they confronted baby sickness and sometimes death. Leading the discussion was fellow trainee Cassidy Lewis, an angular woman who had made it her project to win over the NICU nurses to chaplain service. She even looked like a labor and delivery nurse as she stood next to the blackboard, wearing a NICU fleece, scrubs-like pants, and flat shoes, with a hospital key chain around her neck. Cassidy had a different sort of emotional energy compared to her fellow chaplains; she had a quick impatience, even a bit of a raw edge that was unusual among the warm and gentle group. On another occasion, she admitted as much, telling the group bemusedly that she was in good contact with her anger.

Talking to her fellow chaplains about serving parents in grief, Cassidy was blunt about the challenges they would face. "This is what's unique about labor and delivery patients: when babies die, that's what is different, they never had a chance, they never did anything wrong. Babies are considerably more complicated for people's theodicy [an explanation for why a good God permits the existence of evil] because there was nothing they could have done, so it complicates theodicy. There are none of the things that give hope at the end of life—the life they've lived, the things she's experienced."

Margaret, a middle-aged white woman with three kids of her own, commented that the trauma of these new parents was "so close to my own personal nightmare that it is hard for me to be present in the way I want to. How do I get past that?" "For me it's the opposite," a single woman about ten years younger said. "I have no experience with babies and I'm a little scared of them." Erin Nash mentioned that L&D was a sort of closed unit, that she would be up there sometimes and the nurses would say something like, "Whoa, there's too many people."

Anne Reynolds, the head trainer, then led them in an exercise, asking them to stand up with a ball of yarn. "This could come too close to you if you have your own experience of this," she cautioned. "I do have students come to training having never processed a loss they have had before." The chaplains stood in a circle, tossing a ball of yarn to one another while they said out loud what came to mind when someone they know and care about was going to have a baby. They said things like "I can't wait to meet the baby," "I hope it's a son," "I hope you let me get involved," "Joy," "I know you've been trying for this." Despite the artificiality of the moment, they seemed to lose themselves entirely in the exercise; it felt like they were actually talking to real people who were not there. The yarn got more and more tangled as they tossed the ball back and forth. After a while, Anne stopped the sharing. "OK," she said, "drop the threads." The yarn fell to the floor, looking like a bedraggled heap of abandoned networks.

Anne then directed them to go around the room saying what came to mind when they heard that the baby had died. The response was immediate and visceral—some of the chaplains cried, and one got a tissue—it was very moving to watch. The chaplains reminded each other there could be ambivalence, that we did not know how the parents felt about the birth or death. But the overall mood was deeply sad, and a pall descended on the room. I wondered to myself whether the yarn helped people focus on what they were saying as if it were real; the tangle on the floor had an air of finality.

Sounding boards allow professionals—not just students—to be vulnerable in front of each other, to process difficult moments aloud, and to learn about alternative approaches. For the chaplains, their sessions were an explicit moment to think about relationship, including how they related to one another. "That's just space for them to use," Anne told me. "A big piece of it is also reflecting how they [the chaplain trainees] experience each other, because they [educators] don't really train us for that in divinity school or seminary. They don't really teach us how we're coming off to other people." One chaplain had even spent a meeting coding the way the group interacted, keeping track of critique, support, conflict, and response; Anne showed me the result (see figure 8.1). The group then spent twenty minutes talking about what the chart revealed, Anne told me. The image meant they were able to see "where the lines of sort of communication were going," she said. "This one was pretty even, actually."

Outside of formal training programs like that of the chaplains (and even in some of those), these communities are not at all inevitable, and

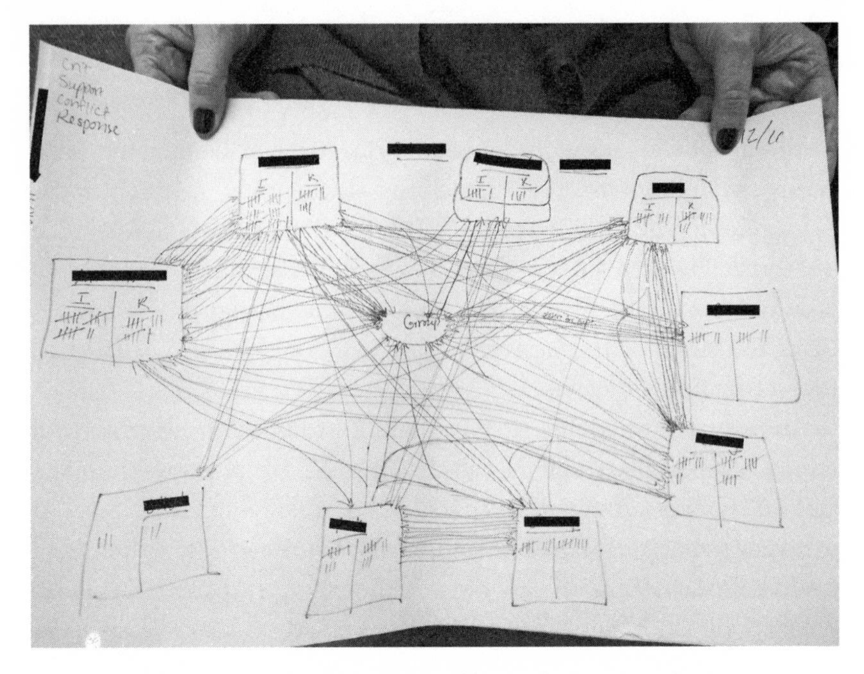

FIGURE 8.1. A chaplain's rendition of group dynamics in discussion.

often require deliberate nurturing and commitment. "The extent to which caregivers are emotionally 'held' within their own organizations is related to their abilities to 'hold' others similarly," writes the organizations scholar William Kahn. Joyce Fletcher, another organizations researcher, called that nurturing the work of "creating team," arguing that "individuals who feel understood, accepted, appreciated, or 'heard' are more likely to extend the same acceptance to others, leading to a kind of group life characterized by . . . a zest for interaction and connection."[7]

Just a few weeks after the chaplains practiced thinking about grief and babies, the exercise became real for them. Margaret came into the tiny sleep lab room from being on call all night, and told the group she had been called to a hospital room where a little boy had just died of cancer. Yet when she got there, she realized she did not know enough about the suffering family to be able to do much of anything helpful. A nurse who knew the family's situation better stepped up and said, "the most beautiful heartfelt thing. The nurse was a better chaplain than I was," Margaret said, with gratitude and regret all at once. "It was so beautiful and really spoke to the heart."

When she had finished, Anne, the group leader, asked gently, "What would it be like to just trust that you are there? What would it feel like to let go of that feeling uncomfortable, that you are not enough?"

Margaret looked at her and repeated her words: "To let go of that . . ."

Hank jumped in. "There's power in just you seeing what you saw, being the witness," he declared. "Those moments of deep, deep grief. Just seeing them validates something, makes it real. Even that—I think there's power in that."

The case had also troubled Cassidy, the chaplain in the NICU fleece, and when it was her turn to share, her characteristic confidence faltered. "What sticks with me when a child dies is not their dead body, I can handle that, even though it feels a little bit callous to say, that is not what haunts me," she said. "What haunts me, what keeps replaying in my mind, is when a mother—there are mother's voices in my mind that I can't get out. This mom was screaming, 'What did he do to deserve this? He's only three. What did I do?'" The memory stayed with her, Cassidy said. "If everything else fades, what continues for me is the women's helpless cries for change, and there's nothing that they can sacrifice to make it different. And I couldn't get her screams out of my mind last night. And I don't know what to do with that, because I can't . . ."

"There's this box of screams, but when another one is added it just brings back all the other ones." Erin, her fellow trainee, responded to her agony. "I wish we all had pillows right now, I wish we could scream into the pillow," she said. "You're holding all these sounds and images, and the toll it takes on you and all of us is worth lamenting too. Where is our space to lament for what seems like an impossible profession sometimes? Next to the Sleeping Lab! I seriously would invite us all to lament as loudly as we could, go out and smack a tree or something, scream, go out and scream at God, or whatever."

Later on, on a brief break, I was in the restroom with Margaret and Cassidy, who were still talking about the little boy's case. "You did something important there, you did what had to be done," Cassidy said. Margaret agreed weakly.

Back in the tiny room, before the group left for the day, they talked about the community they were building. Anne warned them that they may not have the same kind of community in their future jobs that they had forged in their training. "It is hard when and if you move into chaplaincy, it's impossible to rebuild a place like this, a community like this, people who understand what you're doing on a daily basis. Your family can't do it, and it's not their role for you to sit them down and say, 'This is what happened last night.' It's important to find people who can hear you and understand." We might be used to turning to our family for solace, but they could not serve as a sounding board for these kind of experiences, she said, urging them to find other sources of comfort. Otherwise, "it can be really lonely." Workers without sounding boards were on their own, serving perhaps as individual heroes but without the support that could make their paths sustainable.

Sounding boards require support from their organizational settings, so that people have the time and capacity to see each other regularly, to talk about their practice, to develop a history with each other. But they wither in conditions of burnout. After Margaret's session concluded, I ran into her in the parking lot, just before she finally went home. She told me it's really important "to unwind, to process it all so it doesn't build upon itself," so you don't feel like every child on a bike is going to get hit by a car, or every car is not going to wrap itself around a tree. You have to realize it's really extreme, she said. "But sometimes it's hard to do that, under overwork conditions."

At the University Transformational Care clinic, they were explicit about trying to create such a community, particularly because they were relying so much on medical assistants to expand their role. Yet the status differential among the participants could make a discussion feel a bit more like lecturing and a little less like sharing. One Friday morning, I sat in on their weekly clinic meeting. In attendance were physicians, social workers, nurses, and a cadre of former medical assistants who were the "care transformers"; most of the transformers seemed to be Latina and most of the others appeared to be white. At the meeting, the transformers presented the profile of a given patient—who they were, their medical goals, their history—and the group talked about their care, as well as any issues or problems that came up in the week.

The meeting was supposed to create a team, and to provide a site where people could talk about the hard work of connecting. They started the meeting with a meditation, heads bowed as they listened to music. They made plans for a retreat at a winery, and somewhere in the middle they handed around a "team jar," in which people commented on appreciating someone else ("Janine, thank you so much for helping on a patient on Monday who went to Express Care, you're amazing"). As one of the nurses announced at the meeting, "If we don't take care of ourselves, we can't take care of anybody."

But the bulk of the meeting was about a problem that Dorothy, the social worker, sought to solve. It turned out that a care transformer, Luisa, had been stepping out for lunch when she passed by an anxious clinic patient waiting for lab results. The patient was upset, pacing, spiraling out of control, wondering out loud about why the lab results were not forthcoming when her EKG was normal. Luisa interceded with the lab reception to make sure she was seen more quickly, and then sat with her until she got her results a few minutes later.

"I did eat. I took her to the café, and then I went back to work," Luisa reported.

"That was very generous of you," said Dorothy, but in a way that suggested it was a mistake. "She has a history of medical trauma and severe anxiety."

"I understand boundaries are important," Luisa said. "I don't know how to meet her needs and still have boundaries."

But the social worker seemed to blame Luisa for responding. "I want [the patient] to practice saying, 'I have a panic disorder and it would help all of us if I could be seen,'" she said, as she held forth about being an "overfunctioner," apparently meaning someone who helps others more than is good for them. "As a woman, I know I can be an overfunctioner for someone else," Dorothy said, noting how "that's tough behavior to change." Nancy, a clinic physician, chimed in: "Our model of being infinitely accessible is not good for this patient."

While the clinic meeting moved on to discuss other items, Dorothy circled back around to talk again about boundaries for the last ten minutes, while people fidgeted and rustled papers. Luisa's confusion highlighted the difficulty in drawing the boundary between serving and overserving, especially in an institution set up to meet needs that had been forsaken by others.

It was a little hard for me to gauge whether the meeting was sowing community or division, however. On the one hand, the air of reprimand made the meeting seem more like a site of evaluation rather than one of refuge and experimentation. But later on, Catalina told me she found the clinic meetings effective for team building at University Transformational Care. At her old practice in the Central Valley, they had monthly meetings that were much more fraught, she said. "I felt that there was a lot of tension already built up from a lot of issues. But it's like built over a whole month. How do you not explode at that moment, right? Or how do you even bring it up at that moment, you know? You just kind of held it in already for so long. I feel like if they had a lot more, like, team building . . . you can kind of set the guidelines of 'This is the right way to do it,' right?" Perhaps what I had perceived as a reprimand from Dorothy was to the rest of the group an instance of helpful feedback, especially in light of their regular meetings, and the meditation, "team jar," and other ritualistic moments they used to gather themselves together.

"How Do We Systematize Love?": Connective Culture

Relational design, or the way people are positioned vis-à-vis each other in a given organization, can help contribute to a supportive social architecture. But leaders, mentors, and sounding boards are not sufficient on their own; a successful social architecture seemed also to rely on having a shared set of beliefs and practices that reinforced empathic emotional collaboration. Common values, norms, and rituals comprised the connective culture that gathered people under its umbrella.

Bert Juster, the head of the school for boys, was certainly committed to relationship. But he knew that his vision alone was not enough to drag the staff and children into connection. Instead, he was explicit about the importance of not just sharing a philosophy with school personnel, but enacting it through planned practices. "We've actually been talking a lot about how to create systems and structures that can allow us to bring our mission into the day-to-day experience of the school," he said. "Because you can't just wish for it, and you can't just talk a good story about your mission and hope that it shows up in your classroom. You really need to create systems in order to do that." The conversation led to thinking about the aspects of the school that they do well, that they want to be sure to keep for the future.

> We were talking about "What are the things at the school that are like magic? What are the things that we do really well, that we need to not forget as we grow, and as new people come in, we don't want to lose that?" One of the things . . . was love. "How do we systematize love? How do we systematize joy?" So, what we did at that time, was we just brainstormed as a group, all the ways that we were doing love in the school. "What does it look like in the morning? What does it look like in the afternoon? What does it look like in the science class?"

One result of this reflective moment, he said, was to build explicit conversation about love into the admissions process, so that families knew from the start that the school made it a priority, which "weeds out the families who are not right for us." "We just have a school where we talk about love, a lot," Bert said. "You know, if a boy is in trouble the first thing that I say to him is 'You know I love you, right?'" Not every family was open to that kind of language, he said, and they did not have time to win over those who were not convinced already.

Another occasion for "systematizing love" was to create the term "love space," instead of "safe space," to frame how to have difficult conversations. For Bert, the distinction was crucial. "Safe space is all about what you can't say, what you can't do, and a love space is all about saying, 'No, if you have a racist thought in your head you can say it, but you, you gotta own it, yeah.' So 'love space' is all about challenging each other to be the best that you can be, instead of assuming the worst and just preparing fences or barriers. That 'agree to disagree' drives me crazy."

The school enacted "love space" in its "restorative justice circles," in which students in conflict talked openly about their experiences and made plans to alleviate harm. The circles had been crucial for the school, Bert said. "I was thinking how important they've been, for us at least, in having students not actually result in just having shame, or ignoring a transgression or whatever they did to hurt the community, or to hurt another person. Whether or not you're doing a restorative justice circle with just three, or whether you're doing it with the entire grade, or the entire school, it's a way to have that aha moment solidified in repairing your relationship with the community and making it right."

The school's culture had been both challenged and represented at a recent graduation, where they had a nascent tradition that every graduating boy would give a short speech. Bert told the story of an African American boy who stood up to give his speech (about half of the class was non-white, and of that group, about a quarter was African American). The poorest boy in the class, he was also the darkest-skinned boy, Bert said.

> He had to take like, two buses and a subway to get to school every day, [had] a single mom, and at his graduation, he talked about how difficult it was to go to a private school where he was the darkest-skinned African American boy. The title of his speech was "What I Learned from My White School by Being the Darkest-Skinned African American Boy," and it was this beautiful, powerful speech about "I learned this from this person, I learned this from this person, I learned to love my Blackness," but also, I mean, it was bittersweet, there were moments in there where I was like, "Ugh, I wish we could've done better for him." Because one of his colleagues made a racist remark in class one day and he talked about it all.

Even though the boy's speech did not show the school in the best light, just the fact that he gave it reflected the school culture of including every voice, in its authentic moment. It also reflected Bert's emphasis on connection. "The act of me trusting him, to say, 'I believe you, you do your speech.' Even though it might not shed good light on the school, on graduation day, live streamed to grandparents across the country . . ." he chuckled ruefully. "But because of that trust, and because of the relationship that we had developed over time, he learned something in a super authentic way."

Bert thought the speech was also unforgettable for the other boys who were listening, the privileged kids carpooling in from the affluent suburbs. "The power of that moment for the other boys in the class, right? That's a highly relational act," he said. "That wouldn't have happened without relationships." But being able to say his truth in front of the community was also important for the boy who gave the speech, Bert said. He quoted the philosopher Levinas about how "a person cannot learn until they are held hostage by someone else's gaze." Being witnessed in that moment gave the boy something also, Bert said. "If I hadn't allowed him, well, then it was a missed opportunity for all of us to hear his story and he would not have fully integrated it into his being, going back to that self-reflection piece. And [if he were not] held witness, being held hostage by all of our gaze in that moment, he wouldn't actually be able to learn that night." Bert continued, "In order to have a community kind of steeped in love and intimacy, you have to give kids the opportunities to kind of love themselves. And that means understanding the self, it means interrogating the self, and I think you can only do that in relationship with other people." The speech embodied the school's culture—and reflected its emphasis on relationship—even as it also revealed the relational harms that had happened there.

Rituals and routines like graduations and restorative justice circles both reflect and establish culture as it permeates an organization. But culture also reveals itself in how people interpret and experience spontaneous events, particularly those that represent a clash of cultures. A good example is a particularly traumatic event that Catalina narrated to me, about a visit she had made with a patient to a specialist.[8]

One of the remarkable services that University Transformational Care offered was the option to have a "care transformer" accompany patients to appointments with specialists. Catalina's patient had been suffering excruciating back pain, and was dissatisfied with the advice from the neurology clinic, which to date had been simply that she should lose weight. She had asked Catalina to go with her to her next appointment.

"During the visit, her legs, her back were manipulated in ways that were very painful to her, and I'm not sure if the provider sympathized with her pain, you know," Catalina recalled. It was excruciating for her to sit through, but she also felt like the patient was asking her not to say anything in the moment. I asked whether the patient was clearly expressing the pain. "Yes," Catalina replied. "She was crying. Yes." I wondered aloud how Catalina managed that difficult moment, when she may have felt trapped between what she was witnessing and what the patient wanted her to do. She recalled, "I was just sitting there, like, stunned. Yeah, like, wow, she's really in pain and she's crying, and I would ask the patient, 'Are you OK?' [and she would reply,] 'Yes. Yes, I'm fine.' And she would just kind of like, tell me just 'It's fine,' that she's OK. Even though she's crying."

"So it seemed like the patient was saying, 'Don't intercede on my behalf,'" I said to Catalina. "Basically," she agreed. "I mean, I think so. Yeah, it was hard. It was hard seeing that." At the end of the visit, Catalina said, "I left the clinic possibly as upset as she was."

We might not necessarily be surprised that a neurologist would refrain from changing anything about his approach in the presence of a medical assistant; the two occupy distant ends of the spectrum of status and pay, as a neurologist could easily have ten to fifteen years of training on top of a college degree, earning anywhere from five to ten times the salary of a medical assistant. But in her experience, Catalina said, when specialists saw the medical assistant there in the appointment, they would actually treat the patients better. "So that was very unusual to have that provider not respond to your presence," I said. "Exactly," she said. In this visit, she saw firsthand what happened when physicians were indifferent.[9]

I asked her how she handled it after they left. First, she said, she apologized to the patient on behalf of the neurologist. "I told her that I would bring it up to my manager and to the team and see how we can best help her with the situation. And so that's what I did when I got back. And we did get the manager from that clinic, a call." The manager said they would speak to the neurologist about the report. "And hopefully he would be more mindful with other patients as well," Catalina said. "Just because we don't feel it, that doesn't mean they're not."

The incident was traumatic for Catalina, not to mention the patient, because it was shocking to bear witness to the neurologist having so little care for the effect of his manipulations. But it also reflected a collision that illustrates for us an important point: organizational cultures are internal spheres of interaction, with their own rules and routines that may or may not reflect the external environment. In the neurologist's clinical exam room, a different culture prevailed, one in which the patient felt unable to assert herself, in which her discomfort and tears did not count enough to stop the procedure, and in which her witness—Catalina, a "care transformer" but perhaps "only" a medical assistant in the eyes of the neurologist—did not matter enough to improve the connective labor at hand, to make the specialist better able to "see" the patient.

When the visit ended and they returned to University Transformational Care, Catalina's actions threw the contrast between two clinics' cultures into vivid relief. She apologized to the patient on behalf of the specialist, which reflected the responsibility to her that Catalina felt, as well as the priority of relationship: an apology was taking steps toward redressing relational harm. Furthermore, Catalina took the matter up with her own manager and the clinic staff, seeking to make changes on the patient's behalf. Instead of silencing Catalina with observations about the medical hierarchy, the clinic helped her make a report to the manager of the neurologist's practice. We do not know—and Catalina does not know—what ultimately happened ("I believe that the provider was talked to. At least that's what they said they were going to do"), which further reflects the fact that the two clinics were not part of the same solar system, culturally speaking, and that University Transformational Care could not enforce its own cultural tenets on another clinic. But Catalina's outrage, her response, and her employer's support tells us a story about the values embedded in the connective culture at University Transformational Care, how they promote and prioritize relationship, and how they established a social architecture that worked, if only in the micro-environment of the clinic itself.

"This Is Your Two Hours": Material Resources

The third dimension of organizations who were "doing it right" was the distribution of material resources, including the allotment of time and space, the ratios of worker to client, the cognitive load workers faced, the extent of technology and data use, the compensation workers receive—all the resources that underlie a given interaction.

The most important of these is time. As we saw in chapters 5 and 6, the sense of time scarcity was palpable among connective labor practitioners. Outside of exceptional organizations, the only people who felt like they had enough time were those who opted for the "personal service" model of providing connective labor to rich people. Part of what made University Transformational Care so radical was the sense of luxuriant time there. The expanse of time was a feature of their "capitated" model in which they did not have to charge a separate fee for every service. "You know, I tell [the patients] exactly in the beginning of the office visit, 'This is two hours long. This is your two hours,'" Catalina said. "'If you take less, that's really because, you know, things move a lot quicker than they should. But if you want to take up the whole two hours, that is OK with me.' I think after the first visit, it really builds their trust, or the relationship."

We measure time in quantities—an hour here, fifteen minutes there—but it has other characteristics as well. How crowded does it feel? How fast or slow? Sociologist Ben Snyder uses the word "timescapes" to capture some of these qualities. "Each work timescape features the braiding of multiple rhythms of mental and physical energy expenditure, giving the individual worker different experiences of pace, sequence, tempo and articulation," he writes. Many organizations crowd the timescape of connective labor—loading many people onto one practitioner, stacking appointments one after the other with little or no break, requiring many different kinds of thinking and communicating in the same interactive space. These dimensions affect the rhythms of the work as much as the sheer quantity of time.[10]

The complexity of time also explains a certain unevenness in organizational support for connective labor, even among those units with a social architecture that mostly fostered witnessing work. In particular, the chaplains in training spent hours together talking about relationship and how to make connections, which were undoubtedly a priority for them and their supervisors. Yet they also endured a grueling schedule of being on twenty-four-hour call every week and having to spend the morning in class the next day. They did not face the pressure of having to process a certain number of people every fifteen minutes like the primary care physicians, but as we know, this site asked them to log their interactions with patients for three different tracking systems. Away from their supervisors, they talked to me about feeling overwhelmed and burned out, about overwork and exploitation, even as they were surrounded by leaders who prioritized relationship, spent hours cultivating a warm and caring community, and shared a common set of norms and rituals that valued connecting. The complexity of time meant that for the chaplains their experience was partially buoyed by their social architecture, and partially depleted.

Time was not the only dimension that mattered, of course, although it was a vital one. Another important factor was what scholars have called the "attentional load." How many different clients or students or patients did one person maintain, and how many at any one time? Even when workers faced just one patient, how many other things—emails, data entry, scheduling, testing—were they being asked to do at the same time? As we saw in chapter 6, teachers and physicians in particular were asked to balance enormous loads simultaneously; recall Simon, the physician who likened the experience to getting periodically shocked in an experiment. Those organizations whose social architecture supported connective labor took steps to control the attentional load—reducing Catalina's patient roster to six, for example.[11]

Ultimately, organizations offering an alternative social architecture took attention seriously. Not only did that mean limiting how much a worker had to attend to; it also meant treating that attention as if it were precious, a unique catalyst for reflective resonance. Mariah Dreyfus ran several programs for inmates and formerly incarcerated people out of an elite university on the West Coast. One program involved teaching entrepreneurial skills to ex-inmates, with four-month sessions of classes lasting four hours at a time. "So it's a lot of time we're spending together as a group," she said.

But in addition to time, the concentrated attention of multiple others was the program's treasure, she said. Every other week, the entrepreneur-to-be would also meet with a local business executive, as well as two students—one each from local law and business schools. She described to me the slow transformation the process provoked in the men. "So it's like all those people are investing in them," she said. "And what's interesting is that it takes a while for our entrepreneurs to begin to feel comfortable having that much attention on them. Like, 'You mean, you just want to know about what I think? You mean, you just want to be here and invest in my plan? Like we're just going to be talking about what I want to do?'" Shocked by the experience, they asked plaintive questions that revealed just how uncertain they were, how unsure that they even warranted all that focus from others.

It was a sharp contrast to the entitlement that characterized the student body at her elite institution. She described the process of teaching the men, who had most recently been inmates inside a correctional facility, stripped of their individuality: "And we have to—'train' is not the right word—but, like, empower them to think about: 'You need to set the agenda. Like this meeting is about you and what you want to get out of it. What is helpful is to come in with an agenda of what you want to get accomplished.' All that stuff is [a form of] undoing, particularly for those people who have been inside for so long, where they were completely disempowered."

For formerly incarcerated people, that kind of attention, focused on eliciting a plan that would best reflect their desires and skills, was a potent elixir. But the first step was convincing them they deserved it, Mariah said. "Last year I had a guy who had done twenty-six years," she told me. "He's starting a handyman business. He got up, he was working with a professional speaking coach, and at the end the professional speaking coach says, 'The one thing I might gain from you, Jerry, is like the value of your business is what it is. Like why you're totally worth $50 an hour. Like that needs to come across.' And Jerry sheepishly sat there and he said, 'But I got paid nine cents an hour for twenty years. Like, I actually don't believe it myself.'"

I marveled to her that Jerry was able to say that out loud, to which Mariah responded: "Well, that's the magic of the cohort and being in the class together for so long and actually building trust." The time and duration of the program built that trust, which was vital. Research tells us that those who need the most help are often those who feel least entitled to ask, in part because they have often been mistreated by institutions in ways that undermine trust. But it was the people's focused attention that was going to reap the most rewards for the ex-inmates, Mariah thought.[12]

"I'm a strong believer in high touch," she said. "And that can be a faculty member, and for a college student, it could a college advisor, like a college counselor. It could be another student who's like two years ahead of you, who is your person. But I think we all need people. I mean like, 'Who are your people?' And I think this population needs those people just as much as we do." In Mariah's program as for others, the gaze of others became a resource whose distribution—or unexpected concentration—lent power to the interaction.

Finally, it was not that this supportive social architecture was devoid of the scripting and counting that proved so problematic in chapter 6. Instead, these contexts managed to make templates and data needs feel meaningful, incorporating them within and subordinating them to their mandate of relationship. For example, one of Catalina's tasks was to serve as a scribe, which alleviated the burden of the primary care physicians in the University Transformational Care clinic, unlike those who railed against the data entry demands in chapter 6. Yet taking on this work was actually an expansion of the medical assistant's traditional role, so it felt like a bigger responsibility to her; moreover, it enabled her to stay in the exam room while the patient was being seen by the provider, further building her awareness and connection. With her responsible for the EHR, it was much better for the doctor-patient relationship, Catalina said. "I send referrals out, I order labs, I will put in this request, description, everything. I do it here while the provider talks. So that's one benefit because the provider's not looking into here [the computer] and getting lost if they don't really know where to go. Scrambling through different tabs. But because we are here, we do that, we're more familiar with the chart. The provider takes all their time looking at the patient, you know, 'I'm here, I'm talking to you, I'm listening to what you're saying.'"

She actually did not mind the EHR at all, Catalina said, comparing it to the chaos of the charts at her old community clinic.

> You're looking for the last colonoscopy, it will bring you all that patient's colonoscopies, instead of having to go through all those medical records and flipping pages, right? The orders, we're putting them in right away. [When it was] the chart, the doctor's writing down the orders [in the clinic room], then has to come out with the chart, give you the chart or tell you, "Hey, we need this, this, this and that." And then you have to go back and get the lab order, fill that out for the patient, give it to the patient. It's just, I think—

"It's ridiculous," I finished for her. "Yes," she said emphatically. "I mean, I think the electronic medical records is like the best thing." After months of my hearing doctors complain about it, Catalina's words came as a surprise, but they also illustrated a core truth: the problem was not necessarily the technology but how it was used, by whom, and for what purpose. With a scribe and detached from billing, the EHR actually felt like a means to enhance care.

The same was true for templates, which could be ways to connect rather than sources of distance. Bert Juster's teachers had brainstormed a list of the school's core values, and how to "systematize them," for example. "Like for politeness, every day we have a duty station where a teacher stations themselves at the front of school and greets every kid with a handshake or a hug and looks them in the eye. If they don't get the eyes, we make some sort of joke, or cute little 'Oh, give me those blues!' Or whatever, and then to get the kid to look at us. That was one way we've systematized kindness and politeness."

They also standardized the curriculum to some degree, so that every teacher had to think about building ten core questions into what they were teaching, he said. "One of the questions is: How are you introducing the students to art? How are you differentiating for the different kinds of learners in this unit? How are you developing autonomy in the learners? How are you developing a sense of self in the student? So, there are all these essential questions so that when teachers are developing a unit, there is, in every unit, in every class, in every grade, teachers are thinking about social justice and equity issues," Bert said. "Whether that's balancing molecular equations, or it's the US Constitution, because those are important to the school."

Finally, while as a private school it was not mandated to use standardized testing, the school did not shy away from examinations, but used them as a way of figuring out where the kids were, in preparation for them to apply to local high schools. "That's why I think standardization is really awesome because it's just another tool to leverage understanding about the self," Bert said. Systems and standardization were part of the material resources that could contribute to a supportive social architecture, as long as they were subordinated to the priority of relationship.

The Privilege of Supportive Social Architecture

The elite institutions, the independent schools, the private practices—the examples in this chapter stem from places of great privilege. Of course, privilege does not necessarily characterize the people they serve—from the ex-inmates-turned-entrepreneurs to the grief-stricken families to the "darkest-skinned boy" who took "two buses and a subway" to get to his school—but the organizations themselves are a far cry from the community clinic in Appalachia we saw in chapter 5 or Pamela's classroom from chapter 6, whose students came to school from a crime scene. Is a social architecture that supports a full-bodied witnessing only possible in rich institutions? Is such a social architecture a luxury?

The answer is yes and no. Connective labor is labor that involves "high-touch" attention from one human to another, and it resists the productivity gains that other kinds of work demonstrate; chapter 7 showed us some of what that resistance looks like, in stories of misrecognition and rupture. Furthermore, the relentless drive to extract efficiencies in this work is grounded in the perennial hunt for profit in the private sector, and in an environment of scarcity in the public sector, where neoliberal policies have made the welfare state wither. As we have seen in previous chapters, these factors mean that many private and public entities feature a social architecture that is unsustainable or unsustaining.[13]

Moreover, some institutions fostered connective labor by controlling whom they serve and how. Bert Juster talked about an admissions process that "weeded out" families that did not already believe in "love talk" in a boys' school, for example; he said the school did not have time to convert those who did not already believe. Unlike a public institution, his independent school was not required to take all comers nor to bend their social priorities to testing requirements for funding.

Yet the skills of seeing the other, of reaching an emotional understanding of another person with them, are by no means restricted to high earners. Catalina Maldonado is a medical assistant with a high school education; even though her job is expanded considerably, she is still much cheaper to employ than a doctor, even those in primary care who earn the least of all physicians. Indeed, as care scholars have documented, this emotional work is feminized and naturalized, meaning often considered to be part of the essential nature of women, and thus requiring no particular training or compensation. While such work might resist productivity gains, it is also true that the labor costs of managing data analytics and other administrative-focused positions have greatly expanded. Labor costs are a choice, an investment that employers, administrators, and policymakers decide to make in service to their priorities. To be sure, adequate compensation is crucial to support good connective labor—most obvious in its absence, as when exhausted home healthcare aides have to pile on multiple shifts in order to sustain their households—but the capacity to see the other is not reserved for those at the top of the pay ladder. Wages are not what makes this work a luxury.

Furthermore, the offices of University Transformational Care were festooned with posters and charts documenting how much money they were saving the institution (for an example, see figure 8.2). The relationship-focused approach meant that patients were taking better care of themselves, were managing their own conditions better, and were not going to the emergency room. While these gains seem patently clear, research suggests that relationship-focused approaches in other industries are also cost effective.

One problem is that the costs of a social architecture that fosters relationship are often felt at a micro level, local to a school or clinic, while its gains are felt at a macro level, in a given society. Students who are seen, who feel connected to their teachers, stay engaged in school, and we all benefit when kids get a better education. Counselors who achieve a therapeutic alliance see better results in their clients, and we all benefit when people's mental health needs are met. The radical experiment of University Transformational Care was possible in part because the costs and gains were both local—as the university was responsible for the healthcare spending of the target population. The benefits of a connective social architecture are thus bigger than any one organization, and more expansive than any cost-benefit analysis can itemize. Such a scope suggests that public support for policies that can enable organizations to make the right choices would be helpful and warranted.

Beyond the Individual Hero

It is not that the people in this chapter are the only ones who are "doing it right." Talking to all the different kinds of connective labor practitioners for this book—the teachers and therapists, the hairdressers, the medical professionals—I was often struck by the individual heroism of the people involved, how they fought to connect to their clients and students, even when their organizations made it difficult. Yet when connective labor is sustainable, it is ultimately deeply collaborative, reflecting a relational design that prioritizes connection, including a community with adequate support and a collective commitment from the top on down. Organizations like University Transformational Care or the boys' school where Bert worked were unusual oases that reflected different dimensions, enabling not heroism but instead deep emotional collaboration and connection in their relational design, connective culture, and resource distribution.

Of course, each of the elements that construct a social architecture are interrelated. A community of practice is going to reflect the ambient values, while norms and rituals are going to arise in a place that sets aside enough time for people to meet, and a dedicated leader is going to make sure the time and resources exist for connection and empathy. Conversely, any combination of these without one or the other leg of the triad makes it that much harder to do connective labor well. A principal might support the idea of teacher-student relationships, but without the time for teachers to devote to it, or a schoolwide culture that reflects those values, he or she—like the principal of Conrad's school—may as well resort to tacking up a list of twenty-five tips on "How to Connect with Students" in the men's restroom.

A high-touch social architecture may not be a luxury per se, but there is no question that it is rare to see these kinds of contexts, where practitioners are buoyed by time, community, and support as they worked to see the other well. If we continue to prioritize efficiency over relationship, we create a new kind of haves and have-nots: those divided by access to other people's attention. Without addressing the conditions of connective labor, we foster the stratification of human contact.

9

CONCLUSION

Choosing Connection

In 2019, a Dutch supermarket company started reserving some of its checkout lanes for those who wanted to stop and chat with the cashier on their way out. Dubbed "Kletskassa," or "chat checkout," the new lanes were marked by a sign that read in Dutch: "the nicest checkout for when you are not in such a hurry." (See figure 9.1.) The move was a response to widespread loneliness, the store proclaimed on its website. "It is a small gesture," the website noted, "but very valuable, especially in a world that is digitizing and accelerating." Just a few feet over, the digitizing world was accelerating away in the store's self-checkout lanes, where customers scanned, weighed, and bagged their own items, inserting their credit cards and pocketing the receipt, all without talking to a soul.[1]

What does grocery shopping have to do with the humane interpersonal work that is connective labor? Why should we even care about what are often perfunctory interactions? Cashiers work in repetitive, surveilled, scripted jobs—"Paper or plastic?" "Are you a member of our loyalty program?" "Would you like to round up your total to the next whole dollar and donate the difference to [named charity]?"—with some of them under pressure to check customers out as quickly as possible, timed on scans to the minute and reprimanded if they fall below a certain number. Where is the relationship in that?

The answer is that grocery shopping used to be thick with connection, even though we can see only the shadow of it now, like the fossilized

FIGURE 9.1. The chat checkout. Source: Jumbo supermarkten.

footprints indicating where ancient peoples once walked. The proliferation of self-checkout lanes is but the latest stage in a long history of stripping relationship from the experience of grocery shopping, a process that has also devolved more and more control—and work—onto the consumer. A century ago, women used to hand their lists to male grocery clerks, who would retrieve the flour or oranges or cans from high shelves and storerooms, haggle over the prices, and hand the goods over for the shoppers to bring home. "Food buying in the first decades of the twentieth century was an intensely, indeed uncomfortably, social encounter," writes historian Tracey Deutsch. "The exchange was neither abstract nor impersonal, but took place through relentlessly messy social relations. . . . The need to serve women as individuals was both fundamental to and a problem for the operation of grocery stores." Grocers knew their clients intimately; clerks kept close track of shoppers' desires, their habits, and their families, soliciting views and peddling influence.[2]

This "relentlessly messy" daily mix all changed in 1916, when a Memphis chain called Piggly Wiggly first let shoppers loose in the store, introducing the experience of choosing items for themselves, with cashiers at the front to ring up the selection. Whereas before the decision to buy was filtered through the dynamic between consumer and clerk, now, if there was a relationship, it was between consumer and product; Piggly Wiggly is thus also credited with the invention of the brand.[3]

In 2023, when word of the Dutch company's new chat checkouts spread to the social media platform Reddit, the response was split between those heralding the move, and those like user demonblack873, who narrated their own reaction: "blasts through the self-checkout as quickly as possible attempting to avoid eye contact with everyone lest they initiate an unsolicited conversation." "Same," was one response. "Mask and earbuds make this even easier. Some people hate self-checkout, but like then you don't have to talk to people!!" These two commenters are not unique; some 30 percent of supermarket transactions take place in a self-checkout lane in the United States, up from 23 percent in 2019. Furthermore, we might say that as the checkout lane goes, so goes society. The number of single-person households tends to expand when economic times are good; in many countries, the elderly and others choose to live alone when they can. Similarly, when the economy is booming, divorce rates tend not to shrink but to rise, as people split up when they can afford to. We may worry about loneliness and the estrangement of contemporary life in general, but when people choose for themselves, many opt for the efficient, the autonomous, the solitary.[4]

Speaking for myself, I don't view the old grocery clerk with nostalgia; while I might urge my family members to stay away from self-checkout now that I've learned about the psychological benefits of casual social interactions, I am glad I get to pick what I want off the shelves. But while the historical transformation of grocery shopping gained consumers time, money, and control in their daily provisioning, it also hollowed out a site of everyday, local, human-to-human relationship that has not been replaced. Meanwhile for grocery workers, the interactive labor of cashiers has been eroded by ever more counting and scripting, their work inching ever closer to that of machines, until in some cases it is not clear—to themselves or to their customers—how they are that different. Ultimately, the problem is less the cashier's replacement by the self-checkout lane's combination of machine and consumer, and more the corrosion of the human job that makes such a substitution appealing in the first place.[5]

These trends happen not just in the United States. Another Reddit commentator took issue with the good press garnered by the Dutch supermarket's innovation. "What has not been explained so far is that, in the larger cities at least, most Dutch supermarket checkout lines are super-efficient," they noted. "The checkout process is done at top speed with very few words exchanged. The queue behind grows impatient and angry if checkouts are not processed quickly. Dutch commercial life in general values efficiency and profitability over service. . . . Worse than that, most of these cashiers have been, or are being, replaced by scanning devices. At the supermarket near our home, there is one employee watching over nine scanning devices. So I suspect this 'introduction' of 'slow' checkouts may also be for the people who are unable or unwilling to use the scanners." According to this observer, the chat checkout may have been simply the relabeling as "choice" what the store needed to do to manage those unable to adjust to the mechanization of shopping.[6]

The emphasis on speed, efficiency, and profit in grocery work led to the same trends—the scripting and counting, the degradation of human interaction, the replacement by machines—that this book has documented in other occupations as yet still awash in connective labor. If we are surprised today that people might look to grocery stores for relationship, people of tomorrow might be surprised that we ever looked to clinics or classrooms for the same. It may sound hyperbolic, but at one time grocery clerks were that central, that vital; as one observer writes about clerks and shoppers at the time, "he knows their family troubles and family joys."[7]

People have long considered such personal, emotional interactions as somehow beyond the reach of systems that try to count or control them, such as data analytics, automation, or even very detailed manuals scripting worker behavior step-by-step. To many practitioners, connective labor is not very measurable, not very predictable, and not very automatable, and some pundits agree. A *Harvard Business Review* contributor predicts that in an era of increasingly intelligent machines, "the most valuable workers will be hired hearts." *New York Times* columnist Thomas Friedman has argued that in the future, "when machines and software control more and more of our lives, people will seek out more human-to-human connections—all the things you can't download but have to upload the old-fashioned way, one human to another."[8]

But despite these claims, technology is nonetheless gunning for this work. Data collection and analytics have transformed the practice of primary care, teaching, and nursing, as we have heard, leading to burnout rates of sometimes more than 50 percent in these fields. When the ChatGPT-3 bot was released in the fall of 2022, within a few weeks someone announced their company had used it to offer "mental health support to about 4,000 people," by using a bot-and-human pair to compose replies to people's requests for help. It's a fallacy to think these jobs are somehow safe from the data analytics revolution or impervious to what has been called the "AI spring." Moreover, the overlap of these trends with social inequality means that we are approaching a world in which the affluent pay for personal connective labor from less-advantaged others, who might themselves increasingly have to get it delivered by app or AI. We need to take these trends seriously as a wholesale transformation of connective labor, one that has been taking place without a reckoning that allows us to see the risks involved.

In some ways, the "chat checkout" lanes resemble the move by the experimental school chronicled in chapter 3, where students spent all morning with machines, while the school carved out a special role for human "advisors" entrusted with the children's hearts. At best, the Dutch supermarket is thinking about what humans do better than machines—they can chat with the elderly, they make lonely people feel connected—and implementing a division of labor that offers machines for those customers who prioritize efficiency, and people for those who want connection. This is one vision of the future, one that takes seriously the social intimacy that humans create while also making room for automated transactions: we might call it the "humans as feelers" version.

What's at Stake: Belonging and Social Intimacy across Difference

The Dutch store may be focusing on loneliness, but when it comes to connective labor, there is much more than loneliness at stake. While this book has focused on the individual consequences of seeing and being seen, the collective effects are just as profound. We each have a thin connective tissue called fascia that holds our bones, organs, and muscles in place; seeing each other creates the fascia that anchors our social world. The power of connective labor, especially in the mundane quotidian world of everyday commerce and civic society, is in its capacity to knit together communities of disparate souls—in other words, to create belonging.

Belonging is an emotional state with real consequences, whose contours are honed by powerful societal forces shaping who is in and who is out. Scholars tell us that belonging matters for physical and mental well-being; sadly, much of this research is based on what happens in its absence. Researchers have conducted experiments that find that adults who feel excluded are less likely to donate to charity or play games with others, and they give up earlier in a frustrating task. Furthermore, as the writer Sara Ahmed puts it, "how we feel about others is what aligns us with a collective." Belonging makes a community out of individuals.[9]

At first pass, the path to belonging might seem to rely on sameness. My first book was a study of how kids use consumer goods like electronic games or collectibles to assert themselves as part of a community; entitled *Longing and Belonging: Parents, Children, and Consumer Culture* (2009), the book argued that belonging conferred a kind of dignity, the dignity of being a full social citizen. And certainly the children I studied (and their parents) found comfort in sameness, and were concerned about being different—bringing different food from home for lunch, having different toys or clothes or rituals. I found that across communities of considerable inequality in class and race, kids were not so much trying to best each other with claims about what they owned, as they were trying to match what others were claiming, because some items (and experiences, like going to Disneyland or seeing a popular movie) acted like passports for their social worlds.[10]

We might worry about the drive for sameness underlying belonging, in which people forge links with each other due to commonalities they share, with any differences threatening to become the source of exclusion, a bright boundary between "us" and "them." And to be sure, this kind of belonging has a long history and a broad distribution. But in

Longing and Belonging, I also noted that every child—even the most affluent, with the most indulgent parents—experienced being different at some point, lacking the thing that their friends and classmates were talking about, and had to figure out what to do in that terrifying moment (they often just faked it, hoping to hide their dissimilarity).

Those kids came to know viscerally that being different is part of the human condition. Nonetheless, the modern world features all kinds of standardization designed for certain people and for particular bodies. As might be noted by any short person who has tried to click in with a three-point seat belt designed for a different frame, or any parent who has tried to change a baby's diaper on an airplane, we come face-to-face with our "difference" from others whenever who we are or what we need runs up against expectations that are built into our environment. Part of the experience of being Black or gay or disabled means running up against these built-in expectations more often—in a world that feels designed for whiteness, for heterosexuals, or for the able-bodied. And while infrastructure or institutions can deliver potent reminders of these assumptions, they are also reflected in our interactions, as when people presume poverty means a lack of knowledge about nutrition, or misconstrue the impact of illness on a polyamorous couple's commitment. Difference may be the human universal, but some people face a constant struggle to be legible.

This very struggle, however, is what makes the emotional recognition of connective labor so vital: when it works, it can create belonging *across* difference, forging what I have called social intimacy. When people bear effective witness to someone else's experience, they are not saying, "We recognize what you are feeling, but only because we have had the exact same experience." In fact, the act of seeing the other requires that people are different; if everyone were the same, then people would not strive to be seen by others, because everything—cultural products like movies and books, political conversations, institutional realities like buildings and schedules, and, yes, the social interactions that convey prevailing assumptions—would already reflect us. As the actor and writer Anna Deavere Smith writes: "That genuine moment, that 'real' connection, is no small thing." She asks: "Is it a moment that can happen only when we don't know each other, when we have so much to learn about each other that we hang on every breath together? It is hard to find those moments in our culture because we think we know so much about each other. . . . Those moments are in fact all about that which is not predictable."[11]

Through connective labor, we enact respect for the other; across our differences, witnessing conveys that someone is a fellow human being who deserves to be known. In a world of sometimes extraordinary conflict, this is a tall order, to be sure, and as noted in chapter 7, even people whose life work is to see the other can fail at the job. We see the residue of misrecognition all around us like contrails across the sky. The Black Lives Matter movement and the burgeoning racial awakening in the United States are arguably a reflection of the fury Black Americans feel because of what we might call "misrecognition fatigue," protesting the brutality and othering that comes from being profoundly unseen as fellow human beings. Meanwhile a sense of feeling invisible, as captured by films like *Nomadland* (2020) and *Parasite* (2019), animates working-class rage in the US and abroad, and is rife within the social crises of the aptly termed "deaths of despair," suicide and drug- and alcohol-overuse deaths that have radically lowered life expectancy. In the midst of a depersonalization crisis, "being seen" is already in too short supply.

Yet this is not a call to automate connective labor in response to these problems of access and availability. As we have seen, the scripting and counting of that work erodes its social power, until it fails to sustain either practitioners or the people they are charged with seeing. In fact, those trends feed the sense of inevitability about automation, since AI is better than nothing, and nothing is what it can feel like we are getting when connective labor is scripted and standardized. When efficiency and profit are paramount, we're choosing not the future of humans as feelers but a different one altogether.

Humans as Valets

A Silicon Valley engineer once said the real question we face in an AI future is whether, as humans, we choose to be pets or livestock. His point was to emphasize the inevitability of how far computers would

FIGURE 9.2. Students moving scooters out of the way of delivery robots.
Photo Credit: Sean B. Hecht.

exceed humans in intelligence, and to argue that all we can hope for is that AI agents treat us benevolently. "We give pets medical attention, food, grooming, and entertainment," he told a *Wired* reporter. "But an animal that's biting you, attacking you, barking and being annoying? I don't want to go there." While pithy and a bit shocking, his rendition of the digitalized future does not quite capture what actually seems to happen when machines join us at work, a convergence perhaps best summed up by a picture that ran in the *Los Angeles Times* (figure 9.2). When an array of food delivery robots got stuck near a university campus, trapped by electric scooters in their path, students took pains to clear the way.[12] As the *Times* reported, humans hinder robots with pranks and violence, but they also can be found—like those students—scurrying around to make the world smoother so that the machines are able to do their work. Similarly, if you look at the way industrial robots and humans interact in factories, frequently the human's role is to ease the path for robots, which cannot always handle contingencies that arise. If the recent history of robotics is any indication, then, instead of pets or livestock, in a future dominated by AI, humans seem more likely to serve as valets.

What does a future with humans as valets signify for connective labor? We can sketch its contours by looking at how engineers approach that kind of work today: programming for ostensible tasks, such as conveying information about diabetes care, or helping someone understand quantum mechanics, while ignoring the underlying humane and reflective work that makes it possible. Ours is a world flooded by tech "solutions" to problems of access, performance, and standardization, one that touts its cost savings and drives for efficiency, even as it relies on invisible hordes of low-paid workers to train machine learning bots while courting bias and discrimination. If ever AI designers posit a more modest use for their products, the stark inequality of our world means they sometimes replace humans anyway; in 2023 a Mississippi school district reported that students were learning geometry, Spanish, and high school science via a software program, with a human teacher in the next town whom students could try to consult if they got stuck.[13]

I make no claims to be a futurist; I glean what this version looks like by considering contemporary trends in fields like education or banking—in other words, at the proliferation of videos to teach algebra or algorithms to sort credit risk. These and other tech solutions are intimately tied to the scripting and counting that now pervade so many connective labor jobs, as a fever of data gathering applies an industrial logic to even humane interpersonal work.

"The best we can say is that the 'avalanche of numbers' . . . provides a climate of receptivity (or perhaps resignation) to measurement innovations," writes Michael Power, a British scholar who coined the term "audit society." With words written twenty years ago, Power conveyed the quantification of not just baseball but college rankings and nursing home scores. "At worst, measurement systems are fatal remedies . . . by creating incentives to undermine the very activity being measured, making social agents focus on measures themselves as targets to be managed and 'gamed'. Critical data that cannot be readily quantified are marginalized and rendered invisible, and proxy measures end up representing the thing itself."[14]

With connective labor invisible or represented by proxy numbers such as "client satisfaction," with data-gathering regimes like EHRs or K-12 testing impinging on already constrained practitioner time and resources, data analytics has dramatically altered the experience of providing connective labor. In addition to the degradation of the jobs, however, data analytics also acts as an uncompromising translator, converting humane interpersonal work into the numbers that become "algorithmic chow." AI and data analytics go hand in hand in shaping the coming horizon.[15]

If the future of humans as feelers is divided between those who want more connection and those who seek efficiency, the one of humans as valets likely features a different sort of division, one between those who make the videos and apps, and the rest of us who watch them. It resembles what has been called an attention economy, but the problem at its heart is not so much an info glut—"there is too much to pay attention to"—as it is a celebrity system, one that depends on an artificial scarcity of attention stemming from a ruthless appraisal of who is worthy of being seen. Seeing becomes a one-way affair; if it imparts dignity, purpose, and understanding, it does so in one direction, as most of us are relegated to the audience. It is not just the practitioners' job that gets eliminated, then, but also the reflective resonance that they create with the other, in their mutual, interactive accomplishment.[16]

The challenge we face here is that if we are not careful, we will not get to decide between these two or any other futures that are out there, as the mundane selections we make now—about which checkout lane or automated helpline to use, or about whether we comply with data collection regimes at work, or about how we select teachers or therapists or physicians—ultimately make the choice for us about what kind of future we get to have. The two Reddit commenters might not want a world dominated entirely by apps and videos, but the aggregate of their smaller decisions pushes us there, unless we think about and plan for a more connective culture.

A Social Movement for Connection

The COVID-19 pandemic has highlighted the dire importance of public health, the collective system dedicated to protecting the physical well-being of a population. But we also need a collective system dedicated to protecting the *social* well-being of a population, one that takes belonging and recognition as seriously as seat belt use and vaccination rates. We need to fight for and enable what we might call our "social health."[17]

A fight for social health would take place on several fronts. First, we must identify, make space for, and value emotional recognition in our civic spaces, in halls of commerce, at work, and in schools. Care-work scholars tell us that the best way to improve care is through an alliance between consumers and providers aimed at improving working conditions and pay. Anyone who has sought care for their elderly parent, their child, or their relative with disabilities in the United States understands the conundrum at its core: the work is paid far too little for its workers to thrive—certainly far less than its enormous value to those who contract for it—and yet most individuals are hard-pressed to pay more. While the marketplace of care forces these two actors to be on opposite sides of the issue, in actuality their interests are aligned, and both should advocate for state involvement in benefits, training, regulation, and subsidies. The conundrum is common to many connective labor jobs; a social movement for connection will rely on both practitioners and recipients for its army.[18]

For policymakers, program administrators, and insurers, a fight for social health means they must take a hard look at what is being counted, and how the drive for data analytics has distorted a whole range of processes from teaching to offering counsel. These people must help organizations develop the social architecture they need—via relational design, connective culture, and resource distribution—to create the conditions for humane interactions. They need to be fully aware of what people do for each other, and to take responsibility for enabling humans to do so powerfully and well.

For leadership and lay public alike, a connective culture also means resisting the hype of AI, of apps that are sold on convenience and price, of machine learning bots that seem to attain ever more fluency but can never convey to someone that they are seen by another human being, can never create the meaning that humans create together when they are in relationship with each other. We need to be alert to the unintended consequences of inviting automation and AI to pretend to be humans seeing the other, and of all the marveling when they seem to come close—consequences that matter for those who would be seen as well as those who would do the seeing.

At the same time, I am not suggesting we take our hammers to the computers, bots, and agents, but rather that we treat claims about their contributions and capacities with all the skepticism, caution, and anticipation of their likely effects that they deserve. AI may have a place in relieving dementia caregivers, for example, who struggle with their charges' need for constant repetition. Bots such as ChatGPT or later versions may provide guidance in where to go for treatment of social anxiety or depression. A limited role for data analytics might help us know the subjects where students need more help. As Bert Juster, the principal of the boys' school, indicated, standardization has its place, as "just another tool to leverage understanding about the self." But these uses come about when we bend the tool to the priority of relationship, instead of vice versa. Their contributions are far more limited than some of the scenarios casually spun out by pundits and researchers alike that reflect not just a heedless technophilia but also a willful blindness to the precious, fragile, yet powerful effects of humans seeing each other.

If a world dominated by AI and automation is inevitable, if enough people resemble the Reddit commentators in their preference for "blast[ing] through . . . as quickly as possible attempting to avoid eye contact with everyone lest they initiate an unsolicited conversation," then the humans-as-feelers version of the future would be my preference, since it takes connection seriously and carves out space for what humans do for each other, albeit a limited space relegated to the quirky, elderly, or infirm. But when I weigh the different approaches to connective labor that we have seen in this book, I confess I still prefer the "high-touch" social architecture, enacted in chapter 8, that features committed leadership, communities of practice that act as real sounding boards, and ample time for seeing each other. Compared to either future I've elaborated here, those versions seem to me almost utopian. Yet those stories portray real cases happening right now, and so they are possible, if more of us would only insist upon them.

As others have noted, the goal of seeing the other introduces all sorts of problems: Can we ever actually know another human being? Does the process of trying to see the other mean we force them into preexisting labels that actually deny their particularities? How do we set aside our preconceived notions about social categories to make a connection with the human being in front of us? Many of these thorny questions rest on the very real challenges that difference can pose. Yet the people in this book have illustrated that the experience of feeling seen is a profound one, and the practice of seeing another—however imperfectly—is a powerful way to forge a connection.

A deep sense of belonging rests on not just shared sameness but the purposeful integration across difference. We can be helped to connect across difference through seeing each other, and organizations, clinics, schools, and neighborhoods can cultivate the social architecture that enables us to see each other better. Our social health depends upon it.

ACKNOWLEDGMENTS

There are many people to thank, but first I want to mention my appreciation to those who let me into their lives, allowed me to observe them at work, or who met with me to talk for hours about their experiences. Thanks also to those who reached out to their own networks to connect me to others who joined the project. I cannot name you for confidentiality reasons, but I am deeply grateful and hope your experience and perspective are faithfully represented here. Without your openness and generosity, I would not have been able to write this book at all.

I'm thankful for the support for research and writing from the National Science Foundation (Grant No. 1755419), as well as the American Council of Learned Societies, the Center for Advanced Study in the Behavioral Sciences, the Berggruen Institute, the American Sociological Association's Fund for the Advancement of the Discipline, and several sources from within the University of Virginia, including faculty research awards from the College and Graduate School of the Arts and Sciences, and the East Asia Center.

Thanks to these funds, I was able to support a team of graduate and undergraduate students who provided invaluable research and editing help throughout this project, including Nia Baker, Brooke Dinsmore, Hayley Elszasz, Madison Green, Jaime Hartless, Fauzia Husain, Allister Pilar Plater, Patrice Wright, and Shayne Zaslow, as well as Genevieve Charles, Lena Gloeckler, and Finn Trainer. I've also benefited from the transcription assistance of Paula Kamen, Anne Neison, and Vanessa Nielson.

I am also grateful to the writers and scholars and practitioners who read all or portions of this manuscript, or earlier versions in proposal or article form, including three anonymous reviewers for Princeton

University Press. Arlie Hochschild, Christine Williams, Jean Beaman, and Jessica Calarco each read a full draft when doing so was particularly hard work—that messy stage of buried ideas and tangents—and provided the intensive feedback to make it a happier experience for later readers. Viviana Zelizer, Eva Illouz, and Rachel Sherman provided comments on earlier portions of the argument. Carrie Lane, Steve Lopez, Nancy Chodorow, Steve Hoffman, Iddo Tavory, and Max Greenberg each offered helpful reactions or ideas along the way. A special session at the 2022 American Anthropological Association meetings devoted to the concept of connective labor allowed me to hear what some smart scholars found useful about it; I'm deeply appreciative of Carrie Lane's efforts in organizing that panel, and for her comments, as well as those from Ilana Gershon, Anna Eisenstein, Josh Siem, and Lauren Olsen. Special thanks in particular to those friends and family who read an early version of the manuscript in its entirety to provide their own perspective as therapists, teachers, and managers, connective laborers in their own right: particularly Tina Verba, Taryn La Raja, Rosemary Pugh, and Andy Pugh.

At Virginia, I am indebted to my writing group of more than fifteen years, particularly Jennifer Cyd Rubenstein and Denise Walsh, as well as Sarah Milov, Ira Bashkow, and Jennifer Petersen; they challenge and nudge me to better work. The UVa Field Methods Workshop also proved an early incubator for these ideas. At the Center for Advanced Study in the Behavioral Sciences at Stanford, I benefited from the friendship, acute analytic brains, and sometimes singing prowess of several, particularly Terry Maroney, Barry Zuckerman, and Andy Lakoff; Stanford also gave me valuable, transformative time with Abraham Verghese, Mitchell Stevens, Marianne Cooper, and Jill Vialet, among others. At the Berggruen Center in Los Angeles, I was particularly grateful for the camaraderie and support from Christina Dunbar-Hester, Venkatesh Rao, and Nils Gilman.

I'm also thankful for feedback at talks given at the University of British Columbia, Vancouver; the University of Pennsylvania; Stony Brook University; the University of California, Santa Barbara; the University of Southern California; the University of Minnesota Institute for Advanced Study; the Department of Medicine at Stanford University; the New School for Social Research; the Scandinavian Consortium for Organizational Research at Stanford; the Data and Society Institute; Microsoft Research New England; the Labor-Tech Research Network; and UVa's Human-Machine Interaction Group. Portions of the argument were also aired as part of a keynote session at the Conference on Real Work in the Virtual World in Helsinki, Finland; a session on "Artificial Feelings: The Politics and Perceptions of AI," sponsored by the ASA Section on Science, Knowledge, and Technology; the 2017 Global Carework Summit in Lowell, Massachusetts; the 2021 Social Life of Care conference at the University of Cambridge; and a 2021 Sociology of Emotions Section Session on "Emotions and Inequality" at the ASA annual meetings. I appreciate the invitations, and the thoughtful feedback from these audiences.

Iterations of this research were published in the *New Yorker*; *Signs*; *Theory, Culture and Society*; and *American Behavioral Scientist*. I am very grateful to Anthony Lydgate at the *New Yorker*, and to the anonymous reviewers (three at *Signs*, five at *TCS*, and two at *ABS*), as well as to editors Linda Blum, Martha Fineman, and Amber Jamilla Musser (at *Signs*); Rosalind Gill and Lisa Blackman (at *TCS*); and Jeremy Schulz and Barry Wellman (at *ABS*), for their insightful commentary and editing.

I am deeply thankful to Lauren Sharp at Aevitas Creative, whose early curiosity about and commitment to this book helped bring it to publication. And I have nothing but praise and wonder for Meagan Levinson and her team at Princeton University Press; Meagan read many versions of these chapters, noting where the argument was unsupported or the text was "chewy," cheering in the margins at a felicitous phrase. Editing is quintessential connective labor, and Meagan is a deft practitioner. Simply put—and as I have already told her—she is the editor I have always wanted.

Finally, if writing is solitary, the capacity to write is a communal achievement, and I feel lucky to have the family and friends who enable me to do this work I love. Thanks to Tina Verba, Beth Lorey, Suzie Tapson, Janet Legro, and Taryn and Ray La Raja, all of whom have heard me talk about this project for almost a decade, and have responded with their own stories, ideas, chuckles, and tears. Early conversations with Gary and Karen Mueller on a long drive from New Hampshire to Boston, with Janet Legro on long walks with Charlie, and with Hal Movius around a dinner table in Charlottesville, were pivotal. Thanks to those who make my corner of the world a warmer place, including Deborah Lawrence, Wendy Phileo, Miller Susen, Kate Bennis, Nora Brookfield, and Ann Lucas, and the comp team at Capital Rowing Club. Thanks to my mother, Joanne Pugh, and siblings, Jim, Robin, Andy, and Rosemary Pugh, who doggedly read and commented and shared thoughts about this project from its inception. Thanks to my now-adult children, Sophie, Lucy, and Hallie, who grew up with this book, and whose support, advice, and enthusiasm buoy me every day. And most of all, for everything, thanks to Steve.

APPENDIX

"MAYBE WE'RE GOING TO TURN YOU INTO A CHAPLAIN"

Studying Connection

How do we study the connections that people build with others? How do we know when someone sees the other, or when they feel seen? The experience of emotional recognition is one that involves messages sent and received—sometimes verbal ones that we can hear, but other times through bodily gestures as elusive as a nod, a chuckle, or a wrinkle, or, as Wanda might say, a "vibe" or an "energy." It is also deeply internal; together, people work toward a mutual understanding, but they can disagree about whether or not they have achieved it. How do we ascertain what is going on, about a subject that so straddles the boundaries of self and other?

The short answer is: with humility. Psychologists have conducted experiments that demonstrate that people think they understand the other better than they do, and further, that people sometimes feel understood better than they actually are. While these experiments address the doing of connective labor, they also offer up some truths about studying it.[1]

This research was a very large undertaking, stretching across three states and two continents, and involving hundreds of hours of observations and interviews. I was ably assisted by nine graduate students and three undergraduates who contributed to the project at various moments.

With considerable variety in the kind of person we watched and talked to, we did our best to get a deep and broad sense of the work of seeing the other, and how it is shaped by its context, particularly by what we came to think of as its social architecture. And while our grasp is undoubtedly partial, here is the story of how we obtained it.

Who Are the People in This Study?

To gauge how people pursue connection to others as part of their jobs, as well as how they feel about that pursuit, we resorted to two main approaches to learning: in-depth interviews and ethnographic observation. Aided by graduate students, I interviewed 108 people, most of them people who practice, supervise, or analyze connective labor, including therapists, physicians, chaplains, hairdressers, lawyers, salespeople, and police officers. I watched them in doctor's offices and schoolrooms, therapy sessions and squad cars, in California, Virginia, and Massachusetts. I also interviewed people who supervised, evaluated, or automated this work, from principals and program heads to the engineers working with robotics and AI. From the start, the project meant I was immersed with people who were professional feelers, often astute emotional experts with years of practice in seeing the other. Most days, I keenly felt the privilege of being in their midst, trying to ascertain their truth.[2]

I started the research with plans to focus primarily upon three different groups: nurses, therapists, and teachers. I was initially reluctant to include physicians because there are so many different kinds, and specialties like radiology or pathology did not require much, if any, connecting to patients. But doctors essentially insisted that they be a part of the study. As I talked about my research in different venues, at the time of this research—conducted from 2015 to 2020—physicians were so taken with the onset of the electronic health record and the transformations of their jobs due to managed care and evidence-based medicine, that they were grappling with the same issues that I was: what is the magic that practitioners and their charges make together, and what do systems do to that accomplishment? I would sit down to lunch after a talk, and the doctors there would pepper me with questions about what I was finding, or tell me intriguing facts like: "Hospital medical team members answer more than a thousand alarms in one shift." The intensity of their focus drew me in to their dilemmas; primary care physicians became one of the study's three major groups, although I did end up interviewing some nurses.

I also sought out opportunities to talk to people in other fields—beyond the three primary groups—whom I thought would add a unique perspective. I interviewed a handful of people engaged in criminal justice, for example, including police sergeants on the beat, a police chief lauded for his community policing focus in a struggling West Coast city, and the head of a prisoner reentry program. We talked to people in sales, in childcare, and in community organizing. In all, these primary groups included sixty-seven people: twenty-two educators, nineteen medical practitioners (sixteen of whom were physicians), sixteen therapists, and ten others.

I also decided to seek out practitioners who did not have the college degrees, extensive training, or the often-high salaries of those in the three main groups, who frequently benefited from class advantage or professional societies. This group included a wide range of occupations, such as hairdressers, massage therapists, funeral home directors, home healthcare aides, and the often-paid companions known as "sugar babies." My students and I interviewed fifteen practitioners without college degrees.

Interviewees also included some "recipients" of this work. Of course, all who provide connective labor are themselves receivers, as we all have moments in our lives as students, clients, or patients. People also sometimes straddled these roles at the same time, as when the Fellows at the HIV clinic were mentored by the attending physicians there while also seeing their patients. In addition, the ethnographic portion of the study also gave the opportunity to see how particular connecting efforts landed, or were received by the other. Still, while the focus of the study was on those who did the work under different conditions, I sought out a few opportunities to hear detailed accounts from the other side, as it were. In addition, I wanted to hear about potentially negative cases, where people felt particularly unseen or unheard; to that end, I posted a call on Twitter for people to tell me about when they "broke up" with their therapist, advisor, or hairdresser, and interviewed some of the respondents. In sum, seventeen recipients took part in the research.[3]

As the study was focused on the impact of systems on humane interpersonal work, I also sought to interview program administrators, principals, engineers, or researchers whose jobs were to bring order, efficiency, and technology to this work. This group included public school principals, private school founders, and scientists at the cutting edge of AI research; I ended up interviewing 29 people in this category. In total, interviewees included 44 men and 64 women and nonbinary people; ten identified as LGBTQ. Of 108 interviewees, 26 were people of color.[4]

I adopted a range of tactics to recruit people for the study. My goal was to hear from people who varied in the degree to which their connective labor was systematized, in order to contribute what Kristin Luker has called *logical* as opposed to *statistical* generalizability: in other words, to gather not just anybody in these professions but instead those whose situations would enable us to think more deeply about what we cared about. This procedure differed for each occupation. To recruit therapists, for example, I contacted some employed by massive public institutions whose work is subject to intense scrutiny, supervision, and bureaucratic rules, while also contacting those engaged in private practice, who do not accept insurance, and who might have clients for a decade or more. For teachers and principals, I managed to secure access to an email listserv of hundreds of public school principals in a major West Coast city, and also successfully recruited staff from independent schools, including experimental high-tech alternatives.[5]

Interviews were between one and three hours long, with most taking about a hundred minutes. They began with people narrating how they got into this work (to determine whether they thought such work relied on innate skills or "born this way" ideologies), as well as how they were trained. They then walked through a typical day and a typical interaction with clients or recipients. We asked about their experience with various forms of systematization, from checklists to manuals to information technology to automation. Questions probed at their interpretations, such as how they knew they were doing a good job or whom they viewed as their audience, as well as the emotional layers beneath their spoken words, such as the sense of relief or threat that suffused their stories of employer "solutions" when their work seemed overwhelming.[6]

Watching How It Is Done

In addition to hearing how people understood their work, however, we also wanted to see how they did it, and how it was experienced by their patients, clients, and students. I conducted hundreds of hours of observations of clinical visits, classrooms, and conferences, as well as of the training of therapists, doctors, and teachers, in a variant of what Princeton sociologist Matthew Desmond has called "relational ethnography," to "study processes rather than processed people." I was particularly interested in training, as I saw it as an encounter in which practitioners would be forced to articulate the specifics of their approach. A chance to see how even the most artisanal approach to craft intersected with the mandate to pass on their knowledge to others, training struck me as a moment when practitioners brought to the surface any sorts of abstract principles that governed their work.[7]

As part of this effort, I visited public and private classrooms and clinics in Virginia and California, watching practitioners in schools, companies, and hospitals. The project took me to modest clinics and homes with bars on the windows as well as to the lush landscapes of elite universities and hushed carpeted floors of tony practices for the wealthy. My goal was to follow connective labor wherever people would let me see it.

Specifically, I spent six months watching physicians, nurses, and patients in a federally funded HIV clinic, another six months observing a group of relationship-minded physicians as they met to discuss humanistic medicine practices, and a weekend in a horsemanship workshop teaching medical students about the doctor-patient relationship. I shadowed doctors on their rounds and visited experimental clinics. With the help of an undergraduate research assistant, I spent a semester observing a class for masters students seeking to become school counselors;

I also sat in on a semester-long course in adolescent therapy for PhD candidates who were apprentice clinicians, and spent many hours observing videotaped therapy sessions with supervisors giving commentary to clinicians in training; I also attended periodic meetings with the therapy staff of a VA hospital as they discussed common struggles in their practice. I visited the classrooms of a juvenile detention center, of a high-poverty public middle school, and an innovative private school using technology to try to "disrupt" conventional education. I spent six months with a trainee chaplain program that involved shadowing and interviewing the apprentice chaplains in their shifts with patients as well as their classroom work. I also visited labs where engineers were tackling new AI programs and apps in California, Massachusetts, and Japan. I attended conferences discussing AI research, symposia on police reform, and workshops and talks on medical technology, nursing care, and design thinking. I even joined a police officer for twelve hours of ride-along in his squad car, in a distressed Western city with a community policing initiative.

While many of these encounters were in classrooms or labs where I was just one of several, others took place in very small clinics or classrooms where I would try my best to blend in with the surroundings, but where patients or students could not fail to notice my presence. In some cases, they would studiously ignore me—one chaplain, after a long group session in which she disclosed some intensely private turmoil, told me, "I just decided to trust you"—while other times they would tell me jokes or wink my way, clearly aiming some of their interactive juice in my direction. I would smile gently but not really engage, trying to occupy a more low-key presence. I am sure that the dynamics changed in part for some people, but there were also no indicators of major effects.

Still I checked in periodically about my potential impact with my participants, particularly at the places where I spent the most time, such as among the chaplains and at the HIV clinic. When I asked Karl, a Fellow in the clinic, whether my presence had changed anything about his appointments, he answered in the negative. "Not so much," he said. "There are so many observers coming in and out—fellows residents, attendings, psychologists—that we are used to it." We were coming out of a lengthy visit he had just had with Mr. Shiflett, the widower whose appointment was chronicled in chapter 2. "Really the thing that had the most impact [on that appointment] was knowing that I had someone else to take my 9:30 and my 10 a.m., so I could take the time," Karl observed.

These people were, for the most part, real experts, and some days their expertise was brought home to me with force. At one point in my observations, for example, I watched a group therapy session in which the leader was also being observed by his trainer. I filled my field notes with comments about how well he was mirroring the group members, how carefully he reflected their truths. After the session, however, I listened to the trainer as she complained to me about how the trainee had not been doing his job well enough, about how a member of the group had just been going on and on about something not very relevant to the group, and that the trainee had given only supportive comments, nothing challenging, including no notes about the group process, which the trainer told me was the way to bring such soliloquies back to the group, to make it more relevant for everybody. "He [the trainee] has a long way to go," she told me. Her comments floored me, after my notebook had been filled with my own observations of the man's expertise, how thoroughly and competently he bore witness to the other. The trainer's comments made me realize that to them—the professionals—there was a lot more to this work than simply seeing the other, that what I was watching for was but the ground level of a much more complicated edifice.

Leaving the Field

With any qualitative study, one of the primary questions is "How do you know when you are done?" The consensus among researchers is that this moment comes when you have achieved what they call "saturation," or the sense that you know what your interviewees are going to say before they say it. Doing so in this project was a little tricky, because the practitioners would often sound very similar to each other across wildly different contexts—the home healthcare aide and the dermatologist, the funeral home director and the master therapist at the VA. My task was to hear all those similarities—to grapple with the sense of saturation there—and also to be sensitive to their core differences. Still, by the end of five years in the field, I felt I had some purchase on the study, a sense brought home to me one day as I shadowed Erin, the chaplain whose story opens up the book. After a long day together, she turned to me and laughed. “I don’t know, Allison . . . maybe we’re going to turn you into a chaplain.”

Once I left the field, I still had a large amount of field notes and interviews to consider, and analyzing these materials took almost six months. That process involved reading the materials again and again, thinking about themes that repeatedly arose, and writing memos to push at those themes and weigh them against existing knowledge. We used the online program Dedoose to code research materials and write memos. I analyzed the materials of each particular group at a time, coding the transcripts and field notes of all of the therapists, for example, to glean particular insights by occupation before thinking about the practitioners as a whole. I wrote more than a hundred memos—on topics such as “the costs of data primacy,” “ruptures as opportunities,” and “triangle relationships”—some of which then became the building blocks of larger arguments in the book and in articles.

Ethics, Impact, and Managing Vulnerability

In a study of witnessing, I could not help but bear witness, often to moments of considerable vulnerability. This privilege also presented real ethical concerns, which to my mind dictates for the researcher several unarguable obligations. First among these is to do no harm.

There are a number of ways in which I sought to mitigate any harm that might come about from my presence. First, I participated as much as the people—both practitioners and their charges—seemed to want me to, even when it was a little uncomfortable for me. One day, for example, I joined a group of medical students as they learned about doctor-patient relations through working with horses, and while the rather prickly instructor allowed me to be there, she also made sure I was cut to size when she asked me about my negligible riding background or corrected me in an exercise. Later, however, she commented "what a good sport" I was, and mentioned how they were exposing some vulnerabilities in the ring, and that it was good that everybody participated, that having someone just observe would make it harder to be vulnerable, "especially someone with a pad taking notes." The day had featured not just horseback riding but also stories of how important it was to read people's bodies and gestures. One student recounted her own history of medical trauma in a high, warbling voice, pleading with her peers to "just look in their [patients'] faces." While I did not love the experience, and wished somehow I could have just been a fly on the wall, instead of joining small teams to lead the large animals hither and yon, the prickly instructor was not wrong; it was easier on everyone when we were all somehow a bit vulnerable.

Second, particularly in the context of in-depth interviews, I offered my own version of connective labor, to see and reflect the other's truth to the best of my ability. People often commented in interviews about how it "felt like therapy," or "it's fun for me, it's so fun." One pediatrician, holding forth about how she was not "practicing at the top of her license" (because she was being asked to attend to more basic needs than she had trained for), told me, "Everybody has real problems, and what people need is an hour with you. I mean, seriously." She paused and gestured at me. "With a good listener, it doesn't matter what their training [is]." It was a funny moment—because she was both offering me a compliment and at the same time sort of diminishing it as not worth her own time—but I was grateful that she felt listened to.

"Do no harm" also meant stepping back when people seemed to wish it. When I was observing at the HIV clinic, I would check in with doctors in the morning, to see what their schedule was like that day. One day, Isabel, a Fellow, told me that she did not want me to come to her first appointment. "It's not . . . you're lovely . . ." she paused and started again, "It's just that he's a physician and he had a stroke, and he has no filter and his wife cares for him, and he says, 'Just let me die,' and she says something and then he calls her princess, and he's a former physician so he is very controlling about his medical issues, and just managing the two of them can be difficult." Sometimes I stayed out of appointments when practitioners wanted to manage the tough ones on their own.

Other times I stayed out when the clients or patients themselves wanted it, or when others guessed they might. The HIV patients sometimes brought up confidentiality issues that were less acute in other contexts. One day, as I was talking to Christine, the nurse practitioner, about her upcoming schedule, a nurse came into the room and reported that two people who knew each other previously had encountered each other in the waiting room. "One person is upset now," the nurse said. "They want to know what else do we treat, besides HIV, because she doesn't want that other person to know. It just happened," the nurse added. Christine thanked the nurse, and told me her nine-thirty appointment was also "very worried about cconfidentiality—he owns a business here—but that the people after should be fine." I took a step back when it seemed clear that students or clients would be made uncomfortable by my presence.

Finally, sometimes I opted out of my own accord, declining to witness at all when I sensed my presence would intrude unduly. One day, for example, when I was shadowing the chaplain Erin, we learned that a young woman had died unexpectedly of a Tylenol overdose, as the first chapter recounted, and that her husband, aunt, and nephew were gathered in a tiny room next to the hospital elevator. Erin and I discussed whether I should be there, and although Erin was fine with it—and was thinking aloud about how she would introduce me—I decided to stay out. There was just something that felt very wrong with my being in that small space during that moment of intense, surprising grief.

"Doing no harm" also comes into play when we leave the field. While there is a discussion ongoing in sociology about the alleged benefits of "unmasking"—identifying informants and communities involved in research—I have instead taken steps here to provide a measure of confidentiality, such as using pseudonyms and changing some identifying details to obscure the identity of those who participated. As Sarah Mosseri and I have written elsewhere, public recognition can incur real risks, particularly in an era of trolling and doxing and particularly for communities susceptible to social policing. These risks are borne largely, but not solely, by participants, who are much more vulnerable than researchers, and for whom the individual benefits are often less. Their vulnerability, and the gift they offer in their engagement with our research, obligates us to protect them.[8]

The Honor and Urgency of Studying Connective Labor

I always felt lucky to do this research. Most of the time, in most of these contexts, I was fascinated by the process of becoming a connective labor practitioner, of seeing the other and trying to get better at it, and by the question of how people understood what was valuable about the work. I was also riveted by the dilemma of the moment: how some were trying to change or manage it to make it more accessible, more effective, more profitable, but using counting, scripting, and technology to do so. In fact, my research was so gripping on a regular basis, that on the one occasion in five years when I was bored—in a training session watching students role-play with each other—my field notes actually analyzed just why it was boring (I concluded that it was because they were role-playing either fake situations or ones that did not expose themselves in any way—for example, "I'm worried about hiring an event coordinator for the spring fling"). More often, I watched people talk about illnesses, family dynamics, or anxieties in ways that felt real and thereby as compelling, as profound, as watching emotional lives unspool before me.

Furthermore, the project of identifying, defining, and elaborating upon connective labor gained a certain urgency as I saw it being transformed by administrators and engineers heedless of what made it valuable. The impact of systems was everywhere, and if people complained about technology or the demand for data sometimes, they also acted as if they could not imagine a different world. Indeed, more often than not we saw the hegemony of data, technology, and systems—what I call the cultural juggernaut of the industrial model, discussed in chapter 6. One day, Pilar Plater, who at the time was one of my graduate student assistants, was interviewing Paul Giang, a gig worker in San Francisco, about their experience of seeing the other. At the conclusion of their encounter, Paul told her: "So that's why I think AI is never going to substitute completely the human interaction. A big difference between me having this kind of conversation with you in person than if I filled out an Excel spreadsheet with just the answers to those questions there. I'm sure you probably—hopefully you've got more information than if a survey was filled out." Pilar thanked him and got ready to go. "This was really interesting," Paul replied, as they both started to pack up their belongings.

But then it became clear just how much technology was already part of the conversation about interpersonal humane work, how much it was assumed to be where that work was going in the future. Looking back over the interview, Paul said he was actually a bit surprised about where it ended up going. "I was expecting you to just ask me, you know, where . . ." Paul paused. "I honestly thought you guys were going to do an app, you know?"

NOTES

1. Introduction: The Power of Seeing the Other

1. All names and some identifying details have been changed for confidentiality purposes, except where noted.

2. For a review of research on the impact of sensitivity on manager performance, see Reis, Lemay, and Finkenauer, "Toward Understanding Understanding."

3. Estimates of the percent of workers engaged in connective labor are scarce. For the 12 percent figure, see Duffy, Albelda, and Hammonds, "Counting Care Work." The World Economic Forum predicted 97 million "jobs of tomorrow" would be added by 2025, with a greater emphasis on interaction skills. World Economic Forum, "The Future of Jobs Report 2020."

4. Economists have documented the rising importance of socioemotional skills across the US occupational structure. Deming found job growth and wages were tied to social-skill-intensive occupations between 1980 and 2012, and, studying US graduates, found that social skills were a much stronger predictor of job finding success in the 2000s than in the preceding two decades. These findings echoed similar studies in Sweden and the Netherlands. From 2006 to 2016, "feeling" and "thinking" tasks were each responsible for job growth of a little over 4 percent, the "Feeling Economy" researchers found. The use of therapists has remained remarkably stable over the years; the rise in connective labor is the expansion of other forms of work, and its expansion within old jobs. See Deming, "The Growing Importance of Social Skills." For rates of psychotherapy use, see Olfson and Marcus, "National Trends in Outpatient Psychotherapy." The "our finding is important" quote is from Atalay et al., "The Evolution of Work in the United States," 3. For the "Feeling Economy" research, see Huang, Rust, and Maksimovic, "The Feeling Economy." See also Edin et al., "The Rising Return to Non-cognitive Skill"; and Allen, Belfi, and Borghans, "Is There a Rise in the Importance of Socioemotional Skills in the Labor Market?"

5. For a review of patient-clinician relationship effects, see Kelley et al., "The Influence of the Patient-Clinician Relationship on Healthcare Outcomes." For research documenting the impact of the therapeutic alliance, see Horvath and Luborsky, "The Role of the Therapeutic Alliance in Psychotherapy"; Luborsky, Singer, and Luborsky, "Comparative Studies of Psychotherapies"; Budd and Hughes, "The Dodo Bird Verdict"; Sauer et al., "Client Attachment Orientations"; Mikulincer, Shaver, and Berant, "An Attachment Perspective"; and Safran et al., "Therapeutic Alliance Rupture as a Therapy Event." For teacher-student relationships, see Hamre and Pianta, "Early Teacher–Child Relationships"; Myers and Pianta, "Developmental Commentary"; Reyes et al., "Classroom Emotional Climate"; Hattie, *Visible Learning*; Cornelius-White, "Learner-Centered Teacher-Student Relationships Are Effective"; Bergin and

Bergin, "Attachment in the Classroom"; Klem and Connell, "Relationships Matter"; and Roorda et al., "The Influence of Affective Teacher–Student Relationships."

6. *Moneyball* (2004) is Michael Lewis's account of how the underfunded Oakland Athletics baseball team used data analytics to figure out how to seek out undervalued athletes that managed to catapult them to victory. Lewis, *Moneyball.*

7. See, for example, Bowker and Star, *Sorting Things Out*; Alvehus and Spicer, "Financialization as a Strategy of Workplace Control"; Pardo-Guerra, *The Quantified Scholar*; Griffen and Panofsky, "Ambivalent Economizations"; and Diamond, *Making Gray Gold.*

8. Public school systems continue to seek out "teacher-proof" curricula, despite the importance of teacher adaptations to content; see, for example, Taylor, "Replacing the 'Teacher-Proof' Curriculum." Programs—with names like Success for All, Direct Instruction, and Open Court—were particularly pervasive after the No Child Left Behind Act was passed in 2001. See Colt, "Do Scripted Lessons Work—or Not?" The case of therapists at one agency—the Calgary Counselling Center—being required to administer "outcome questionnaires" and "session ratings" to their clients is discussed in Goldberg, Babins-Wagner, and Miller, "Nurturing Expertise at Mental Health Agencies." Within four months of implementation of the policy, 40 percent of the therapists resigned, the authors recount. "Staff developed clinical dashboards that summarized each therapist's clinical outcomes in comparison with the average outcomes from all of the therapists at CCC. The dashboards are given to staff every 4 months as a tool to gauge their own effectiveness" (206). After seven years, the agency reported, outcomes demonstrated small improvements. See also Goldberg et al., "Creating a Climate for Therapist Improvement."

9. For legitimacy, see Glenn, "Creating a Caring Society"; for standards as important hedges against incompetence, see Reich, *Selling Our Souls*, and Leidner, *Fast Food, Fast Talk.* For standards as protection against discrimination, see Roberts, *Fatal Invention.* How they protect against demanding working conditions is chronicled in Leidner, "Emotional Labor in Service Work." The "world of gray sameness" is in Timmermans and Epstein, "A World of Standards."

10. Fraser, "Contradictions of Capital and Care," has argued that a struggle over "a key set of social capacities: those available for birthing and raising children, caring for friends and family members, maintaining households and broader communities, and sustaining connections more generally" is central to contemporary capitalism. "Globalizing and propelled by debt, this capitalism is systematically expropriating the capacities available for sustaining social connections," she contends (116). My use of the phrase "social architecture" diverges from organization scholars such as Jane Dutton, who applies the concept instead to "the amalgam of social networks, values, and routines that structure an organization and that constrain and enable individual action"—what I would call culture. Instead, I think the term is actually better applied to a more capacious concept, to capture all the dimensions that enable or impede connective labor. As chapters 5 and 8 explore, these include people, culture, and material resources. See Dutton et al., "Explaining Compassion Organizing."

11. For creativity, see Reich, "Disciplined Doctors," and Verghese, "The Importance of Being." For autonomy, see Leidner, *Fast Food, Fast Talk*; Abbott, *The System of Professions*; and Haug, "A Re-examination." For alienation, see Hochschild, *The Managed Heart.* For "industrial objects," see Friedson, *Profession of Medicine*; and for demoralization, see Rupert and Morgan, "Work

Setting and Burnout." Others who write about the dehumanizing consequences include Fox, *Medical Uncertainty Revisited*, and White, "Reason, Rationalization, and Professionalism."

Regarding the more than 50 percent of primary care physicians reporting burnout, see Tolentino et al., "What's New in Academic Medicine"; for the evidence linking it to use of the EHR, see Shanafelt et al., "Relationship between Clerical Burden." A 2017 federal study found "more than half of primary care physicians report feeling stressed because of time pressures and other work conditions." Nearly a third said they needed 50 percent more time for physical exams and follow-up care. See the Agency for Health Care Quality and Research, "Physician Burnout." For stress in teachers and nurses, see Gallup, "State of American Schools." One study of more than three million nurses found that a stressful work environment was the number-one factor for those who left their jobs (almost five hundred thousand); see Shah et al., "Prevalence of and Factors Associated with Nurse Burnout."

12. For the size of the health-app market, see Liquid State, "4 Digital Health Trends," and Matheny et al., "Artificial Intelligence in Health Care." The 38.2 billion valuation of 2021 was reported in by the market research firm Grand View Research, "mHealth Apps Market Size."

13. Lucy Suchman (personal communication) is one such critic, cautioning against relying too much on the hype about what AI can accomplish, and noting the underlying politics to that hype, ranging from immigration hostility to concerns about research budgets. See also Nutt, "The Woebot Will See You Now." *The Daily* podcast featuring ChatGPT was entitled "Did Artificial Intelligence Just Get Too Smart?" and aired December 16, 2022.

14. When asked what the dangers of AI were, the ChatGPT-3 bot replied with these three problems, as well as the loss of human autonomy; its reply is telling, since it reflects what it was able to absorb from what has been written before. For bias and stereotyping in AI, see Benjamin, *Race after Technology*; Noble, *Algorithms of Oppression*; and Eubanks, *Automating Inequality*. For tracking and surveillance, see Zuboff, *The Age of Surveillance Capitalism*. For AI's impact on job disruptions, see Pasquale, *New Laws of Robotics*.

15. The Queen's quote is found in Acemoglu and Robinson, *Why Nations Fail*, 182–83. See also Frey and Osborne, "The Future of Employment."

16. Autor, Levy, and Murnane, "The Skill Content," complicated the story of cognitive and manual tasks with a question of how routine the tasks are. Scholars have expressed some frustration about the arbitrary nature of what counts as "routine"; for example, Benzell et al., "Identifying the Multiple Skills," found that leadership and cooperation were most important for "skill-biased technical change," with leadership and cooperation most important for wage growth, and empathy most vital for employment growth.

17. For "wealth work," see Autor, *Work of the Past*. For a report on Utah's virtual preschool, see Bowles, "An Online Preschool Closes a Gap." See also Dwyer, "The Care Economy?," and Ticona and Mateescu, "Trusted Strangers."

18. See Ticona, Mateescu, and Rosenblat, "Beyond Disruption."

19. I write about this "colliding intensification" in Pugh, "Emotions and the Systematization of Connective Labor." Scholars debate the impact of these developments. One group offers sophisticated analyses of the great variety of systems and how people use them, but their studied neutrality about their impact can underestimate the sometimes-inadvertent impediments that such systems can impose on people's capacity to conduct relationship. Another group argues

that the two systems create each other; when therapists "see" clients, for example, they translate the chaotic wildness of human idiosyncrasy into recognizable thoughts and emotions, just like the depersonalizing checklists, manuals, and apps do. As a result, emotions become commodities, or, to use Eva Illouz's term, "emodities." A third group worries about how the collision contributes to exploitation of workers, arguing that emotional attachment between worker and client is either a myth obscuring corrosive inequalities, a distraction from difficult working conditions, or a vulnerability, the root of an 'emotional hostage effect' that employers exploit to keep workers at bad jobs. See Illouz, "Introduction"; Zelizer, "How I Became a Relational Economic Sociologist"; Reich, *Selling Our Souls*; and England, "Emerging Theories of Care Work." See also Hochschild, *The Managed Heart*; Lopez, "Emotional Labor and Organized Emotional Care"; Diamond, *Making Gray Gold*; and Glenn, *Forced to Care*.

20. Much of what Surgeon General Murthy describes addresses the importance of being seen. "To be at home is to be known," he writes. "It is to be loved for who you are." See Murthy, *Together*, xviii. The "headless horseman" quote is from Claude Fischer's blog "Made in America"; see "Loneliness Epidemic." See also Fischer, "The 2004 GSS Finding of Shrunken Social Networks." The Japanese official title is the Minister for Loneliness and Isolation. The quote on "the problem undergirding many of our other problems" is found in Brooks, "The Blindness of Social Wealth." See also Case and Deaton, *Deaths of Despair*; Allen, "America Is in a Great Pulling Apart"; McCoy and Press, "What Happens When Democracies Become Perniciously Polarized?"; and Bollyky, Kickbusch, and Petersen, "The Trust Gap."

21. Loneliness is as dangerous as obesity; Holt-Lunstad is the source of the widely quoted number "fifteen cigarettes a day," which I could not find in any academic publication, although her published work does equate it with light smoking. See Holt-Lunstad, Smith, and Layton, "Social Relationships and Mortality Risk"; Holt-Lunstad, Smith, and Baker, "Loneliness and Social Isolation." See also Newman, Lohman, and Newman, "Peer Group Membership"; Baumeister et al., "Social Exclusion"; Pressman et al., "Loneliness"; and Twenge et al., "Social Exclusion Decreases Prosocial Behavior." The 2023 review was Wang et al., "A Systematic Review and Meta-Analysis." The "food" quote comes from Baumeister and Leary, "The Need to Belong," 498.

22. In blogs, articles, and books, for more than a decade Claude Fischer has been asserting the persistent connectedness of Americans. The quote about the "total volume of personal contact" comes from a post entitled "Overcoming Distance and Embracing Place." The work on Americans' new types of associations can be found in Paxton and Rap, "Does the Standard Voluntary Association Question," which compared the GSS survey (used by many researchers to chronicle the decline of American association) with groups that are listed on the platform Meetup.com. "We estimate that the GSS may be missing about half of the types of groups that appear on Meetup," the authors conclude. See also Fischer and Durham, "Forms of Group Involvement"; and Fischer, *Still Connected*.

23. The predilection of people to confide in others who are available, rather than necessarily particularly intimate or knowledgeable, was established by Small, "Weak Ties," and Small, *Someone to Talk To*; see also Furman, *Facing the Mirror*. The British study varied how customers interacted with their baristas, finding that those who engaged in conversation ended up with more positive feelings, a greater sense of belonging, and a higher satisfaction about their experience; see Sandstrom and Dunn, "Is Efficiency Overrated?" The "gold standard" of loneliness metrics,

drawn from a larger battery that comprises the UCLA loneliness scale, is a combination of three measures: "I feel left out," "I feel isolated," and "I lack companionship." See Hughes et al., "A Short Scale for Measuring Loneliness." The influential study of weak ties is Granovetter, "The Strength of Weak Ties." See also Sandstrom and Dunn, "Social Interactions and Well-Being," and Dunn et al., "Misunderstanding the Affective Consequences."

24. A review of the costs and benefits of personalized medicine concluded that it generated "modest health benefits" at a high cost; see Vellekoop et al., "The Net Benefit of Personalized Medicine." For findings of personalization's "slightly positive effect on student learning," see Pane et al., "How Does Personalized Learning Affect Student Achievement?" The *Forbes* piece is Morgan, "The 20 Most Compelling Examples."

25. Harry Braverman, in *Labor and Monopoly Capital* (1998 [1974]), analyzed the way managerial control broke down the autonomy of labor across a range of jobs, expanding upon Marx's concept of alienated labor and commodity fetishism, although he did not write much about jobs that would be considered connective labor.

26. The emotion management component is what Arlie Hochschild called "emotional labor" in 1983. See Hochschild, *The Managed Heart*, and Hochschild, "Emotion Work, Feeling Rules, and Social Structure."

27. In *Playing to the Crowd* (2018), Nancy Baym coins the term "relational labor" to describe how musicians feel pressure to reveal of themselves to connect to fans; Hochschild, in *The Managed Heart*, has written about how flight attendants feel pressure to control their emotions for pay.

28. The word "labor" is also useful here, I think, because it invokes the conditions, culture, or climate where it takes place. A few others have coined related terms—for example, Baym's "relational labor", Joyce Fletcher's "relational practice" to describe how engineers build collaborative teams to do their work, Viviana Zelizer's "relational labor" to describe how people use economic tools to establish and convey relationship ties—but these are too specific to capture the broader phenomenon at hand. For emotional intelligence, see Elfenbein and Ambady, "Predicting Workplace Outcomes"; and Mayer, Salovey, and Caruso, "Emotional Intelligence." See also Baym, *Playing to the Crowd*; Fletcher, *Disappearing Acts*; and Zelizer, *The Purchase of Intimacy*.

29. Among philosophers, Levinas suggested that we are all captured by the "gaze of the other"; see Levin, *The Philosopher's Gaze*. As defined by Hochschild, emotion work "requires one to induce or suppress feeling in order to sustain the outward countenance that produces the proper state of mind in others" (*The Managed Heart*, 7); she reserved the term "emotional labor" for that work conducted for a wage, cautioning that workers risk feeling alienated from their very selves when they control emotions for their employers. While initially a very specific term for the work people do to manage their own feelings, "to create a publicly observable facial and bodily display . . . that produces the proper state of mind in others," emotional labor has since expanded to mean all kinds of uses for emotions at work. As I use it here, connective labor is one of many ways that people use feelings at work, which might also include enlisting allies, negotiating sympathy, assessing attachment, or other emotional practices. See Hochschild, *The Outsourced Self*, and Hochschild, *So How's the Family?*. See also Sherman, "Caring or Catering?"; Lopez, "Emotional Labor and Organized Emotional Care"; Kelly, *Disability Politics and Care*; and Pugh, "Connective Labor as Emotional Vocabulary."

30. Adia Harvey Wingfield has written about the emotional expectations that entrap Black men at work, who are constrained from expressing anger in white-dominated workplaces. See Wingfield, "Are Some Emotions Marked 'Whites Only'?" Judith Butler, in *Bodies That Matter* (2011), explores and questions the demand—inherent in notions of recognition—that one must be legible to others.

31. As the anthropologist C. Jason Throop explains, "this saying is pejoratively used to refer to people for whom it is possible to tell immediately what they are thinking or feeling." In his words, "You just look at them and know if they are sad or angry. . . . In allowing one's inner conditions to manifest directly in one's external forms of expression, an individual is thus comparable to a papaya, and as such clearly marked as failing to approximate the virtues of self-governance, concealment, and secrecy." The case of the Yap Islanders and their dismissive view of ripening papayas, so easily seen, was reported in Throop, "On the Problem of Empathy." Throop notes that there are four components to thinking about empathy culturally, and one of these he dubs "appropriateness/possibility," that is, whether it is even suitable to "[seek] out or [demonstrate] knowledge of another's internal states in particular contexts" (406).

32. The history of empathy is reported in Lanzoni, *Empathy*. Mesquita, *Between Us*, explores the cultural variation of emotions like empathy.

33. For work finding that some do not want to be seen, see Blume Oeur, "Recognizing Dignity," and *Black Boys Apart*, as well as Phillippo, "'You're Trying to Know Me.'" Psychologists report that not everybody wants to be understood; see Reis, Lemay, and Finkenauer, "Toward Understanding Understanding." By some measures, as many as 20 percent of people have what is called an "avoidant" attachment style, preferring to hold off intimacy and close connections with others, according to Hirsch and Clark, "Multiple Paths to Belonging." Given this, the experience of being seen may not bring about physical and mental health benefits to the same degree for all. In a 2022 conference paper at a panel organized around the connective labor concept, Ilana Gershon and Anna Eisenstein explore how it intertwines with boundaries for nannies working during the pandemic; see Gershon and Eisenstein, "'Saying No and Staying Flexible.'"

34. Eva Illouz has written about the extensive use of emotional skills and techniques at work and home. See Illouz, *Cold Intimacies*. Researchers estimate the US investment in empathy education could be as low as $21 billion annually, and as high as $47 billion. See Krachman and LaRocca, "The Scale of Our Investment." A review of research on SEL programs and empathy is in Jones and Doolittle, "Social and Emotional Learning." For more about the prevalence of psychometric testing, see Alloway and Cissel, "Psychometric Testing." See also Illouz, "Towards a Post-Normative Critique." The quote "a voice of an innate primary nature" is from page 200.

35. Charles King defines Herzensbildung in his 2019 account of modern anthropology. See King, *Gods of the Upper Air*, 30.

2. The Value of Connecting

1. There is extensive research on the role of empathy in physician-patient relations. In one widely cited article, Larson and Yao argue that clinical empathy is emotional labor, focusing on the physicians' efforts to gin up the appropriate feeling; Vinson and Underman agree but add the organizational factors that make that more likely. Such empathy is a form of seeing the other, and sometimes it does indeed have to be actively stoked, particularly under conditions of burn-

out or overload. I'd argue the connective labor concept enables us to focus on the connections such work relies on and also creates, in other words, (1) its reciprocal qualities and (2) its social outcomes. See Larson and Yao, "Clinical Empathy as Emotional Labor," and Vinson and Underman, "Clinical Empathy as Emotional Labor."

2. Randall Collins writes about the emotional energy generated by coordinated action with others, and argues that it helps bolster institutional stability, in an argument that blends Goffman with Durkheim; Fine, "Review," 1288, suggests he develops the theory but "neglects in practice . . . an interaction that generates solidarity and symbols of group membership." These interactions include the ones I focus on in this book; the shared humanity that connective labor conveys helps create social intimacy and belonging. See Collins, *Interaction Ritual Chains.*

3. The authors write: "The Other is recognized as that which resists its precise constitution, and which nevertheless persists in being present." See Lenay, "'You Never Fail to Surprise Me,'" 393. See also Schilbach et al., "Toward a Second-Person Neuroscience."

4. Collins, *Interaction Ritual Chains,* argues that interactions need four characteristics to generate emotional energy and solidarity: sharing people's bodily co-presence, making a barrier to outsiders, creating a mutual focus, and sustaining a common mood. Chapter 4 discusses how practitioners accomplish these and other aspects of connective labor as craft.

5. Anthropologists have long recognized the emotional dimensions of resonance; see Wikan, *Resonance.* Describing resonance as "empathic vibrations," Messeri, "Resonant Worlds," 133, argues that it can lead to deeper understanding "beyond words and patterns and toward structures of feeling and affect." Yet recent work in the sociology of culture suggests resonance is a dynamic process, one that goes beyond simply sharing a set of common strings. Transcending the alignment of existing worldviews, cultural resonance is about offering something new to something old, scholars assert; see McDonnell, Bail, and Tavory, "A Theory of Resonance." Applying this model to connective labor, it suggests that for it to resonate, the cultural handshake between recognizer and recognized confirms something they think they know with a new image or revelation. In Pugh and Mosseri, "Trust-Building versus 'Just Trust Me,'" we argue for a "textured model of resonance," in which a resonant idea has room for multiple meanings, and offers not resolution so much as the capacity to unearth complex emotions that might haunt a given cultural object. In the connective labor context, then, this idea would suggest that a powerful empathic reflection offers recognition in several dimensions, at least some of which haunt the seen.

6. For the phrase "wild, silky part," see Oliver, *A Poetry Handbook,* 8.

7. With all the heated judgment mothers face today, I wish we could bring the concept of "good-enough mothering" back into common parlance. Psychologists share a broad consensus that mirroring or attunement—a form of emotional recognition—is important for infant development; see Stern, *The Interpersonal World of the Infant.* For good-enough mothering, see Winnicott, *Playing and Reality*; Benjamin, *Shadow of the Other,* is a good example of a more contemporary psychoanalytic account about how infant recognition shapes adult intimacy. Warren Poland has written compellingly about the adult therapists' witnessing; see Poland, "The Analyst's Witnessing and Otherness."

8. Sociologist Michèle Lamont reconciles these positions by considering recognition the cultural dimension of inequality, including material inequalities. Recognition may be more widely available to reflect "growing consensus about the equal worth of social groups," and yet she notes

it is distributed unevenly, as burgeoning inequalities within supposedly meritocratic societies create "recognition gaps," leading to rising resentment among those at the bottom. For the point that recognition has become newly relevant in contemporary times, see Taylor, "The Politics of Recognition"; for philosophers' work on the importance of recognition for social groups, see Fraser and Honneth, *Redistribution or Recognition?*. For the argument that recognition gaps stem from neoliberal cultural scripts, see Lamont et al., *Getting Respect*, a cross-cultural comparison to document available responses to stigmatization, which they regard as recognition's opposite.

9. Axel Honneth, *The Struggle for Recognition*, and Charles Taylor, "The Politics of Recognition," were among the more influential in this conversation. Honneth's typology initially featured love, rights, and solidarity, with the latter broken down into "emotional support," "cognitive respect," and "social esteem." The "emotional support" in the earlier writings was part of the socialization process, and as in psychoanalytic accounts, generated "one's underlying trust in one's self" (133). His later writings, for example, Honneth, "The Point of Recognition," distinguished simply among love, rights, and esteem; see McBridge and Seglow, "Introduction."

10. Emotional recognition is complex, with plenty of room for missteps. Relationships can sour when workers are constrained by a social architecture that impedes a full witnessing of the other, via efficiency metrics, overwork, and routinization. Critical perspectives might do better to take aim at the working conditions of relationships rather than their inherent dangers. When the mutuality of connective labor falters, however, it is sometimes because practitioners just get it wrong in their witnessing, reflecting back to someone an image that is inaccurate, misjudged, or unfair. Social inequalities figure prominently here: as we'll see in chapter 7, because witnessing involves seeing the other, there are serious moments of misrecognition that can arise due to racism, sexism, and other systems of thinking that create what Patricia Hill Collins called "controlling images," stereotypes that get in the way of perceptions of the individual. Furthermore, the time crunch accompanying the rise of data primacy bears implications for inequality: the less time workers have for their witnessing presence, the more they rely on cognitive shortcuts like stereotyping or scripting that can actually lead to harmful misrecognitions. For the impact of the powerful's pejorative images, see Fanon, *Black Skin, White Masks*. For the dangers of misrecognition, see Fleming, Lamont, and Welburn, "African Americans Respond to Stigmatization." For work demonstrating groups that view recognition as an invasion of privacy or threat, see Blume Oeur, "Recognizing Dignity," and Phillippo, "'You're Trying to Know Me.'" The notion that someone can be read "wrongly" should not be understood to mean that there are stable, authentic selves just waiting to be discovered or expressed. Instead, people's selves, or particular versions of their selves, are created in interaction, in part through recognition, and these variants can change with different people, in different contexts or at different times. For "controlling images," see Collins, *Black Feminist Thought*.

11. See Fraser, "Contradictions of Capital and Care"; the quote "the coin of love" is on page 102. See also Illouz, *Cold Intimacies*.

12. There is also evidence that not being heard—or interacting with someone who is trying not to be—is stressful. A different team of researchers from Stanford examined what happens when people suppress emotion. Similarly to the Finnish study, they took some biological measures of two people while they were talking, but the difference here was that both people first watched an intense short film with "graphic footage" of the aftermath of the nuclear bombs

dropped on Japan, after which researchers directed one participant to suppress their emotions entirely in the ensuing conversation. Those people who were told to hold a conversation while suppressing their emotions had increased blood pressure compared to the control group who were never issued that directive. The biggest increase, however, was actually demonstrated in those who listened to people suppressing their own emotions. See Butler et al., "The Social Consequences." The Finnish study was reported in Peräkylä et al., "Sharing the Emotional Load."

13. For studies of how those who have skill in reading others' emotions are more effective at work, see DiMatteo et al., "Sensitivity to Bodily Nonverbal Communication"; and Elfenbein et al., "Reading Your Counterpart." For studies of children who can decode nonverbal meanings, see Halberstadt and Hall, "Who's Getting the Message?"; Izard et al., "Emotion Knowledge"; and Elfenbein and Ambady, "Predicting Workplace Outcomes." For the improved workplace climates linked to manager sensitivity, see Johnson and Indvik, "Organizational Benefits." Research about leaders being viewed as better when they understand others can be found in Bechler and Johnson, "Leadership and Listening"; Johnson and Bechler, "Examining the Relationship"; and Kluger and Zaidel, "Are Listeners Perceived As Leaders?" For a review of studies emphasizing the impact of feeling understood, see Reis, Lemay, and Finkenauer, "Toward Understanding Understanding." For research on patient compliance and its link to feeling understood, see Street et al., "How Does Communication Heal?"; for the impact of feeling understood on therapy clients, see Ackerman and Hilsenroth, "A Review of Therapist Characteristics"; and Pocock, "Feeling Understood."

14. Research suggests what really matters is not so much the accuracy of that understanding, but rather whether others simply *feel* that they are being understood. And as it happens, people think they are more understood than they actually are, at least according to research with romantic partners. A host of different assumptions go into this belief: they think how they feel inside is more perceptible than it is, that their partners are more similar than they are, and that other people are paying closer attention to them than they do. In one telling study, researchers found that how much specific knowledge someone's partner demonstrated about them bore almost no relation to how much they felt understood. "Feeling that one understands one's partner and is understood by one's partner is unrelated to actually knowing one's partner and being known by one's partner," the researchers reported. Still, as another study concluded: "people generally understand others, at least to some degree." For the broader impacts of feeling understood, see Cooley, *Human Nature and the Social Order*. For the links between belongingness and improved physical and mental health, see May, "When Recognition Fails." For the positive impact of self-disclosure in social interactions, see Mehl et al., "Eavesdropping on Happiness." For the issue of (non)reciprocity, see Reis and Shaver, "Intimacy as an Interpersonal Process." For the study linking feeling understood to life satisfaction, see Oishi et al., "Feeling Understood"; the quote "people cannot survive" is on page 490.

15. For the dangers working conditions pose to relationships, regarding efficiency metrics, see Diamond, *Making Gray Gold*; regarding overwork, see Rodriquez, *Labors of Love*; and regarding routinization, see Leidner, *Fast Food, Fast Talk*. For the notion that workers find relationships rewarding, see Baugher, "Pathways through Grief"; Black, "Moral Imagination"; Foner, *The Caregiving Dilemma*; Sherman, "Caring or Catering?"; and regarding home healthcare workers in particular, see Stacey, "Finding Dignity in Dirty Work." To the latter we can add the

voluminous carework scholarship on the dangers of "love rhetoric," which scholars contend obscures the skill involved in such work, suppresses recognition of care as a collective responsibility, justifies the exclusion of care work from labor regulation, and enables exploitation and abuse. See Ray and Qayum, *Cultures of Servitude*; Romero and Pérez, "Conceptualizing the Foundation"; see also Pugh, "Connective Labor as Emotional Vocabulary." For Hochschild's early work on emotion work and emotional labor, see *The Managed Heart*. For research on working conditions that contribute to emotional estrangement, see for example Debesay et al., "Dispensing Emotions"; Lopez, "Emotional Labor and Organized Emotional Care"; and Ray and Qayum, *Cultures of Servitude*.

16. Academics have come up with other related concepts, among them "affective labor" and "intimate labor." Both of these overlap but are not the same as connective labor, which as emotional recognition I intend something more specific. Affective labor has been another influential attempt to capture the way contemporary capitalism colonizes the realm of sentiment. Hardt and Negri define affective labor as immaterial labor centrally involving the creation and manipulation of affects, "generally associated with human contact, with the actual presence of another, but that contact can be either actual or virtual" (*Empire*, 293). The breadth of this definition is striking: as defined thusly, the affective laborer is not just the nurse or flight attendant but also a model on a billboard or an actress in a movie, eliciting desire or shock at one remove. Indeed, all work has affective dimensions, they argue, making it difficult to distinguish, as Gill and Pratt note, "between the hospice nurse and the backroom computer programmer" ("In the Social Factory?," 15). Writes Nathanson, "Affect is biology whereas emotion is biography" (*Shame and Pride*, 50). While this distinction seems a bit overdrawn, the role of the "conceptual frame" in shaping emotions is what I think makes emotions particularly apt here. Meanwhile, intimate labor is an almagam of care, domestic, and sex work, but its inclusion of "bodily and household upkeep" means it is not focused on the reflective attunement of connective labor. For intimate labor, see Boris and Parreñas, *Intimate Labors*. Some writers have described affective labor in service to intimate encounters, such as how call center employees work to generate proximity with callers; see Mankekar and Gupta, "Intimate Encounters."

17. Work documenting status and authority benefits include, for example, Froyum, "'They Are Just Like You and Me'"; Craciun, "The Cultural Work of Office Charisma"; Wolkomir and Powers, "Helping Women and Protecting the Self"; Rogers, "'Helping the Helpless Help Themselves'"; and Deeb-Sossa, "Helping the 'Neediest of the Needy.'"

18. See Kayal, "Healing Homophobia," and Kayal, *Bearing Witness*. See also Allahyari, *Visions of Charity*, and Black, "Moral Imagination in Long-Term Care Workers," 302.

19. Recent studies have argued that people exaggerate the degree to which they are able to put themselves in someone else's shoes; see Eyal, Steffel, and Eppley, "Perspective Mistaking." Yet the researchers noted that perspective-taking, while not always accurate, yielded other benefits—such as increased empathy, a sense of connection, and an inclination to cooperate—and that when people were able to converse with the other, and simply ask them their views, their understanding improved. While scholars have been able to identify people who are strongly empathetic—who report feeling the emotions that they perceive in other people—they have not been able to find that these people are particularly accurate in the emotions they perceive. See Ickes et al., "On the Difficulty"; and Zaki, Bolger, and Ochsner, "It Takes Two." Because of this

disjuncture, researchers have given up the "dispositional approach" and abandoned the hunt for "accurate empathic perceivers," instead seeking to identify particular situations and relationships that generate empathic accuracy; see Zaki, Bolger, and Ochsner, "It Takes Two," 399; Stinson and Ickes, "Empathic Accuracy." The implications of this research are important, because they suggest that we focus on settings rather than individuals, that instead the "social architecture" could have significant impact on how well people can bear witness to each other; see chapters 5 and 8. Indeed, this pushes back a bit on the emphasis upon skill development in some sociological writings; while a minimal level of skill in reading the other is surely critical, this research suggests the environment in which the interactions take place could be consequential.

20. Kolb, "Sympathy Work," investigates how people find their moral identities in reflecting work, and the struggle when the emotional dynamic does not mesh with their self-concepts.

3. The Automation Frontier

1. Erin Cech writes about how the widespread cultural emphasis on "follow[ing] your passion" feeds both overwork among white-collar workers and inequality. See Cech, *The Trouble with Passion.*

2. Arlie Hochschild once documented what she called the "commodity frontier," the moving line that reflected how people divided their family life into more or less meaningful, with certain tasks and moments—driving kids to soccer, say, or organizing their birthday party—designated as open to outsourcing. See Hochschild, *The Commercialization of Intimate Life.*

3. With regard to education, for example, researchers have found that limited forms of online schooling can be as effective as face-to-face instruction. These findings were largely restricted to those students enrolled in traditional schools who take one or two courses online to supplement learning, however. Several national reviews report that when comparing full-time virtual instruction—the kind of distance-learning that millions of students experienced during the pandemic—to that which happens in a classroom, the online students performed significantly worse. A joint Stanford-Harvard project recently announced that students lost more than half a year's worth of learning in math and a quarter of a year in reading due to the pandemic. See Fahle et al., "Local Achievement Impacts of the Pandemic." For effective face-to-face instruction, see Rice, "A Comprehensive Look"; for the national report comparing full-time virtual and in-person instruction, see Molnar et al., "Virtual Schools in the U.S."

4. Some social robots in the form of seals and dogs provide social support for the elderly, which has been shown to improve both physical and mental health. There is also some research showing robots can improve kids' motivation, engagement, and curiosity; they enjoy the novelty and can be less anxious about making mistakes than in front of a human teaching assistant. Some studies report a significant drop-off in learning after a week, however—presumably when the novelty wears off. The review of therapy chatbot benefits, which found high satisfaction but also the inappropriate response to depression, was conducted by Vaidyam et al., "Chatbots and Conversational Agents in Mental Health." The story of Tessa the chatbot is told in Wells, "An Eating Disorders Chatbot Offered Dieting Advice." The phrasing "early stages of maturity" comes from Kueper et al., "Artificial Intelligence and Primary Care Research," 256. One cancer center spent $62 million on Watson before abandoning it; the story of Watson's rise and fall is

captured in Lohr, "Whatever Happened to IBM's Watson?" For studies of robots in therapeutic contexts with children, see Dautenhahn, "Roles and Functions"; Dautenhahn and Werry, "Towards Interactive Robots"; and Sharkey and Sharkey, "The Crying Shame." Extensive work has looked at the use of robots with the elderly; see, for example, Banks, Willoughby, and Banks, "Animal-Assisted Therapy"; Melson et al., "Robotic Pets in Human Lives"; and Waytz, Cacioppo, and Epley, "Who Sees Human?" For a review of socially assistive robots in education, see Papadopoulos, "A Systematic Review." See also Miner et al., "Key Considerations."

5. Byambasuren, "Prescribable mHealth Apps," includes the meta-review; the quotation about "low quality" is found on page 9.

6. The bot's passing the licensing exam was reported in Kung et al., "Performance of ChatGPT." Ayers and colleagues found the chatbot gave more empathic responses to questions posted on the Reddit social media forum, as detailed in Ayers et al., "Comparing Physician and Artificial Intelligence Chatbot Responses."

7. In this chapter, I continue to use pseudonyms for most of those quoted, except when they have previously appeared in print (as in a previous *New Yorker* article; see Pugh, "Automated Health Care Offers Freedom from Shame") or when they state it as their preference. In these cases, I note where I use their real name, as in the case of Lucy Suchman.

8. I wrote about this case for the *New Yorker*; see Pugh, "Automated Health Care Offers Freedom from Shame." Timothy Bickmore is his real name.

9. For more information about Bickmore's virtual nurse experiment, see Bickmore et al., "Automated Promotion of Technology Acceptance."

10. See Thompson, "Time, Work-Discipline, and Industrial Capitalism."

11. See Bowles, "An Online Preschool."

12. For the benefit of the doubt that humans give machines, see studies by Turkle, "Authenticity in the Age of Digital Companions," and Reeves and Nass, *The Media Equation*. The "roboticists have learned" quotation comes from Turkle, *Alone Together*, 20.

13. The quotation comes from Sharkey and Sharkey, "A Crying Shame."

14. For more information about the AI kiosks at the border, see Breselor and Higginbotham, "Welcome."

15. For how businesses cultivate relationships with customers, see Gutek and Welsh, *The Brave New Service Strategy*.

16. For the effect of class on the customization of learning by elementary school students, see Calarco, *Negotiating Opportunities*.

17. The research here is robust, leading some to investigate the phenomenon of "overtrusting robots" (Aroyo et al., "Overtrusting Robots"). For the finding that people disclose more about their financial problems in an online survey, see Morris and Kennedy, "Personal Finance Questions"; for blood donors, see Weisband and Kiesler, "Self-Disclosure on Computer Forms"; for children's sadness, see Lucas et al., "It's Only a Computer"; see also Kang and Gratch, "Virtual Humans." For privacy concerns of confessing to computers, see Miner et al., "Key Considerations." As Burrell and Fourcade argue in "The Society of Algorithms," the coding elite consolidates power by "disrupting" the professional's monopoly power to judge others, on the basis that "human reasoning is inadequate, even the expert decision-making of high-status professionals" (10). See also Pasquale, *New Laws of Robotics*.

18. See Chui, Manyika, and Miremadi, "Four Fundamentals of Workplace Automation."

19. The same spring I visited the Silicon Valley school, just thirty miles away a public school resorted to Khan Academy videos to teach math. A 2023 *Washington Post* report told of a Mississippi school district using videos in light of a teacher shortage. See Balingit, "A Teacher Shortage." For "deskilling," see Braverman, *Labor and Monopoly Capital.*

20. I also discuss this material in Pugh, "Emotions and the Systematization of Connective Labor."

21. Lily Irani notes that those who employed Mechanical Turk workers often characterized the work they outsourced as menial or rote and highlighted the algorithmic techniques for managing workers; they did both to appeal to venture capitalists, who valued "tech companies" over "labor companies." See Irani, "Difference and Dependence."

22. I also discuss this material in Pugh, "Constructing What Counts as Human at Work." Having to prove they are human bore some similarities to the task of current call center employees, whom scholars have noted have to prove to customers that they are not foreign, in essence, that they are more valued humans; see Poster, "Who's On the Line?"

4. How to Be a Human: Connective Labor as Artisanal Practice

1. Among those who have called interpersonal work the last human job are Friedman, "From Hands to Heads to Hearts," and Gershon, "The Automation-Resistant Skills."

2. The quotation about "interaction" comes from Schilbach et al., "Toward a Second-Person Neuroscience." Chanda Prescod-Weinstein has challenged the phrase "dark matter," arguing that the material is actually invisible. See Prescod-Weinstein, *The Disordered Cosmos.*

3. There is a large research tradition itemizing the practices of workers who deploy relationships, voluminous even if we focus on just three primary fields: the therapeutic alliance, teachers' "rapport" with students, and the physician-patient relationship. Scholars identify a very wide range of practices that constitute relationship-building, and there is no shortage of lists of such practices, such as "knowledge, trust, loyalty, and regard" in Eveleigh et al., "An Overview of 19 Instruments," or "respect, trust, caring, and cohesiveness" in Downey, "Recommendations for Fostering Educational Resilience." To some degree, the lists vary by occupation: researchers suggest positive regard or affection more frequently for teachers and therapists, for example, while urging teachers and physicians to consider the outside lives of their students or patients; time is clearly more centrally a problem for teachers and physicians, while touch is available only to some practitioners. One of the most comprehensive recent efforts to identify relevant strategies that "foster physician humanism and connection with patients" led to the Presence 5, a list of practices compiled by a team at Stanford Medical School; see Zulman et al., "Practices to Foster Physician Presence." Team members undertook a literature review, observed primary care, and interviewed varied practitioners—including ten clinicians, and thirty nonmedical participants such as realtors, yoga instructors, and managers, many of them connective labor workers—to come up with an initial list of thirty-one practices; three rounds of review by "expert panels" culled and combined these to arrive at five: (1) prepare with intention, (2) listen intently and completely, (3) agree on what matters most, (4) connect with the patient's story, and (5) explore emotional cues. Overall, these lists are often part analysis, part recommendation; the Presence

5, like other lists, is designed very explicitly as an intervention that practitioners should adopt (and was published in the highly visible *JAMA*). The lists do tend to ignore the external factors that impede or enable such efforts, such as the social architecture outlined in chapters 5 and 8. Thus they inadvertently make it seem like connection is simply up to the workers' choice or ability, and not subject to the potential impact of contextual factors that are likely consequential—such as, in the case of physicians, a time famine and the press for efficiency, the use of the EHR and other managerial dictates, and the culture of professional medicine, with its emphases on objectivity, evidence, and rationality. In light of the pressures in many occupations to standardize and commodify connective labor, this chapter focuses on those aspects of the work that defy its easy systematization.

4. In *Interaction Ritual Chains*, Collins points out crucial accomplishments of rituals: in addition to the mutual focus and the shared mood, they also make a barrier to outsiders through their co-presence.

5. The multiple uses of emotion here go beyond emotion management to their deployment as sensory tools and their experience as signals; in this framing, I rely upon Cottingham and Erickson's notion of emotion practice, in which "emotions emerge from and act upon a mindful body that is dynamically structured, collective yet individual" (192). See Cottingham and Erickson, "The Promise of Emotion Practice." See also Cottingham, *Practical Feelings*.

6. Mariana Craciun has written about how professionals use emotions in supportive, didactic, and inductive ways; while therapists in different traditions might all use a kind of attunement to sense the other, they justify and explain their reliance differently depending on their "epistemic community," and its relationship to insurers and other objectifying stakeholders concerned with evidence, standards, and quantification. See Craciun, "Emotions and Knowledge in Expert Work."

7. Alexandra Vinson documents how this naming work is learned by physicians in medical education. See Vinson, "'Constrained Collaboration.'"

8. Hochschild calls this work "emotional labor"; see *The Managed Heart*.

9. Raka Ray and Seemin Qayum coined the term "love rhetoric" to capture the way domestic workers and the families who employ them talk about their relations; Mary Romero and Nancy Pérez have argued that the use of that rhetoric in research threatened to depoliticize and deskill caring labor. In a 2023 piece in the journal *Signs*, I review these debates and urge scholars to turn toward and not away from emotions, arguing that the concept of connective labor is a way to incorporate more nuance, including to emphasize the use of feelings as a tool, into how we talk about sentiment in care work. See Pugh, "Connective Labor as Emotional Vocabulary." See also Ray and Qayum, *Cultures of Servitude*; Romero and Pérez, "Conceptualizing the Foundation"; and Saraceno, "Social Inequalities in Facing Old-Age Dependency."

10. Collaboration also meant putting the patient or client or student at the center of the interaction, instead of the worker exerting their authority to define the encounter top-down. Across many fields, practitioners told me about the importance of centering the other, their words echoing each other and suggesting the presence of a certain cultural zeitgeist. This zeitgeist was never more apparent than when listening to proponents of competing approaches to therapy—cognitive behavioral therapy (CBT) and more interpretive, emotion-based approaches such as interpersonal therapy or attachment-based therapy. Therapists from both traditions held forth

to me that theirs was the truly collaborative approach. With both pointing fingers at the other about which style was more hierarchical, it was clear that the notion of being collaborative and client-centered was culturally dominant. When Martina Verba, a practicing therapist and friend, read this section, she took issue with the opposition between collaboration and hierarchy, noting that "there is a hierarchy in a relationship in which there is a helper and one receiving help; that is part of what makes the relationship work, but within it there can be collaboration." That seems a wise observation; the opposition I make here is how it was presented to me by practitioners of either approach, but I suspect they would agree that the distinction is overdrawn.

11. See, for example, Safran et al., "Therapeutic Alliance Rupture"; and Lewis, "Repairing the Bond."

12. Vinson explores the "constrained collaboration" of physicians and patients negotiating physician authority amid a movement for the patient empowerment. See Vinson, "'Constrained Collaboration.'"

13. Elizabeth Bernstein argues that the flexible commercialization of postindustrial capitalism has led to the rise of "bounded authenticity": "the sale and purchase of authentic emotional and physical connection" (154). While she was referring to sexual intimacy and changes in the meaning and conduct of prostitution, her point applies to connective labor as well; these workers struggled with twin demands for connection and for limits. See Bernstein, "Bounded Authenticity."

14. Research suggests that people like it when robots make small mistakes; see Shiomi, Nakagawa, and Hagita, "Design of a Gaze Behavior." See also Salem et al., "To Err Is Human(-Like)."

5. The Social Architecture of Connective Labor

1. For studies of "state empathy," which neurologists tell us can be activated in different situations, see Jackson, Meltzoff, and Decety, "How Do We Perceive"; Rameson, Morelli, and Lieberman, "The Neural Correlates of Empathy"; Batson et al., "Empathy, Attitudes, and Action"; Stel and Vonk, "Mimicry in Social Interaction"; and Clark, Robertson, and Young, "'I Feel Your Pain.'"

2. Health care provides one fitting example; Light, "The Rhetorics and Realities," offers a beautiful summary of the development of American health care, and the costs and benefits of different regimes of medicine. Regarding the weakness of the community health clinic model in contemporary US, he writes: "Those doing real community health care are at the periphery of the corporatized health care system, trying to survive it, like Finland treading warily for years at the edge of the Soviet empire" (128); The "crumbs from the table" quotation is from page 13.

3. Extensive research documents the pitfalls of routinization and overload. Overload means that residents are often schooled in reducing empathy in service to efficiency, according to Szymczak and Bosk, who write: "Sarah's criticism of Peter for being 'too empathetic' is reminiscent of the process involved in 'training for detached concern' . . . except that now emotional detachment is a tool of efficiency—being emotionally invested in a patient takes too much time" ("Training for Efficiency," 350). Sometimes practitioners must choose between adhering to systemic rules about staffing and efficiency, and their perceived care obligations. Szymczak et al. quote one physician who described a first-year resident who stayed over time. "People call him inefficient. It's really because he's doing what we all should be doing, but there was not time to do it, and he'd be there two hours late. I'm not talking fifteen minutes. He would call that doctor,

he would call the family member, the kind of things that people wanna believe doctors would do for them, and he did them all" ("To Leave or to Lie?," 363). For an in-depth analysis of the problem of overload and its potential reform, see Kelly and Moen, *Overload.*

4. Jessica Calarco's book *Holding It Together,* expanding on her observation, is forthcoming from Portfolio Press. For the care deficit, see Hochschild, *The Commercialization of Intimate Life.*

5. This "cost disease" exerts a downward force on wages, especially for what Blinder called "ordinary personal service jobs (such as cutting hair and teaching elementary school)," as opposed to "luxury personal-service jobs (such as plastic surgery and chauffeuring) ("Offshoring," 123–24).

6. Jenna worked for a public entity, a county hospital system that accepted the state version of Medicare; her employer may not be concerned with "profit" per se, but anemic support from the state meant it still had to contain labor costs by loading its providers with large patient rosters.

7. As noted in chapter 1, unlike organizational scholar Jane Dutton, I use the term "social architecture" to capture an overarching concept of how an organization shapes human connection based on not just connective cultures but also its material resources, including time, space, and technology. In my view this idea is similar to how architecture configures the structure of a building. See Dutton et al., "Explaining Compassion Organizing."

8. For research regarding the links between burnout and high workloads, see Maslach, Schaufeli, and Leiter, "Job Burnout"; for emotion work, see Zapf et al., "Emotion Work and Job Stressors"; and for emotional dissonance, see Kenworthy et al., "A Meta-Analytic Review."

9. The American Medical Association is funding a "joy in medicine" program to reduce physician burnout; see https://www.ama-assn.org/system/files/joy-in-medicine-roadmap.pdf. A tradition of humanistic medicine focuses on relationships as a source of physician resilience; see Serwint and Stewart, "Cultivating the Joy of Medicine." Lilius, "Recovery at Work," 573, explores the sustaining, as opposed to depleting, quality of relationships.

10. This troika resembles the three kinds of organizations that scholars have found shape nursing care. As Grant et al. note, "nurses' styles of care may vary depending on whether their hospitals were founded to address the physical suffering of the indigent, provide special services to wealthy clientele, or bring health care costs under control for the middle class" ("Affirming Selves," 196). See also Reich, *Selling Our Souls.*

11. Scholars have found that medical practices can have different cultures, varying on their degree of collegiality, emphasis on autonomy, and focus on profit maximization, and that these differences affect the kind of medical care they provide. In particular, researchers found that "without adequate support systems, a culture of autonomy may have a detrimental effect on the quality of patient care." See Shackelton, "Does the Culture of a Medical Practice," 102.

12. Research confirms that advantaged patients demonstrate considerable "cultural health capital," securing advantages in obtaining care and treatment through an attentive "vigilance." See Shim, "Cultural Health Capital"; Gengler, "'I Want You to Save My Kid!'"; Gage-Bouchard, "Culture, Styles of Institutional Interactions, and Inequalities."

13. Of course the word "corporate" lumps together many different kinds of entities, from small businesses or practices to large ones, from founder-led to those run by private equity or publicly traded firms. Still, these settings often share the similarities I outline here.

14. Extensive research documents the pervading logic of customer satisfaction in private schools and well-resourced suburban public ones. See Calarco, "Avoiding Us versus Them," and Lewis-McCoy, *Inequality in the Promised Land*. Relational cultures can dictate the place of emotion; judges, for example, are largely governed by a professional code that includes a "script of judicial dispassion" enjoining them from emotional display. Maroney chronicles the conflict between the ideal of the dispassionate judge and the everyday reality of judges' feelings; see Maroney, "The Persistent Cultural Script of Judicial Dispassion."

15. Ticona, a scholar at the University of Pennsylvania's Annenberg School, studies the impact of digital platforms for care work in a forthcoming book; see also Flanagan, "Theorising the Gig Economy"; McDonald, Williams, and Mayes, "Means of Control"; Ticona, Mateescu, and Rosenblat, "Beyond Disruption"; and Fetterolf, "It's Crowded at the Bottom." In 2019, Care .com had 11.5 million registered worker profiles in the US and $192 million in revenue (Ticona, "Red Flags, Sob Stories, and Scams"). Steven Vallas and Juliet Schor note how platform-based work exposes workers to an "evaluative infrastructure" that can inhibit autonomy, even in the absence of hierarchical controls; see Vallas and Schor, "What Do Platforms Do?"

16. Carrie Lane coined the term "a company of one" to refer to the self-conscious branding required of American workers generally that she discovered in a study of the unemployed. The concept is apt for gig economy workers as well. See Lane, *A Company of One*.

17. GroupCare is a pseudonym, like almost all of the names in this book.

18. For how "personal lifestyle workers" manage the distinction between being a professional and being a servant, see Sherman, "Caring or Catering?"

6. Systems Come for Connective Labor

1. Extensive research corroborates Nathan's narrative here. In a study of residents in training, for example, Szymczak and Bosk write: "The residents in our study were responsible for approximately the same number of patients as residents of an earlier era—12 to 14—yet are expected to care for them in a shorter amount of time in a crowded technological environment that is more tightly controlled by formal organizational protocols (see Note 3). The interaction of these factors makes the workday feel frenzied and unpredictable for residents" ("Training for Efficiency," 349).

2. This process is called Taylorization, named for Frederick Taylor's practice of breaking down a given job's component parts into smaller tasks that are then siphoned off, some of them to be done by cheaper labor. The infiltration of this trend into a given industry has varied, depending in part on the power of workers to fend it off (through unions or through professional codes and credos) or on the relative power of clients to insist upon personalized, artisanal labor. Therapists have had uneven exposure to the drive for efficiency, for example, with some relatively free of that imperative and others who work in government settings or in "disrupted" startups having to hew closely to productivity measures. For the classic analysis of Taylorization and its impact on the "deskilling" of work, see Braverman, *Labor and Monopoly Capital*. LaTonya Trotter has argued that nurse practitioners actually take on more complexity in their care work than physicians, because they address medical, social, and organizational problems often brought on by their patients' poverty, a more comprehensive approach than that

of physicians, who are more likely to limit themselves to the strictly medical. See Trotter, *More than Medicine*.

3. The account of how a checklist transformed catheter-related bloodstream infections was dangerously oversimplified in the popular press, argue Bosk and his colleagues, who write: "Indeed, it would be a mistake to say there was one 'Keystone checklist': there was not a uniform instrument, but rather, more than 100 versions." Instead of a story about how a checklist solved the problem, the example actually "models how to achieve results in wider contexts: recruit advocates within the organisation, keep the team focused on goals, create an alliance with central administration to secure resources, shift power relations, create social and reputational incentives for cooperating, open channels of communications with units that face the same challenges, and use audit and feedback." See Bosk et al., "Reality Check for Checklists," 445. For the impact of checklists on surgical and ER outcomes, see Hales et al., "Development of Medical Checklists." For the BATHE script, see Lieberman and Stuart, "The BATHE Method."

4. In "Training for Efficiency," Szymczak and Bosk argue that contemporary hospitals train residents for "efficiency," and residents respond with a skepticism about the system, which translates into workarounds that customize standardization that gets in their way. The result could be hazardous, they write. "Residents direct their resentment to a complex set of demands that are embedded in an opaque system and respond to these barriers to efficiency by working around them. Workarounds may well have negative unintended consequences. Things get dropped in the space between the formal and informal ways of doing things. . . . Clean supplies stored in a locker, not in the always-locked-and-difficult-to-access sterile central processing unit, could become contaminated and lead to a hospital-acquired infection" (355). Researchers confirm that standardization is often customized in practice: see, for example, Timmermans and Epstein, "A World of Standards"; Espeland and Stevens, "A Sociology of Quantification"; Bowker and Star, *Sorting Things Out*. In *Metrics at Work*, Christin documented how organizational logics and cultural ideas shape the way metrics are interpreted and enacted. Leidner, "Emotional Labor in Service Work," explores how scripting can protect workers from demanding or chaotic situations or clients. The research into the "high-value practices" and how they used "standing orders and protocols," was reported in Simon et al., "Exploring Attributes of High-Value Primary Care."

5. Mariana Craciun compares CBT analysts and psychotherapists and how they talk about metrics, evidence, and standards. See Craciun, "Emotions and Knowledge in Expert Work." See also Zeavin, *The Distance Cure*.

6. For the review, see Budd and Hughes, "The Dodo Bird Verdict."

7. Budd and Hughes, "The Dodo Bird Verdict," 516.

8. For the large study of the Crisis Text Hotline, see Althoff, Clark, and Leskovec, "Large-Scale Analysis." For research documenting that scripting threatens innovation, see Reich, "Disciplined Doctors," and Verghese, "The Importance of Being"; for its detrimental impact on autonomy, see Leidner, *Fast Food, Fast Talk*; Abbott, *The System of Professions*; Haug, "A Re-examination." For the argument that scripting alienates workers from their own feeling, see Hochschild, *The Managed Heart*; for research demonstrating that it demoralizes workers, see Rupert and Morgan, "Work Setting."

9. For the capacity of surveys to downplay the uniqueness of the particular case, see Fox, *Medical Uncertainty Revisited*, and White, "Reason, Rationalization, and Professionalism."

10. The survey then takes a turn toward reasons the person might have to live, asking about religion, family members left behind, and possible sources of help. Some of them conclude with a "contract" for the person to sign with the therapist, pledging to seek out other means of coping and to refuse to act on urges to kill themselves during a set period.

11. Lutfey, "On Practices of 'Good Doctoring,'" found that practitioners adopted a range of different personas—"educators, detectives, negotiators, salesmen, cheerleaders and policemen"—in their attempt to manage compliance among diabetes patients.

12. Charles II died in 1685 at age fifty-four. He suffered from chronic gout, and in his last few days, physicians bled him three times (removing thirty-four ounces of blood); modern analyses suggest the bleeding contributed to his dehydration, which would have led to his "acute renal insufficiency . . . acute posterior reversible encephalopathy syndrome, [and a] cerebral oedema." See Aronson, "When I Use a Word."

13. For the argument that counting lends legitimacy to service work, see Glenn, "Creating a Caring Society"; and Levy, "Relational Big Data," 73.

14. Brayne, *Predict and Surveil*, 101. See Beer, *Metric Power*. For how measurement exacerbates race and class inequalities, see Benjamin, *Race after Technology*; Eubanks, *Automating Inequality*; Noble, *Algorithms of Oppression*; and O'Neil, *Weapons of Math Destruction*.

15. Math professor Michael Thaddeus has documented how Columbia University implausibly ascended the *US News* rankings; see http://www.math.columbia.edu/~thaddeus/ranking/investigation.html. The "comfortably crowd out" is from Beer, *Metric Power*, 184. Some scholars argue that counting can improve equity; Vale and Perkins, for example, call for doctors to do more documentation of what are called the "social determinants of health"—nonmedical social factors such as poverty or shift work—so that they take those factors into account when treating patients. Yet, as they note, standardizing how doctors do so "could even constitute a loss of social data if it displaces clinicians' local practices." See Vale and Perkins, "Discuss and Remember," 7. For the "I know my students by data" quote, see Valli and Buese, "The Changing Roles of Teachers." For the capacity of measurement to change the very thing being measured, see Espeland and Sauder, "Rankings and Reactivity."

16. Verghese, "The Importance of Being," 1927.

17. For data on the percent of physicians using an EHR, see the Office of the National Coordinator for Health Information Technology, "Office-Based Physician." See also Sinsky et al., "Allocation of Physician Time." Regarding the more than 50 percent of primary care physicians reporting burnout, see Tolentino et al., "What's New in Academic Medicine"; for the evidence linking it to use of the EHR, see Shanafelt et al., "Relationship between Clerical Burden." For an in-depth account of the EHR rollout in the US, see Schulte and Fry, "Death by 1,000 Clicks."

18. See Hochschild, *The Second Shift*.

19. See Mausethagen, "A Research Review."

20. This account echoes the story of a Canadian therapy practice that implemented satisfaction surveys; 40 percent of the therapists in the practice quit. See Goldberg, Babins-Wagner, and Miller, "Nurturing Expertise at Mental Health Agencies."

21. I also discuss this material in Pugh, "Emotions and the Systematization of Connective Labor."

22. One could certainly argue that the fist-to-five check-in produces what Eva Illouz has called "emodities." For SEL curriculum as a way of rationalizing emotions, see Gillies, "Social and Emotional Pedagogies"; for how they create neoliberal subjects, see Wilce and Fenigsen, "Emotion Pedagogies"; and Williamson, "Psychodata." For "emodities," see Illouz, "Introduction." These tactics happen not just in education; Nik Brown has analyzed the use of hope indices by oncology practitioners to apply scales "to grade, rank, score and rate a subject's affective orientation to their future" ("Metrics of Hope," 127).

23. For "more than half," see Agency for Health Care Quality and Research, "Physician Burnout."

24. See Marken and Agrawal, "K-12 Workers"; Corlette et al., "Impact of the COVID-19 Pandemic"; The Commonwealth Fund, "Stressed Out and Burned Out"; Jefferson et al., "GP Wellbeing."

25. Chochinov, "Dignity and the Essence of Medicine."

7. Connecting across Difference: The Power and Peril of Inequality

1. "Motivational interviewing" is an inquiry-based mode of talking to people about changes they might want to make, and is increasingly widespread in contemporary medicine. See Lundahl et al., "Motivational Interviewing"; and Lutfey, "On Practices of 'Good Doctoring.'"

2. African American women endure shocking rates of maternal mortality and other pregnancy- and childbirth-related complications; most research has targeted practitioners' implicit bias or patient vulnerabilities as prominent causes. The connective labor perspective helps decode what might be happening in the exam room, as the physician-patient relationship becomes the site of misrecognition by providers and its shattering impact on women of color, among a host of potential factors producing such acute racial discrepancies in maternal health outcomes. The doctor's judgment here is particularly common for mothers; as Elliott and Bowen write in "Defending Motherhood," "there is a long history of elites scrutinizing the mothering practices of immigrants, poor women and women of color." The "weathering" phenomenon has been explored by Geronimus, "The Weathering Hypothesis." For more details of this vignette and the impact of doctor-patient relationships on Black women's reproductive health, see Wright, "Affective Burdens."

3. For "controlling images," see Collins, *Black Feminist Thought*.

4. The "vital human need" quote comes from Charles Taylor, "The Politics of Recognition," 26. Judith Butler, in *Bodies That Matter* (2011), decried the procrustean boxes of recognition's demand for legibility.

5. Blume Oeur examines "being unknown" in "Recognizing Dignity," as well as his award-winning book *Black Boys Apart* (2018). Phillippo, "'You're Trying to Know Me,'" reports the results of a study of disadvantaged high school students and personalized relationships; the quote "trying to know my business" is from page 453, and "You're here for science, for math, and you're trying to know me" is from page 442. See also Groark, "Social Opacity."

6. Foucault thought pastoral power combined two kinds of power, a fostering of self-governing subjects and a disciplining surveillance, making it both individualizing and totalizing, a blend he called "tricky." The quote "guidance through knowledge of people's secrets" is from

Cook and Brunton, "Pastoral Power," 546. Foucault's statement that the blend was "tricky" is in Foucault, "The Subject and the Power," 782; see also Martin and Waring, "Realising Governmentality." For "the point is that not everything is bad," see Foucault, *Ethics*, 256.

7. In education, this phenomenon has been called the compensatory-complementary paradox, wherein positive teacher-student relationships are more common among white and middle-class students, but have a more intense positive impact among Black, Latinx, and working-class students. For an explanation of the paradox, see Dinsmore, "Relational Cultures, Inequality and Belonging." Research showing less advantaged students benefiting more from connection includes Olsson, "The Role of Relations"; Muller, "The Role of Caring"; Erickson, McDonald, and Elder, "Informal Mentors and Education"; Crosnoe, Johnson, and Elder, "Intergenerational Bonding in School"; and Lewis et al., "Con cariño"; but see Malecki and Demaray, "Social Support as a Buffer." In medical research, parallel findings are reported by Shi, "The Impact of Primary Care"; Shi, Green, and Kazakova, "Primary Care Experience"; Shi et al., "Vulnerability"; and Atlas et al., "Patient–Physician Connectedness."

8. See Pugh, "Connective Labor as Emotional Vocabulary."

9. The devastating psychological impact of being misrecognized by the powerful was explored by Fanon in *Black Skin, White Masks*. This is what Harvard sociologist Michèle Lamont has dubbed stigmatization: the attachment of particular disadvantaged identities with negative judgments. The exploration of stigmatization is captured in Fleming, Lamont, and Welburn, "African Americans Respond to Stigmatization."

10. For research into disproportionate school discipline, see Losan and Skiba, *Suspended Education*. Victor Rios writes about the span of punitive surveillance across home and school for youth of color; see Rios, *Punished*. See also Dinsmore and Pugh, "The Paradox of Constrained Well-Being." Elijah Anderson has written about the concept of "white space"; see Anderson, *Black in White Space*. Patrice Wright researches the emotional toll of being Black in white spaces; see Wright, "Affective Burdens."

11. Scholars have found that effects of systematization, such as compressed time, "team" approaches to health care (otherwise known as "fragmented care"), and measurement regimes contribute to the use of stereotypes, or controlling images. See Diamond-Brown, "The Doctor-Patient Relationship"; O'Neil, *Weapons of Math Destruction*.

12. See Stepanikova, "Racial-Ethnic Biases," 338. See also Johnson et al., "The Impact of Cognitive Stressors"; and Burgess, "Are Providers More Likely to Contribute."

13. See Roberts, *Torn Apart*; Roberts, "Child Protection as Surveillance." See also Seim, "Stretched Thin." For discretionary power, see Brayne, *Predict and Surveil*.

14. I have written elsewhere about the different kinds of information that interviews generate. Hank's stories here did have the feel of the honorable, as he told me how cherished he was in the low-income community. Still, his criticism of the savior model and how he inhabited it did not put him in the best light, and his account of moving through various jobs revealed a certain humility, tapping into a visceral level of despair that laid the groundwork for transforming his approach to others. See Pugh, "What Good Are Interviews."

15. We should not overstate the "choice" Betty had to decline, however, given the systemic context in which she had to work for two different agencies just to muster enough hours to live on, where her skill and experience did not do much to change the relatively flat wage structure,

and other artifacts of broad structural inequalities. While Betty may have been theoretically able to refuse individual clients, to some degree the difficult clients reflected social problems, the entitlement enacted by recalcitrant patients or demanding adult children, combined with the financial insecurity that felt to some workers like coercion. She could not refuse all of them, and so she learned how to accommodate their expectations. Gendered subordination soaked through to the cultural ground, shaping the options and constraining individual power, before these workers even stepped up to "choose."

16. Rachel Sherman has analyzed the way personal investors and other lifestyle workers finesse the threat of being treated like a servant by their wealthy employers by demanding they be considered professionals instead; see Sherman, "Caring or Catering?"

17. See Ray and Qayum, *Cultures of Servitude.*

18. See, for example, Romero and Pérez, "Conceptualizing the Foundation"; and Saraceno, "Social Inequalities in Facing Old-Age Dependency." Elizabeth Bernstein's concept of "bounded authenticity" is also relevant here, and her argument that humane interpersonal work—like the sex work she analyzed—often involves producing a version of authentic emotional connection, but within limits often imposed by market exchange. See Bernstein, *Temporarily Yours.*

8. Doing It Right: Building a Social Architecture That Works

1. For the factors that improve teacher-student relationships, see Bingham and Sidorkin, *No Education without Relation*; Day, "School Reform and Transitions"; Fredriksen and Rhodes, "The Role of Teacher Relationships"; Mausethagen, "A Research Review"; and Valli and Buese, "The Changing Roles of Teachers."

2. For the "moral, intellectual and social skills required," see Schlechty, *Schools for the Twenty-First Century*, xix.

3. See Hersey, Blanchard, and Johnson, *Management of Organization Behavior.* Bass wrote about the vision that transformational leaders offer; see Bass, "From Transactional to Transformational Leadership."

4. For the 25 percent figure of the principal's impact, see Marzano, Waters, and McNulty, *School Leadership That Works.* One study of career assistant principals noted they cultivated an ethic of care; "they see nuances in people's efforts at good performance and acknowledge them, they recognize the diverse and individual qualities in people and devise individual standards of expectation, incentives and rewards," in other words, they perform connective labor as leaders. See Marshall et al., "Caring as Career," 282. For the influential review, see Hallinger, "Leading Educational Change," 339. Indeed, as one review concluded, "quantitative findings suggested that principals' relationship-oriented behaviors may mediate the effect of teachers' emotional regulation ability on their emotional wellbeing"; see Berkovich and Eyal, "Educational Leaders and Emotions," 144. See also Cai, "Can Principals' Emotional Intelligence Matter."

5. Experiential learning is vital for clinical practice, including emotional practice, reports Grant and Kinman, "Emotional Resilience." A study of nurses found that a mentoring intervention enabled them to shift from students to professionals and do so with the enhanced capacity for empathic reflection of the other. See Sergeant and Laws-Chapman, "Creating a Positive Workplace Culture." One study compared the practical experience of training teachers, clergy,

and therapists, finding that each used role-plays or actual practice in their training. The different occupations had different strengths and weaknesses; for example, therapists did not get trained in how to manage groups, and could have benefited from teachers' extensive training there, while teachers did not get trained in how to manage resistance among their students, and could have used what psychologists received. See Grossman et al., "Teaching Practice." Mentoring is widely considered important for career achievement, but when formal networks are segregated by gender they can be damaging for women's careers; see Williams, *Gaslighted*.

6. Later, Andy told me that when a patient said they were too dry, it was not always reliable. "Some people can say they're too dry because it's a medication that they're taking that makes their mouth dry. They're really not dry, it's just that perception of their mouth being dry. And the other findings on our exam that shows that they actually might be fluid overloaded," Andy said. "The worst thing you could do then [is] take a bunch of fluids today and then they end up in the hospital later that day. So you've got to be right. And heart failure patients are ones where it can go sour pretty quickly. And you want to make sure because you know you wouldn't want to make that mistake."

7. For organizations that "hold," see Kahn, "Caring for the Caregivers." Research suggests that peer support groups inside and outside the workplace and supervision that offers "feedback, technical assistance, and support . . . enable caregivers to reconnect to their own experiences and offset their emotional withdrawals from themselves as well as others." Fletcher argues that the skill in creating team relied on "a type of cognitive complexity . . . the capacity to freely and wholeheartedly engage with another's subjectivity (i.e., drawing out other's ideas), being able to acknowledge and affirm that reality while maintaining and being in touch with one's own" ("Relational Practice," 174).

8. Organizational cultures matter in providing the frame through which people interpret particular feelings, and the scripts for responding to them. As Dutton, Workman, and Hardin, "Compassion at Work," put it, they "encourage people to pay attention to particular kinds of feelings, provide frames for making sense of the sufferer's and focal actors' situations, and provide scripts for certain kinds of actions" (291). They provide what Grant, Morales, and Sallaz have called "pathways to meaning" that also inform worker identities, which can shape their style of care, and how they respond to situational cues and calls to standardize their tasks. See Grant, Morales, and Sallaz, "Pathways to Meaning."

9. Information on the mean salaries of medical assistants versus neurologists is provided by the US Bureau of Labor Statistics' Occupational Employment and Wage Statistics program; see https://www.bls.gov/oes/.

10. For "each work timescape," see Snyder, *The Disrupted Workplace*, 14. Wajcman analyzes the cultural imperative for optimizing productivity through efficient time management in Wajcman, "How Silicon Valley Sets Time," and Wajcman, *Pressed for Time*.

11. Studies have shown that "attentional demands affect emotional responses," that people's capacity for empathy depends on how much else they are asked to attend to; see Dutton, Workman, and Hardin, "Compassion at Work"; and Dickert and Slovic, "Attentional Mechanisms."

12. For research on the relationship of inequality to help-seeking, see Calarco, "'I Need Help!'"

13. Lopez's study of nursing homes that had adopted new initiatives to "humanize" their relations with their frail elderly patients found that even as some parts of the facilities—for example, the activity department—emphasized a "person-centered" approach, the (much-larger) nursing department continued treating the residents as objects, because "incentives for delivering efficient care, however, pull nursing homes in the opposite direction." See Lopez, "Culture Change and Shit Work."

9. Conclusion: Choosing Connection

1. The Netherlands is one of those countries that has identified loneliness as an important social problem; it holds a "Week against Loneliness" every fall and maintains an elaborate website promoting social integration activities and knowledge. Information about the Dutch checkout lanes can be found on the company website https://nieuws-jumbo-com/jumbo-geeft-startschot-voor-opening-200-kletskassas/. The Reddit comments can be found at https://www.reddit.com/r/MadeMeSmile/comments/107gqi2/a_dutch_supermarket_chain_introduced_slow/; accessed January 16, 2023.

2. The grocery history can be found in Deutsch, *Building a Housewife's Paradise.*

3. In doing so, the Piggly Wiggly store also inadvertently launched the development of brands; see Ross, "The Surprising Way." The self-checkout transaction figures come from FMI, the Food Industry Association (see https://www.fmi.org/our-research/supermarket-facts), and from Meyersohn, "Nobody Likes Self-Checkout."

4. The "mask and earbuds" comment was posted by "captain_duckie" on January 9, 2022. See https://www.reddit.com/r/MadeMeSmile/comments/107gqi2/comment/j3nu1xp/. See also Hellerstein and Morrill, "Booms, Busts, and Divorce."

5. Of course, it is not necessarily a loss if one realm of "relentlessly messy" daily relationship vanished in this manner, as long as others opened up so that people were effectively seen somewhere. Just like the churn of jobs that characterizes the US economy, it is possible that we experience a churn of connecting, in which new domains to connect pop up as old ones disappear. To be sure, any new domains must include more virtual ones; online gaming, for example, represents a vast new expanse of interaction connecting people across large distances and social categories. Sometimes those connections are deep and/or meaningful. See, for example, Perry et al., "Online-Only Friends."

6. This comment was posted by egocrat on January 9, 2023; see https://www.reddit.com/r/MadeMeSmile/comments/107gqi2/comment/j3obfgp/, accessed January 16, 2023.

7. The quote is from Deutsch, *Building a Housewife's Paradise,* 33.

8. The "hired hearts" quote is from Seidman, "From the Knowledge Economy to the Human Economy." The "upload the old-fashioned way" quote is from a 2017 op-ed: see Friedman, "From Hands to Heads to Hearts."

9. I also wrote about this in a blog post for a group of Oslo researchers in January 2023; see https://uni.oslomet.no/belong/2023/01/10/on-belonging-sameness-and-difference/. While a number of researchers have divided the emotional dimension (the "personal, intimate, feeling of being 'at home'") from its political sources ("the discursive resource which constructs, claims, justifies, or resists forms of socio-spatial inclusion/exclusion"), I think we should resist such an

impulse, as it cordons off the realm of sentiment to an intimate individual sphere devoid of politics. Instead, the two dimensions entwined is how most of us encounter belonging; personal, intimate feelings arise from the personal, intimate experience of power, of being included or excluded because of the meanings ascribed to particular categories and identities enacted in daily life. These quotes come from Antonsich, "Searching for Belonging," 645. See Newman, Lohman, and Newman, "Peer Group Membership"; Baumeister et al., "Social Exclusion Impairs Self-Regulation"; and Ahmed, "Collective Feelings," 27.

10. Scholars who have worried about belonging rooted in sameness include Antonsich, "Searching for Belonging"; May, *Connecting Self to Society*; and Yuval-Davis, "Belonging and the Politics of Belonging."

11. The quotes are from Smith, *Talk to Me*.

12. The "pets or livestock" question, and the quotes, are from engineer Anthony Levandowski. See Harris, "Inside the First Church of Artificial Intelligence." The *Los Angeles Times* picture ran in conjunction with a story documenting how people both help and hinder robot work; see White, "Kicks, Pranks, Dog Pee."

13. On the problem of algorithmic bias in credit scoring, see Pasquale, *The Black Box Society*. The story of the Mississippi school district was reported in Balingit, "A Teacher Shortage."

14. The quotes come from Power, "Counting, Control and Calculation," 774–75.

15. The term "algorithm chow" was coined by Sally Applin, in Applin, "They Sow, They Reap."

16. For the attention economy, see Davenport and Beck, "Getting the Attention You Need."

17. Low-income people and people of color, among others, suffer from a "recognition gap" derived from their perceived lower status, according to Lamont, who argues that the cultural apparatus that Americans use to recognize others is too limited to neoliberal conceptions of merit and financial achievement. See Lamont, "Addressing Recognition Gaps."

18. See, for example, Cranford, *Home Care Fault Lines*.

Appendix. "Maybe We're Going to Turn You into a Chaplain": Studying Connection

1. For work on why people feel more understood than they are, see Reis, Lemay, and Finkenauer, "Towards Understanding Understanding."

2. Graduate students assisted in about 10 percent of the interviews, 5 percent of the observations, and 15 percent of the data entry, coding, and analysis. Allister Pilar Plater and Jaime Hartless conducted interviews, while Hayley Elszasz, Jaime Hartless, Fauzia Husain, Allister Pilar Plater, and Shayne Zaslow assisted with secondary research and analysis. Undergraduate student Genevieve Charles also conducted some ethnographic observations. Other students helped with some secondary research.

3. My future research focuses more intently upon these connective labor recipients.

4. These subtotals do not add up to 108 because some of these groups overlap.

5. See Luker, *Salsa Dancing into the Social Sciences*.

6. Interviews were conducted with an eye to gathering four kinds of information, as explored in Pugh, "What Good Are Interviews?": the honorable (revealing what people think

is most admirable), the schematic (revealing the frames and metaphors that people use to think with), the visceral (revealing the emotional gut of how people understand their experiences), and meta-feelings (revealing the distance between how someone feels and how they ought to feel).

7. See Desmond, "Relational Ethnography," 548.

8. These comments regarding the dangers of unmasking form part of the argument in an article I published with Sarah Mosseri. See Pugh and Mosseri, "Trust-Building versus 'Just Trust Me.'"

REFERENCES

Abbott, Andrew. *The System of Professions: An Essay on the Division of Expert Labor*. Chicago: University of Chicago Press, 1988.

Acemoglu, Daron, and James A. Robinson. *Why Nations Fail: The Origins of Power, Prosperity, and Poverty*. New York: Crown Publishing, 2013.

Ackerman, Steven J., and Mark J. Hilsenroth. "A Review of Therapist Characteristics and Techniques Positively Impacting the Therapeutic Alliance." *Clinical Psychology Review* 23, no. 1 (2003): 1–33.

Agency for Health Care Quality and Research. "Physician Burnout." Publication: 17-M018–1-EF. 2017. https://www.ahrq.gov/sites/default/files/wysiwyg/professionals/clinicians-providers/ahrq-works/impact-burnout.pdf. Accessed July 6, 2022.

Agrawal, Pramila, Changchun Liu, and Nilanjan Sarkar. "Interaction between Human and Robot." *Interaction Studies* 9, no. 2 (2008): 230–57.

Ahmed, S. "Collective Feelings: Or, the Impressions Left by Others." *Theory, Culture & Society* 21, no. 2 (2004): 25–42.

Allahyari, Rebecca Anne. *Visions of Charity: Volunteer Workers and Moral Community*. Berkeley: University of California Press, 2000.

Allen, Danielle. "America Is in a Great Pulling Apart: Can We Pull Together?" *Washington Post*, January 31, 2023. https://www.washingtonpost.com/opinions/2023/01/31/danielle-allen-american-democracy-renovation-series/.

Allen, Jim, Barbara Belfi, and Lex Borghans. "Is There a Rise in the Importance of Socioemotional Skills in the Labor Market? Evidence from a Trend Study among College Graduates." *Frontiers in Psychology* 11 (2020): 1710.

Alloway, Tracey Packiam, and Heather Cissel. "Psychometric Testing in the Workplace." *The Score* (newsletter of Division 5 of the American Psychological Association). April 2017. https://www.apadivisions.org/division-5/publications/score/2017/04/psychometric-testing. Accessed January 30, 2023.

Althoff, Tim, Kevin Clark, and Jure Leskovec. "Large-Scale Analysis of Counseling Conversations: An Application of Natural Language Processing to Mental Health." *Transactions of the Association for Computational Linguistics* 4 (2016): 463–76.

Alvehus, Johan, and André Spicer. "Financialization as a Strategy of Workplace Control in Professional Service Firms." *Critical Perspectives on Accounting* 23, no. 7–8 (2012): 497–510.

Anderson, Elijah. *Black in White Space: The Enduring Impact of Color in Everyday Life*. Chicago: University of Chicago Press, 2022.

Antonsich, Marco. "Searching for Belonging: An Analytical Framework." *Geography Compass* 4, no. 6 (2010): 644–59.

Applin, Sally. "They Sow, They Reap: How Humans Are Becoming Algorithm Chow." *Medium*, May 13, 2017.

Aronson, Jeffrey K. "When I Use a Word . . . Counterfactual Medical History: The Death of Charles II." *BMJ* 377:01345 (May 27, 2022).

Aroyo, Alexander M., Jan De Bruyne, Orian Dheu, Eduard Fosch-Villaronga, Aleksei Gudkov, Holly Hoch, Steve Jones, et al. "Overtrusting Robots: Setting a Research Agenda to Mitigate Overtrust in Automation." *Paladyn, Journal of Behavioral Robotics* 12, no. 1 (2021): 423–36.

Atalay, Enghin, Phai Phongthiengtham, Sebastian Sotelo, and Daniel Tannenbaum. "The Evolution of Work in the United States." *American Economic Journal: Applied Economics* 12, no. 2 (2020): 1–34.

Atlas, Steven J., Richard W. Grant, Timothy G. Ferris, Yuchiao Chang, and Michael J. Barry. "Patient–Physician Connectedness and Quality of Primary Care." *Annals of Internal Medicine* 150, no. 5 (2009): 325–35.

Autor, D. H. *Work of the Past, Work of the Future*. Working Paper No. 25588; 2019, 49. https://www.nber.org/papers/w25588.

Autor, David H., Frank Levy, and Richard J. Murnane. "The Skill Content of Recent Technological Change: An Empirical Exploration." *Quarterly Journal of Economics* 118, no. 4 (2003): 1279–1333.

Ayers, John W., Adam Poliak, Mark Dredze, et al. "Comparing Physician and Artificial Intelligence Chatbot Responses to Patient Questions Posted to a Public Social Media Forum." *JAMA Intern Med*. April 28, 2023.

Balingit, Moriah. "A Teacher Shortage So Acute That Students Are Expected to Learn Without One." *Washington Post*, January 19, 2023. https://www.washingtonpost.com/education/2023/01/19/teacher-shortage-mississippi/.

Banks, Marian R., Lisa M. Willoughby, and William A. Banks. "Animal-Assisted Therapy and Loneliness in Nursing Homes: Use of Robotic versus Living Dogs." *Journal of the American Medical Directors Association* 9, no. 3 (2008): 173–77.

Barbaro, Michael, et al. "Did Artificial Intelligence Just Get Too Smart?" *New York Times*, December 16, 2022. https://www.nytimes.com/2022/12/16/podcasts/the-daily/chatgpt-openai-artificial-intelligence.html.

Bass, Bernard M. "From Transactional to Transformational Leadership: Learning to Share the Vision." *Organizational Dynamics* 18, no. 3 (1990): 19–31.

Batson, C. Daniel, Johee Chang, Ryan Orr, and Jennifer Rowland. "Empathy, Attitudes, and Action: Can Feeling for a Member of a Stigmatized Group Motivate One to Help the Group?" *Personality and Social Psychology Bulletin* 28, no. 12 (2002): 1656–66.

Baugher, John Eric. "Pathways through Grief to Hospice Volunteering." *Qualitative Sociology* 38, no. 3 (2015): 305–26.

Baumeister, Roy F., C. Nathan DeWall, Natalie J. Ciarocco, and Jean M. Twenge. "Social Exclusion Impairs Self-Regulation." *Journal of Personality and Social Psychology* 88, no. 4 (2005): 589.

Baumeister, Roy F., and Mark R. Leary. "The Need to Belong: Desire for Interpersonal Attachments as a Fundamental Human Motivation." *Psychological Bulletin* 117, no. 3 (1995): 497–529.

Baym, Nancy K. *Playing to the Crowd: Musicians, Audiences, and the Intimate Work of Connection.* New York: NYU Press, 2018.

Bechler, Curt, and Scott D. Johnson. "Leadership and Listening: A Study of Member Perceptions." *Small Group Research* 26, no. 1 (1995): 77–85.

Beer, David. *Metric Power.* London: Palgrave Macmillan, 2016.

Benjamin, Jessica. *Shadow of the Other: Intersubjectivity and Gender in Psychoanalysis.* New York: Routledge, 2013.

Benjamin, Ruha. *Race after Technology: Abolitionist Tools for the New Jim Code.* Medford, MA: Polity Press, 2019.

Benzell, Seth Gordon, Erik Brynjolfsson, Frank MacCrory, and George Westerman. "Identifying the Multiple Skills in Skill-Biased Technical Change." Working Paper, 2019. https://ide.mit.edu/wp-content/uploads/2019/08/Identifying-the-Multiple-Skills-in-SBTC-8-2-19.pdf. Accessed February 4, 2023.

Bergin, Christi Ann, and David A. Bergin. "Attachment in the Classroom." *Educational Psychology Review* 21, no. 2 (2009): 141–70.

Berkovich, Izhak, and Ori Eyal. "Educational Leaders and Emotions: An International Review of Empirical Evidence 1992–2012." *Review of Educational Research* 85, no. 1 (2015): 129–67.

Bernstein, Elizabeth. "Bounded Authenticity and the Commerce of Sex." In Eileen Boris and Rhacel Salazar Parreñas, *Intimate Labors: Cultures, Technologies, and the Politics of Care,* 148–65. Stanford, CA: Stanford University Press, 2010.

Bernstein, Elizabeth. *Temporarily Yours.* Chicago: University of Chicago Press, 2007.

Bickmore, Timothy, Laura Vardoulakis, Brian Jack, and Michael Paasche-Orlow. "Automated Promotion of Technology Acceptance by Clinicians Using Relational Agents." *Intelligent Virtual Agents conference (IVA).* 2013.

Bingham, Charles Wayne, and Alexander M. Sidorkin. *No Education without Relation.* Vol. 259. New York: Peter Lang, 2004.

Black, Helen K. "Moral Imagination in Long-Term Care Workers." *OMEGA-Journal of Death and Dying* 49, no. 4 (2004): 299–320.

Blinder, Alan. "Offshoring: The Next Industrial Revolution?" *Foreign Affairs* 85, no. 2 (March/April 2006): 113–28.

Blume Oeur, Freeden. *Black Boys Apart: Racial Uplift and Respectability in All-Male Public Schools.* Minneapolis: University of Minnesota Press, 2018.

Blume Oeur, Freeden. "Recognizing Dignity: Young Black Men Growing Up in an Era of Surveillance." *Socius* (2016): 2. https://doi.org/10.1177/2378023116633712.

Bollyky, Thomas J., Ilona Kickbusch, and Michael Bang Petersen. "The Trust Gap: How to Fight Pandemics in a Divided Country." *Foreign Affairs,* January 30, 2023. https://www.foreignaffairs.com/united-states/trust-gap-fight-pandemic-divided-country.

Boris, Eileen, and Rhacel Salazar Parreñas. *Intimate Labors: Cultures, Technologies, and the Politics of Care.* Stanford, CA: Stanford University Press, 2010.

Bosk, Charles L., Mary Dixon-Woods, Christine A. Goeschel, and Peter J. Pronovost. "Reality Check for Checklists." *The Lancet* 374, no. 9688 (2009): 444–45.

Bowker, Geoffrey C., and Susan Leigh Star. *Sorting Things Out: Classification and Its Consequences.* Cambridge, MA: MIT Press, 2000.

Bowles, Nellie. "An Online Preschool Closes a Gap but Exposes Another." *New York Times*, A1, July 7, 2019.

Braverman, Harry. *Labor and Monopoly Capital: The Degradation of Work in the Twentieth Century*. New York: NYU Press, 1998 (1974).

Brayne, Sarah. *Predict and Surveil: Data, Discretion, and the Future of Policing*. New York: Oxford University Press, 2020.

Breselor, Sara, and Adam Higginbotham. "Welcome to the United States Border." *Wired*, February 21, 2013 (2): 90.

Brooks, David. "The Blindness of Social Wealth." *New York Times*. April 16, 2018.

Brown, Nik. "Metrics of Hope: Disciplining Affect in Oncology." *Health* 19, no. 2 (2015): 119–36.

Budd, Rick, and Ian Hughes. "The Dodo Bird Verdict—Controversial, Inevitable and Important: A Commentary on 30 Years of Meta-analyses." *Clinical Psychology & Psychotherapy: An International Journal of Theory & Practice* 16, no. 6 (2009): 510–22.

Burgess, Diana J. "Are Providers More Likely to Contribute to Healthcare Disparities under High Levels of Cognitive Load? How Features of the Healthcare Setting May Lead to Biases in Medical Decision Making." *Medical Decision Making* 30, no. 2 (2010): 246–57.

Burrell, Jenna, and Marion Fourcade. "The Society of Algorithms." *Annual Review of Sociology* 47 (July 2021): 213–37.

Butler, Emily A., Boris Egloff, Frank H. Wilhelm, Nancy C. Smith, Elizabeth A. Erickson, and James J. Gross. "The Social Consequences of Expressive Suppression." *Emotion* 3, no. 1 (2003): 48–67.

Butler, Judith. *Bodies That Matter: On the Discursive Limits of Sex*. New York: Routledge, 2011.

Byambasuren, Oyungerel, Sharon Sanders, Elaine Beller, and Paul Glasziou. "Prescribable mHealth Apps Identified from an Overview of Systematic Reviews." *NPJ Digital Medicine* 1, no. 1 (2018): 12.

Cai, Qijie. "Can Principals' Emotional Intelligence Matter to School Turnarounds?" *International Journal of Leadership in Education* 14, no. 2 (2011): 151–79.

Calarco, Jessica McCrory. "Avoiding Us versus Them: How Schools' Dependence on Privileged 'Helicopter' Parents Influences Enforcement of Rules." *American Sociological Review*, 85, no. 2 (2020): 223–46.

Calarco, Jessica McCrory. "'I Need Help!': Social Class and Children's Help-Seeking in Elementary School." *American Sociological Review* 76, no. 6 (2011): 862–82.

Calarco, Jessica McCrory. *Negotiating Opportunities: How the Middle Class Secures Advantages in School*. Oxford University Press, 2018.

Calarco, Jessica McCrory. *Holding It Together: How Women Became America's Social Safety Net*. New York: Portfolio Press, forthcoming.

Case, Anne, and Angus Deaton. *Deaths of Despair and the Future of Capitalism*. Princeton, NJ: Princeton University Press, 2022.

Cech, Erin. *The Trouble with Passion: How Searching for Fulfillment at Work Fosters Inequality*. Berkeley: University of California Press, 2021.

Chochinov, Harvey Max. "Dignity and the Essence of Medicine: The A, B, C, and D of Dignity Conserving Care." *BMJ* 335, no. 7612 (2007): 184–87.

Christin, Angèle. *Metrics at Work: Journalism and the Contested Meaning of Algorithms*. Princeton, NJ: Princeton University Press, 2020.

Chui, Michael, James Manyika, and Mehdi Miremadi. "Four Fundamentals of Workplace Automation." *McKinsey Quarterly*. November 2015.

Clark, Malissa A., Melissa M. Robertson, and Stephen Young. "'I Feel Your Pain': A Critical Review of Organizational Research on Empathy." *Journal of Organizational Behavior* 40, no. 2 (2019): 166–92.

Collins, Patricia Hill. *Black Feminist Thought: Knowledge, Consciousness, and the Politics of Empowerment*. New York: Routledge, 1990.

Collins, Randall. *Interaction Ritual Chains*. Princeton, NJ: Princeton University Press, 2004.

Colt, Sarah. "Do Scripted Lessons Work—or Not?" "Making School Work with Hedrick Smith." 2005. https://www.pbs.org/makingschoolswork/sbs/sfa/lessons.html.

The Commonwealth Fund. "Stressed Out and Burned Out: The Global Primary Care Crisis: Findings from the 2022 International Health Policy Survey of Primary Care Physicians." 2022. https://www.commonwealthfund.org/publications/issue-briefs/2022/nov/stressed-out-burned-out-2022-international-survey-primary-care-physicians. Accessed December 19, 2022.

Cook, Catherine, and Margaret Brunton. "Pastoral Power and Gynaecological Examinations: A Foucauldian Critique of Clinician Accounts of Patient-Centred Consent." *Sociology of Health & Illness* 37, no. 4 (2015): 545–60.

Cooley, Charles Horton. *Human Nature and the Social Order*. New York: Routledge, 2017 (1902).

Corlette, Sabrina, Robert Berenson, Erik Wengle, Kevin Lucia, and Tyler Thomas. "Impact of the COVID-19 Pandemic on Primary Care Practices." *The Urban Institute*. 2021. https://www.urban.org/sites/default/files/publication/103596/impact-of-the-covid-19-pandemic-on-primary-care-practices.pdf. Accessed December 19, 2022.

Cornelius-White, Jeffrey. "Learner-Centered Teacher-Student Relationships Are Effective: A Meta-analysis." *Review of Educational Research* 77, no. 1 (2007): 113–43.

Cottingham, Marci D. *Practical Feelings: Emotions as Resources in a Dynamic Social World*. New York: Oxford University Press, 2022.

Cottingham, Marci D., and Rebecca J. Erickson. "The Promise of Emotion Practice: At the Bedside and Beyond." *Work and Occupations* 47, no. 2 (2020): 173–99.

Craciun, Mariana. "The Cultural Work of Office Charisma: Maintaining Professional Power in Psychotherapy." *Theory and Society* 45, no. 4 (2016): 461–83.

Craciun, Mariana. "Emotions and Knowledge in Expert Work: A Comparison of Two Psychotherapies." *American Journal of Sociology* 123, no. 4 (2018): 959–1003.

Cranford, Cynthia J. *Home Care Fault Lines: Understanding Tensions and Creating Alliances*. Ithaca, NY: Cornell University Press, 2020.

Crosnoe, Robert, Monica K. Johnson, and Glen H. Elder. "Intergenerational Bonding in School: The Behavioral and Contextual Correlates of Student-Teacher Relationships." *Sociology of Education* 77 (2004): 60–81.

Dautenhahn, Kerstin. "Roles and Functions of Robots in Human Society: Implications from Research in Autism Therapy." *Robotica* 21, no. 4 (2003): 443–52.

Dautenhahn, Kerstin, and Iain Werry. "Towards Interactive Robots in Autism Therapy." *Pragmatics and Cognition* 12, no. 1 (2004): 1–35.

Davenport, Thomas H., and John C. Beck. "Getting the Attention You Need." *Harvard Business Review* 78, no. 5 (2000): 118–26.

Day, Christopher. "School Reform and Transitions in Teacher Professionalism and Identity." *International Journal of Educational Research* 37, no. 8 (2002): 677–92.

Debesay, Jonas, Ivan Harsløf, Bernd Rechel, and Halvard Vike. "Dispensing Emotions: Norwegian Community Nurses' Handling of Diversity in a Changing Organizational Context." *Social Science & Medicine* 119 (2014): 74–80.

Deeb-Sossa, Natalia. "Helping the 'Neediest of the Needy': An Intersectional Analysis of Moral-Identity Construction at a Community Health Clinic." *Gender & Society* 21, no. 5 (2007): 749–72.

Deming, David J. "The Growing Importance of Social Skills in the Labor Market." *Quarterly Journal of Economics* 132, no. 4 (2017): 1593–1640.

Department of Health Care Access and Information. "Mental and Behavioral Health Diagnoses in Emergency Department and Inpatient Discharges by Healthy Places Index Ranking." State of California. 2023. https://hcai.ca.gov/visualizations/mental-and-behavioral-health-diagnoses-in-emergency-department-and-inpatient-discharges-by-healthy-places-index-ranking/#additional-information. Accessed May 19, 2023.

Desmond, Matthew. "Relational Ethnography." *Theory and Society* 43, no. 5 (2014): 547–79.

Deutsch, Tracey. *Building a Housewife's Paradise: Gender, Politics, and American Grocery Stores in the Twentieth Century*. Raleigh: University of North Carolina Press, 2010.

Diamond, Timothy. *Making Gray Gold: Narratives of Nursing Home Care*. Chicago: University of Chicago Press, 2009.

Diamond-Brown, Lauren. "The Doctor-Patient Relationship as a Toolkit for Uncertain Clinical Decisions." *Social Science & Medicine* 159 (2016): 108–15.

Dickert, Stephan, and Paul Slovic. "Attentional Mechanisms in the Generation of Sympathy." *Judgment and Decision Making* 4, no. 4 (2009): 297–306.

DiMatteo, M. Robin, Howard S. Friedman, and Angelo Taranta. "Sensitivity to Bodily Nonverbal Communication as a Factor in Practitioner-Patient Rapport." *Journal of Nonverbal Behavior* 4 (1979): 18–26.

Dinsmore, Brooke. "Relational Cultures, Inequality and Belonging: Race, Class, and Teacher-Student Relationships at Two U.S. High Schools." PhD dissertation, University of Virginia, 2023.

Dinsmore, Brooke, and Allison J. Pugh. "The Paradox of Constrained Well-Being: Childhood Autonomy, Surveillance and Inequality." *Sociological Forum* 36, no. 2 (2021): 448–70.

Downey, Jayne A. "Recommendations for Fostering Educational Resilience in the Classroom." *Preventing School Failure* 53 (2008): 56–63.

Duffy, Mignon, Randy Albelda, and Clare Hammonds. "Counting Care Work: The Empirical and Policy Applications of Care Theory." *Social Problems* 60, no. 2 (2013): 145–67.

Dunn, Elizabeth W., Jeremy C. Biesanz, Lauren J. Human, and Stephanie Finn. "Misunderstanding the Affective Consequences of Everyday Social Interactions: The Hidden Benefits of Putting One's Best Face Forward." *Journal of Personality and Social Psychology* 92 (2007): 990–1005.

Dutton, Jane E., Monica C. Worline, Peter J. Frost, and Jacoba Lilius. "Explaining Compassion Organizing." *Administrative Science Quarterly* 51, no. 1 (2006): 59–96.

Dutton, Jane, Kristina Workman, and Ashley Hardin. "Compassion at Work." *Annual Review of Organizational Psychology and Organizational Behavior* 1 (2014): 277–304.

Dwyer, Rachel E. "The Care Economy? Gender, Economic Restructuring, and Job Polarization in the US Labor Market." *American Sociological Review* 78, no. 3 (2013): 390–416.

Edin, Per-Anders, Peter Fredriksson, Martin Nybom, and Bjorn Ockert. "The Rising Return to Non-cognitive Skill." IZA Discussion Paper No 10914. 2017. https://dx.doi.org/10.2139/ssrn.3029784.

Elfenbein, Hillary Anger, and Nalini Ambady. "Predicting Workplace Outcomes from the Ability to Eavesdrop on Feelings." *Journal of Applied Psychology* 87, no. 5 (2002): 963–71.

Elfenbein, Hillary Anger, Maw Der Foo, Judith White, Hwee Hoon Tan, and Voon Chuan Aik. "Reading Your Counterpart: The Benefit of Emotion Recognition Accuracy for Effectiveness in Negotiation." *Journal of Nonverbal Behavior* 31 (2007): 205–23.

Elliott, Sinikka, and Sarah Bowen. "Defending Motherhood: Morality, Responsibility, and Double Binds in Feeding Children." *Journal of Marriage and Family* 80, no. 2 (2018): 499–520.

England, Paula. "Emerging Theories of Care Work." *Annual Review of Sociology* 31, no. 1 (2005): 381–99.

Erickson, Lance D., Steve McDonald, and Glen H. Elder. "Informal Mentors and Education: Complementary or Compensatory Resources?" *Sociology of Education* 82, no. 4 (2009): 344–67.

Espeland, Wendy Nelson, and Michael Sauder. "Rankings and Reactivity: How Public Measures Recreate Social Worlds." *American Journal of Sociology* 113, no. 1 (July 2007): 1–40.

Espeland, Wendy Nelson, and Mitchell L. Stevens. "A Sociology of Quantification." *European Journal of Sociology/Archives Européennes de Sociologie* 49, no. 3 (2008): 401–36.

Eubanks, Virginia. *Automating Inequality: How High-Tech Tools Profile, Police, and Punish the Poor*. New York: St. Martin's Press, 2018.

Eveleigh, Rhona M., Esther Muskens, Hiske van Ravesteijn, Inge van Dijk, Eric van Rijswijk, and Peter Lucassen. "An Overview of 19 Instruments Assessing the Doctor-Patient Relationship: Different Models or Concepts Are Used." *Journal of Clinical Epidemiology* 65, no. 1 (2012): 10–15.

Eyal, Tal, Mary Steffel, and Nicholas Epley. "Perspective Mistaking: Accurately Understanding the Mind of Another Requires Getting Perspective, Not Taking Perspective." *Journal of Personality and Social Psychology* 114, no. 4 (2018): 547.

Fahle, Erin, Thomas J. Kane, Tyler Patterson, Sean F. Reardon, and Douglas O. Staiger. "Local Achievement Impacts of the Pandemic." *Education Recovery Scorecard*. October 28, 2022. https://educationrecoveryscorecard.org/wp-content/uploads/2022/10/Education-Recovery-Scorecard_Key-Findings_102822.pdf.

Fanon, Frantz. *Black Skin, White Masks*. New York: Grove Press, 1967.

Fetterolf, Elizabeth. "It's Crowded at the Bottom: Trust, Visibility, and Search Algorithms on Care.com." *Journal of Digital Social Research* 4, no. 1 (2022): 49–72.

Fine, Gary Alan. "Review of Interaction Ritual Chains." *Social Forces* 83, no. 3 (2005): 1287–88.

Fischer, Claude. "Loneliness Epidemic: An End to the Story?" *Made in America*. August 24, 2018. https://madeinamericathebook.wordpress.com/2018/08/24/loneliness-epidemic-an-end-to-the-story/.

Fischer, Claude. "Overcoming Distance and Embracing Place: Personal Ties in the Age of Persistent and Pervasive Communication." *Made in America*. December 20, 2021. https://madein

americathebook.wordpress.com/2021/12/20/overcoming-distance-and-embracing-place-personal-ties-in-the-age-of-persistent-and-pervasive-communication/.

Fischer, Claude, and Xavier Durham. "Forms of Group Involvement: Alternatives to the Standard Question." *Sociological Perspectives* 65, no. 4 (2002): 661–83.

Fischer, Claude S. *Still Connected: Family and Friends in America Since 1970*. New York: Russell Sage Foundation, 2011.

Fischer, Claude S. "The 2004 GSS Finding of Shrunken Social Networks: An Artifact?" *American Sociological Review* 74, no. 4 (2009): 657–69.

Flanagan, Frances. "Theorising the Gig Economy and Home-Based Service Work." *Journal of Industrial Relations* 61, no. 1 (2019): 57–78.

Fleming, Crystal M., Michèle Lamont, and Jessica S. Welburn. "African Americans Respond to Stigmatization: The Meanings and Salience of Confronting, Deflecting Conflict, Educating the Ignorant and 'Managing the Self.'" *Ethnic and Racial Studies* 35, no. 3 (2012): 400–17.

Fletcher, Joyce K. *Disappearing Acts: Gender, Power, and Relational Practice at Work*. Cambridge, MA: MIT Press, 2001.

Fletcher, Joyce K. "Relational Practice: A Feminist Reconstruction of Work." *Journal of Management Inquiry* 7, no. 2 (1998): 163–86.

Foner, Nancy. *The Caregiving Dilemma: Work in an American Nursing Home*. Berkeley: University of California Press, 1995.

Foucault, Michel. *Ethics: Subjectivity and Truth; The Essential Works of Foucault 1954–1984*, Vol. I, trans. R. Hurley. Harmondsworth: Penguin Books, 1994.

Foucault, Michel. "The Subject and the Power." In *Michel Foucault: Beyond Structuralism and Hermeneutics*, edited by H. Dreyfus and P. Rabinow, 208–26. Chicago: University of Chicago Press, 1982.

Fox, Renee. "Medical Uncertainty Revisited." In *Handbook of Social Studies in Health and Medicine*, edited by Albrecht, Gary L., Ray Fitzpatrick, and Susan C. Scrimshaw, 409–25. Thousand Oaks, CA: Sage Publications, 2000.

Fraser, Nancy. "Contradictions of Capital and Care." *New Left Review* 100 (2016): 99–117.

Fraser, Nancy, and Axel Honneth. *Redistribution or Recognition?: A Political-Philosophical Exchange*. London: Verso, 2006.

Fredriksen, K., and J. Rhodes. "The Role of Teacher Relationships in the Lives of Students." *New Directions for Youth Development* (2004): 45–54. doi:10.1002/yd.90.

Frey, Carl Benedikt, and Michael Osborne. "The Future of Employment: How Susceptible Are Jobs to Computerisation?" 2013. Accessed February 4, 2023. http://www.oxfordmartin.ox.ac.uk/downloads/academic/The_Future_of_Employment.pdf.

Friedman, Thomas L. "Opinion: From Hands to Heads to Hearts." *New York Times*. January 4, 2017. https://www.nytimes.com/2017/01/04/opinion/from-hands-to-heads-to-hearts.html.

Friedson, Elliot. *Profession of Medicine: A Study of the Sociology of Applied Knowledge*. Chicago: University of Chicago Press, 1988.

Froyum, Carissa. "'They Are Just Like You and Me': Cultivating Volunteer Sympathy." *Symbolic Interaction* 41, no. 4 (2018): 465–87.

Furman, Frida Kerner. *Facing the Mirror: Older Women and Beauty Shop Culture*. New York: Routledge, 1997.

Gage-Bouchard, Elizabeth A. "Culture, Styles of Institutional Interactions, and Inequalities in Healthcare Experiences." *Journal of Health and Social Behavior* 58, no. 2 (2017): 147–65.

Gallup. "State of America's Schools: The Path to Winning Again in Education." 2014. https://www.gallup.com/services/178769/state-america-schools-report.aspx.

Gengler, Amanda M. "'I Want You to Save My Kid!': Illness Management Strategies, Access, and Inequality at an Elite University Research Hospital." *Journal of Health and Social Behavior* 55, no. 3 (2014): 342–59.

Geronimus, Arline T. "The Weathering Hypothesis and the Health of African-American Women and Infants: Evidence and Speculations." *Ethnicity & Disease* (1992): 207–21.

Gershon, Ilana, and Anna Eisenstein. "'Saying No and Staying Flexible Can Coexist': Utilizing Connective Labor to Assert Personal Autonomy as a Nanny During the Pandemic." Conference paper, American Anthropological Association meetings, Seattle, 2022.

Gershon, Livia. "The Automation-Resistant Skills We Should Nurture." BBC. 2017. https://www.bbc.com/worklife/article/20170726-the-automation-resistant-skills-we-should-nurture.

Gill, Rosalind, and Andy Pratt. "In the Social Factory?: Immaterial Labour, Precariousness and Cultural Work." *Theory, Culture & Society* 25, no. 7–8 (2008): 1–30.

Gillies, Val. "Social and Emotional Pedagogies: Critiquing the New Orthodoxy of Emotion in Classroom Behaviour Management." *British Journal of Sociology of Education* 32, no. 2 (2011): 185–202.

Glenn, Evelyn Nakano. "Creating a Caring Society." *Contemporary Sociology* 29, no. 1 (2000): 84–94.

Glenn, Evelyn Nakano. *Forced to Care: Coercion and Caregiving in America.* Cambridge, MA: Harvard University Press, 2010.

Goldberg, Simon B., Robbie Babins-Wagner, and Scott D. Miller. "Nurturing Expertise at Mental Health Agencies." In *The Cycle of Excellence: Using Deliberate Practice to Improve Supervision and Training*, edited by Tony Rousmaniere, Rodney Goodyear, Scott D. Miller, and Bruce E. Wampold, 199–217. Hoboken, NJ: John Wiley and Sons, 2017.

Goldberg, Simon B., Robbie Babins-Wagner, Tony Rousmaniere, Sandy Berzins, William T. Hoyt, Jason L. Whipple, Scott D. Miller, and Bruce E. Wampold. "Creating a Climate for Therapist Improvement: A Case Study of an Agency Focused on Outcomes and Deliberate Practice." *Psychotherapy* 53, no. 3 (2016): 367.

Grand View Research. "mHealth Apps Market Size, Share & Trends Analysis Report by Type (Fitness, Medical), by Region (North America, Europe, Asia Pacific, Latin America, Middle East & Africa), and Segment Forecasts, 2022–2030." https://www.grandviewresearch.com/industry-analysis/mhealth-app-market. Accessed January 16, 2023.

Granovetter, Mark S. "The Strength of Weak Ties." *American Journal of Sociology* 78 (1973): 1360–80.

Grant, Don, Alfonso Morales, and Jeffrey J. Sallaz. "Pathways to Meaning: A New Approach to Studying Emotions at Work." *American Journal of Sociology* 115, no. 2 (2009): 327–64.

Grant, Don, Rebecca J. Erickson, Beth Duckles, and Christine Sheikh. "Affirming Selves through Styles of Care: When, How, and Why Hospital Workers Craft Different Patient-Centered Cultures." *Social Problems* 63, no. 2 (1 May 2016): 180–202.

Grant, Louise, and Gail Kinman. "Emotional Resilience in the Helping Professions and How It Can Be Enhanced." *Health and Social Care Education* 3, no. 1 (2014): 23–34.

Griffen, Zachary, and Aaron Panofsky. "Ambivalent Economizations: The Case of Value Added Modeling in Teacher Evaluation." *Theory and Society* 50, no. 3 (2021): 515–39.

Groark, Christina J., and Robert B. McCall. "Community-Based Interventions and Services." In *Rutter's Child and Adolescent Psychiatry*, edited by Michael Rutter, Dorothy Bishop, Daniel S. Pine, Stephen Scott, Jim Stevenson, Eric Taylor, and Anita Thapar, 971–88. Malden, MA: Wiley Blackwell, 2008.

Groark, Kevin P. "Social Opacity and the Dynamics of Empathic In-Sight among the Tzotzil Maya of Chiapas, Mexico." *Ethos* 36, no. 4 (2008): 427–48.

Grossman, Pamela, Christa Compton, Danielle Igra, Matthew Ronfeldt, Emily Shahan, and Peter Williamson. "Teaching Practice: A Cross-Professional Perspective." *Teachers College Record* 111, no. 9 (2009): 2055–2100.

Gutek, Barbara A., and Theresa M. Welsh. *The Brave New Service Strategy: Aligning Customer Relationships, Market Strategies and Business Structures*. New York: Amacom, 2000.

Halberstadt, Amy G., and Judith A. Hall. "Who's Getting the Message? Children's Nonverbal Skill and Their Evaluation by Teachers." *Developmental Psychology* 16, no. 6 (1980): 564.

Hales, Brigette, Marius Terblanche, Robert Fowler, and William Sibbald. "Development of Medical Checklists for Improved Quality of Patient Care." *International Journal for Quality in Health Care* 20, no. 1 (2008): 22–30.

Hallinger, Philip. "Leading Educational Change: Reflections on the Practice of Instructional and Transformational Leadership." *Cambridge Journal of Education* 33, no. 3 (2003): 329–52.

Hamre, Bridget K., and Robert C. Pianta. "Early Teacher–Child Relationships and the Trajectory of Children's School Outcomes through Eighth Grade." *Child Development* 72, no. 2 (2001): 625–38.

Hardt, Michael, and Antonio Negri. *Empire*. Cambridge, MA: Harvard University Press, 2000.

Harris, Mark. "Inside the First Church of Artificial Intelligence." *Wired*. November 15, 2017. https://www.wired.com/story/anthony-levandowski-artificial-intelligence-religion/.

Hattie, John. *Visible Learning: A Synthesis of Over 800 Meta-Analyses Relating to Achievement*. London: Routledge, 2008.

Haug, Marie. "A Re-examination of the Hypothesis of Physician Deprofessionalization." *The Milbank Quarterly* 66 (1988): 48–56.

Hellerstein, Judith K., and Melinda Sandler Morrill. "Booms, Busts, and Divorce." *The BE Journal of Economic Analysis & Policy* 11, no. 1 (2011).

Hersey, Paul, Kenneth Blanchard, and Dewey Johnson. *Management of Organizational Behavior: Leading Human Resources*. 8th ed. Upper Saddle River, NJ: Prentice Hall, 2001.

Hirsch, Jennifer L., and Margaret S. Clark. "Multiple Paths to Belonging That We Should Study Together." *Perspectives on Psychological Science* 14, no. 2 (2019): 238–55.

Hochschild, Arlie Russell. *The Commercialization of Intimate Life: Notes from Home and Work*. Berkeley: University of California Press, 2003.

Hochschild, Arlie Russell. "Emotion Work, Feeling Rules, and Social Structure." *American Journal of Sociology* 85, no. 3 (1979): 551–75.

Hochschild, Arlie Russell. *The Managed Heart: Commercialization of Human Feeling*. Berkeley: University of California Press, (1983) 2012.

Hochschild, Arlie Russell. *The Outsourced Self: What Happens When We Pay Others to Live Our Lives for Us*. New York: Metropolitan Books, 2012.

Hochschild, Arlie Russell. *The Second Shift*. New York: Penguin Books, 1989.

Hochschild, Arlie Russell. *So How's the Family?: And Other Essays*. Berkeley: University of California Press, 2013.

Holt-Lunstad, Julianne, Timothy B. Smith, and J. Bradley Layton. "Social Relationships and Mortality Risk: A Meta-analytic Review." *PLoS Medicine* 7, no. 7 (2010): e1000316.

Holt-Lunstad, Julianne, Timothy B. Smith, Mark Baker, Tyler Harris, and David Stephenson. "Loneliness and Social Isolation as Risk Factors for Mortality: A Meta-analytic Review." *Perspect Psychol Sci* 10 (2015): 227–37.

Honneth, Axel. "The Point of Recognition: A Rejoinder to the Rejoinder." In *Redistribution or Recognition: A Political-Philosophical Exchange*, edited by N. Fraser and A. Honneth, 237–68. London: Verso, 2003.

Honneth, Axel. *The Struggle for Recognition: The Moral Grammar of Social Conflicts*. Cambridge, MA: MIT Press, 1996.

Horvath, Adam O., and Lester Luborsky. "The Role of the Therapeutic Alliance in Psychotherapy." *J Consult Clin Psychol* 68 (1993): 651–73.

Huang, Ming-Hui, Roland Rust, and Vojislav Maksimovic. "The Feeling Economy: Managing in the Next Generation of Artificial Intelligence (AI)." *California Management Review* 61, no. 4 (2019): 43–65.

Hughes, Mary Elizabeth, Linda J. Waite, Louise C. Hawkley, and John T. Cacioppo. "A Short Scale for Measuring Loneliness in Large Surveys: Results from Two Population-Based Studies." *Research on Aging* 26, no. 6 (2004): 655–72.

Ickes, William, Ann Buysse, Hao Pham, Kerri Rivers, James R. Erickson, Melanie Hancock, Joli Kelleher, and Paul R. Gesn. "On the Difficulty of Distinguishing 'Good' and 'Poor' Perceivers: A Social Relations Analysis of Empathic Accuracy Data." *Personal Relationships* 7, no. 2 (2000): 219–34.

Illouz, Eva. *Cold Intimacies: The Making of Emotional Capitalism*. Malden, MA: Polity, 2007.

Illouz, Eva. "Introduction: Emodities or the Making of Emotional Commodities." In *Emotions as Commodities*, edited by Eva Illouz, 1–29. New York: Routledge, 2017.

Illouz, Eva. "Towards a Post-Normative Critique." In *Emotions as Commodities*, edited by Eva Illouz, 197–213. New York: Routledge, 2017.

Irani, Lilly. "Difference and Dependence among Digital Workers: The Case of Amazon Mechanical Turk." *South Atlantic Quarterly* 114, no. 1 (2015): 225–34.

Izard, Carroll, Sarah Fine, David Schultz, Allison Mostow, Brian Ackerman, and Eric Youngstrom. "Emotion Knowledge as a Predictor of Social Behavior and Academic Competence in Children at Risk." *Psychological Science* 12, no. 1 (2001): 18–23.

Jackson, Philip L., Andrew N. Meltzoff, and Jean Decety. "How Do We Perceive the Pain of Others? A Window into the Neural Processes Involved in Empathy." *Neuroimage* 24, no. 3 (2005): 771–79.

Jefferson, Laura, Su Golder, Claire Heathcote, Ana Castro Avila, Veronica Dale, Holly Essex, Christina van der Feltz Cornelis, Elizabeth McHugh, Thirimon Moe-Byrne, and Karen Bloor. "GP Wellbeing during the COVID-19 Pandemic: A Systematic Review." *British Journal of General Practice* 72, no. 718 (2022): e325–e333.

Johnson, Pamela R., and Julie Indvik. "Organizational Benefits of Having Emotionally Intelligent Managers and Employees." *Journal of Workplace Learning* (1999).

Johnson, Scott D., and Curt Bechler. "Examining the Relationship between Listening Effectiveness and Leadership Emergence: Perceptions, Behaviors, and Recall." *Small Group Research* 29, no. 4 (1998): 452–71.

Johnson, Tiffani J., Robert W. Hickey, Galen E. Switzer, Elizabeth Miller, Daniel G. Winger, Margaret Nguyen, Richard A. Saladino, and Leslie R. M. Hausmann. "The Impact of Cognitive Stressors in the Emergency Department on Physician Implicit Racial Bias." *Academic Emergency Medicine* 23, no. 3 (2016): 297–305.

Joint Message from the UK and Japanese Loneliness Ministers. Gov.uk. June 17, 2021. https://www.gov.uk/government/news/joint-message-from-the-uk-and-japanese-loneliness-ministers.

Jones, Stephanie M., and Emily J. Doolittle. "Social and Emotional Learning: Introducing the Issue." *The Future of Children* 27, no. 1 (2017): 3–11.

Kahn, William A. "Caring for the Caregivers: Patterns of Organizational Caregiving." *Administrative Science Quarterly* (1993): 539–63.

Kang, Sin-Hwa, and Jonathan Gratch. "Virtual Humans Elicit Socially Anxious Interactants' Verbal Self-Disclosure." *Computer Animation and Virtual Worlds* 21, no. 3–4 (2010): 473–82.

Kayal, Philip M. *Bearing Witness: Gay Men's Health Crisis and the Politics of AIDS*. New York: Routledge, 2018.

Kayal, Philip M. "Healing Homophobia: Volunteerism and 'Sacredness' in AIDS." *Journal of Religion and Health* 31, no. 2 (1992): 113–28.

Kelley, John M., Gordon Kraft-Todd, Lidia Schapira, Joe Kossowsky, and Helen Riess. "The Influence of the Patient-Clinician Relationship on Healthcare Outcomes: A Systematic Review and Meta-Analysis of Randomized Controlled Trials." *PLoS ONE* 9, no. 4 (2014): e94207.

Kelly, Christine. *Disability Politics and Care*. Vancouver: University of British Columbia Press, 2016.

Kelly, Erin L., and Phyllis Moen. *Overload: How Good Jobs Went Bad and What We Can Do About It*. Princeton, NJ: Princeton University Press, 2021.

Kenworthy, Jared, Cara Fay, Mark Frame, and Robyn Petree. "A Meta-Analytic Review of the Relationship between Emotional Dissonance and Emotional Exhaustion." *Journal of Applied Social Psychology* 44, no. 2 (2014): 94–105.

King, Charles. *Gods of the Upper Air: How a Circle of Renegade Anthropologists Reinvented Race, Sex and Gender in the Twentieth Century*. New York: Doubleday, 2019.

Klassen, Robert M., Nancy E. Perry, and Anne C. Frenzel. "Teachers' Relatedness with Students: An Underemphasized Component of Teachers' Basic Psychological Needs." *Journal of Educational Psychology* 104, no. 1 (2012): 150–65.

Klem, Adena M., and James P. Connell. "Relationships Matter: Linking Teacher Support to Student Engagement and Achievement." *Journal of School Health* 74 (2004): 262–73.

Kluger, Avraham N., and Keren Zaidel. "Are Listeners Perceived as Leaders?" *International Journal of Listening* 27, no. 2 (2013): 73–84.

Kolb, Kenneth H. "Sympathy Work: Identity and Emotion Management among Victim-Advocates and Counselors." *Qualitative Sociology* 34 (2011): 101–19.

Krachman, Sara Bartolino, and Bob LaRocca. "The Scale of Our Investment in Socio-Emotional Learning." Working paper. Transforming Education, Boston. 2017. https://transformingeducation.org/wp-content/uploads/2017/10/Inspire-Paper-Transforming-Ed-FINAL-2.pdf. Accessed December 1, 2022.

Kristof, Nicholas. "Let's Wage a War on Loneliness." *New York Times*. November 9, 2019. https://www.nytimes.com/2019/11/09/opinion/sunday/britain-loneliness-epidemic.html.

Kueper, Jacqueline K., Amanda L. Terry, Merrick Zwarenstein, and Daniel J. Lizotte. "Artificial Intelligence and Primary Care Research: A Scoping Review." *The Annals of Family Medicine* 18, no. 3 (2020): 250–58.

Kung, Tiffany H., Morgan Cheatham, Arielle Medinilla, ChatGPT, Czarina Sillos, Lorie De Leon, Camille Elepano, et al. "Performance of ChatGPT on USMLE: Potential for AI-Assisted Medical Education Using Large Language Models." *medRxiv* (2022): 2022–12. https://www.medrxiv.org/content/10.1101/2022.12.19.22283643v2. Accessed January 22, 2023.

Lamont, Michèle. "Addressing Recognition Gaps: Destigmatization and the Reduction of Inequality." *American Sociological Review* 83, no. 3 (2018): 419–44.

Lamont, Michèle, Graziella Moraes Silva, Jessica Welburn, Joshua Guetzkow, Nissim Mizrachi, Hanna Herzog, and Elisa Reis. *Getting Respect*. Princeton, NJ: Princeton University Press, 2016.

Lane, Carrie M. *A Company of One*. Ithaca, NY: Cornell University Press, 2011.

Lanzoni, Susan. *Empathy: A History*. New Haven, CT: Yale University Press, 2018.

Larson, Eric B., and Xin Yao. "Clinical Empathy as Emotional Labor in the Patient-Physician Relationship." *JAMA* 293, no. 9 (2005): 1100–1106.

Leidner, Robin. "Emotional Labor in Service Work." *The ANNALS of the American Academy of Political and Social Science* 561 (1999): 81–95.

Leidner, Robin. *Fast Food, Fast Talk: Service Work and the Routinization of Everyday Life*. Berkeley: University of California Press, 1993.

Lenay, Charles, John Stewart, Marieke Rohde, and Amal Ali Amar. "'You Never Fail to Surprise Me': The Hallmark of the Other; Experimental Study and Simulations of Perceptual Crossing." *Interaction Studies* 12, no. 3 (2011): 373–96.

Levin, David Michael. *The Philosopher's Gaze: Modernity in the Shadows of Enlightenment*. Berkeley: University of California Press, 1999.

Levy, Karen E. C. "Relational Big Data." *Stanford Law Review* Online 66 (2013): 73.

Lewis, James L., Robert K. Ream, Kathleen M. Bocian, Richard A. Cardullo, Kimberly A. Hammond, and Lisa A. Fast. "Con Cariño: Teacher Caring, Math Self-Efficacy, and Math Achievement among Hispanic English Learners." *Teachers College Record* 114, no. 7 (2012): 1–42.

Lewis, Jerry M. "Repairing the Bond in Important Relationships: A Dynamic for Personality Maturation." *American Journal of Psychiatry* 157, no. 9 (2000): 1375–78.

Lewis, Michael. *Moneyball: The Art of Winning an Unfair Game*. W. W. Norton and Company, 2004.

Lewis-McCoy, R. L'Heureux. 2014. *Inequality in the Promised Land*. Stanford, CA: Stanford University Press.

Lieberman III, Joseph A., and Marian R. Stuart. "The BATHE Method: Incorporating Counseling and Psychotherapy into the Everyday Management of Patients." *Primary Care Companion to the Journal of Clinical Psychiatry* 1, no. 2 (1999): 35.

Light, Donald W. "The Rhetorics and Realities of Community Health Care: The Limits of Countervailing Powers to Meet the Health Care Needs of the Twenty-First Century." *Journal of Health Politics, Policy and Law* 22, no. 1 (1997): 105–45.

Lilius, Jacoba M. "Recovery at Work: Understanding the Restorative Side of 'Depleting' Client Interactions." *Academy of Management Review* 37, no. 4 (2012): 569–88.

Liquid State. "4 Digital Health Trends (Apps) to Consider for 2018." *Liquid State*. https://liquid-state.com/digital-health-app-trends-2018/.

Lohr, Steven. "Whatever Happened to IBM's Watson?" *New York Times*. July 16, 2021. https://www.nytimes.com/2021/07/16/technology/what-happened-ibm-watson.html. Accessed February 3, 2023.

Lopez, Steven H. "Culture Change and Shit Work: Empowering and Overpowering the Frail Elderly in Long-Term Care." *American Behavioral Scientist* 58, no. 3 (2014): 435–52.

Lopez, Steven H. "Emotional Labor and Organized Emotional Care: Conceptualizing Nursing Home Care Work." *Work and Occupations* 33, no. 2 (2006): 133–60.

Losan, Daniel, and Russell Skiba. *Suspended Education: Urban Middle Schools in Crisis*. Los Angeles: The Civil Rights Project, 2010.

Luborsky, Lester, Barton Singer, and Lise Luborsky. "Comparative Studies of Psychotherapies: Is It True That Everyone Has Won and All Must Have Prizes?" *Archives of General Psychiatry* 32, no. 8 (1975): 995–1008.

Lucas, Gale M., Jonathan Gratch, Aisha King, and Louis-Philippe Morency. "It's Only a Computer: Virtual Humans Increase Willingness to Disclose." *Computers in Human Behavior* 37 (2014): 94–100.

Luker, Kristin. *Salsa Dancing into the Social Sciences: Research in an Age of Info-Glut*. Cambridge, MA: Harvard University Press, 2009.

Lundahl, Brad, Teena Moleni, Brian L. Burke, Robert Butters, Derrik Tollefson, Christopher Butler, and Stephen Rollnick. "Motivational Interviewing in Medical Care Settings: A Systematic Review and Meta-analysis of Randomized Controlled Trials." *Patient Education and Counseling* 93, no. 2 (2013): 157–168.

Lutfey, Karen. "On Practices of 'Good Doctoring': Reconsidering the Relationship between Provider Roles and Patient Adherence." *Sociology of Health & Illness* 27, no. 4 (2005): 421–47.

Malecki, Christine Kerres, and Michelle Kilpatrick Demaray. "Social Support as a Buffer in the Relationship between Socio-economic Status and Academic Performance." *School Psychology Quarterly* 21 (2006): 375–95.

Mankekar, Purnima, and Akhil Gupta. "Intimate Encounters: Affective Labor in Call Centers." *positions: East Asia Cultures Critique* 24, no. 1 (2016): 17–43.

Marken, Stephanie, and Sangeeta Agrawal. "K-12 Workers Have Highest Burnout Rate in U.S." Gallup Poll. 2022. https://news.gallup.com/poll/393500/workers-highest-burnout-rate.aspx. Accessed May 27, 2023.

Maroney, Terry A. "The Persistent Cultural Script of Judicial Dispassion." *Calif. L. Rev.* 99 (2011): 629.

Marshall, Catherine, Jean A. Patterson, Dwight L. Rogers, and Jeanne R. Steele. "Caring as Career: An Alternative Perspective for Educational Administration." *Educational Administration Quarterly* 32 (1996): 271–94.

Marti, Patrizia, et al. "Engaging with Artificial Pets." In *ACM International Conference Proceeding Series* 132 (2005): 99–106.

Martin, Graham P., and Justin Waring. "Realising Governmentality: Pastoral Power, Governmental Discourse and the (Re)constitution of Subjectivities." *Sociological Review* 66, no. 6 (2018): 1292–1308.

Marzano, Robert J., Timothy Waters, and Brian A. McNulty. *School Leadership That Works: From Research to Results*. Alexandria, VA: Association for Supervision and Curriculum Development, 2001.

Maslach, Christina. *Burnout: The Cost of Caring*. Englewood Cliffs, NJ: Prentice-Hall, 1982.

Maslach, Christina, Wilmar B. Schaufeli, and Michael P. Leiter. "Job Burnout." *Annual Review of Psychology* 52, no. 1 (2001): 397–422.

Matheny, Michael, S. Thadaney Israni, Mahnoor Ahmed, and Danielle Whicher. "Artificial Intelligence in Health Care: The Hope, the Hype, the Promise, the Peril." Washington, DC: National Academy of Medicine, 2019.

Mausethagen, Sølvi. "A Research Review of the Impact of Accountability Policies on Teachers' Workplace Relations." *Educational Research Review* 9 (2013): 16–33.

May, Vanessa. *Connecting Self to Society: Belonging in a Changing World*. Macmillan International Higher Education, 2013.

May, Vanessa. "When Recognition Fails: Mass Observation Project Accounts of Not Belonging." *Sociology* 50, no. 4 (2016): 748–63.

Mayer, John D., Peter Salovey, and David R. Caruso. "Emotional Intelligence: Theory, Findings, and Implications." *Psychological Inquiry* 15, no. 3 (2004): 197–215.

McBride, Cillian, and Jonathan Seglow. "Introduction: Recognition: Philosophy and Politics." *European Journal of Political Theory* 8, no. 1 (2009): 7–12.

McCoy, Jennifer, and Benjamin Press. "What Happens When Democracies Become Perniciously Polarized?" *Carnegie Endowment for International Peace*. January 18, 2022. https://carnegieendowment.org/2022/01/18/what-happens-when-democracies-become-perniciously-polarized-pub-86190. Accessed January 31, 2023.

McDonald, Paula, Penny Williams, and Robyn Mayes. "Means of Control in the Organization of Digitally Intermediated Care Work." *Work, Employment and Society* 35, no. 5 (2021): 872–890.

McDonnell, Terence E., Christopher A. Bail, and Iddo Tavory. "A Theory of Resonance." *Sociological Theory* 35, no. 1 (2017): 1–14.

McKinsey Global Institute. "A Future That Works: Automation, Productivity and Employment." McKinsey and Company. January 2017. http://www.mckinsey.com/~/media/McKinsey/Global%20Themes/Digital%20Disruption/Harnessing%20automation%20for%20a%20 future%20that%20works/MGI-A-future-that-works-Full-report.ashx. Accessed January 15, 2017.

Mears, Ashley. "Working for Free in the VIP: Relational Work and the Production of Consent." *American Sociological Review* 80, no. 6 (2015): 1099–1122.

Mehl, Matthias R., Simine Vazire, Shannon E. Holleran, and C. Shelby Clark. "Eavesdropping on Happiness." *Psychological Science* 21 (2010): 539–41.

Melson, Gail F., Peter H. Kahn Jr., Alan Beck, and Batya Friedman. "Robotic Pets in Human Lives: Implications for the Human–Animal Bond and for Human Relationships with Personified Technologies." *Journal of Social Issues* 65, no. 3 (2009): 545–67.

Mesquita, Batja. *Between Us: How Cultures Create Emotions*. New York: W. W. Norton, 2022.

Messeri, Lisa. "Resonant Worlds: Cultivating Proximal Encounters in Planetary Science." *American Ethnologist* 44, no. 1 (2017): 131–42.

Meyersohn, Nathaniel. "Nobody Likes Self-Checkout: Here's Why It's Everywhere." *CNN Business*, July 10, 2022. https://www.cnn.com/2022/07/09/business/self-checkout-retail/index.html. Accessed January 16, 2023.

Mikulincer, Mario, Phillip R. Shaver, and Ety Berant. "An Attachment Perspective on Therapeutic Processes and Outcomes." *Journal of Personality* 81, no. 6 (2013): 606–16.

Miner, Adam S., Nigam Shah, Kim D. Bullock, Bruce A. Arnow, Jeremy Bailenson, and Jeff Hancock. "Key Considerations for Incorporating Conversational AI in Psychotherapy." *Frontiers in psychiatry* 10 (2019): 746.

Molnar, Alex, Gary Miron, Najat Elgeberi, Michael K. Barbour, Luis Huerta, Sheryl R. Shafer, and Jennifer K. Rice. "Virtual Schools in the U.S." Boulder: National Education Policy Center, 2019. http://nepc.colorado.edu/publication/virtual-schools-annual-2019.

Morgan, Blake. "The 20 Most Compelling Examples of Personalization." *Forbes*. March 29, 2021. https://www.forbes.com/sites/blakemorgan/2021/03/29/the-20-most-compelling-examples-of-personalization/?sh=9b3a15771b16. Accessed January 29, 2023.

Morris, George E., and Courtney Kennedy. "Personal Finance Questions Elicit Slightly Different Answers in Phone Surveys Than Online." Pew Research Center: FactTank News in the Numbers. 2017. https://www.pewresearch.org/fact-tank/2017/08/04/personal-finance-questions-elicitslightly-different-answers-in-phone-surveys-than-online/.

Muller, Chandra. "The Role of Caring in the Teacher-Student Relationship for At-Risk Students." *Sociological inquiry* 71, no. 2 (2001): 241–55.

Murthy, Vivek H. *Together: The Healing Power of Human Connection in a Sometimes Lonely World*. New York: Harper Wave, 2020.

Myers, Sonya S., and Robert C. Pianta. "Developmental Commentary: Individual and Contextual Influences on Student–Teacher Relationships and Children's Early Problem Behaviors." *Journal of Clinical Child & Adolescent Psychology* 37, no. 3 (2008): 600–8.

Nathanson, Donald L. *Shame and Pride: Affect, Sex, and the Birth of the Self*. New York: W. W. Norton, 1994.

Newman, Barbara M., Brenda J. Lohman, and Philip R. Newman. "Peer Group Membership and a Sense of Belonging: Their Relationship to Adolescent Behavior Problems." *Adolescence* 42, no. 166 (2007).

Noble, Safiya Umoja. *Algorithms of Oppression*. New York: New York University Press, 2018.

Nutt, Amy Ellis. "The Woebot Will See You Now: The Rise of Chatbot Therapy." *Washington Post*. December 3, 2017. https://www.washingtonpost.com/news/to-your-health/wp/2017/12/03/the-woebot-will-see-you-now-the-rise-of-chatbot-therapy/?utm_term=.461b9bb35bf6. Accessed December 4, 2017.

Office of the National Coordinator for Health Information Technology. "Office-Based Physician Electronic Health Record Adoption." 2019. *Health IT Quick-Stat* #50.

Oishi, Shigehiro, Sharon A. Akimoto, Joo Ree K. Richards, and Eunkook M. Suh. "Feeling Understood as a Key to Cultural Differences in Life Satisfaction." *Journal of Research in Personality* 47, no. 5 (2013): 488–91.

Olfson, Mark, and Steven C. Marcus. "National Trends in Outpatient Psychotherapy." *American Journal of Psychiatry* 167, no. 12 (2010): 1456–63.

Oliver, Mary. *A Poetry Handbook*. New York: Houghton Mifflin Harcourt, 1994.

Olsson, Elin. "The Role of Relations: Do Disadvantaged Adolescents Benefit More from High-Quality Social Relations?" *Acta Sociologica* 52, no. 3 (2009): 263–86.

O'Neil, Cathy. *Weapons of Math Destruction: How Big Data Increases Inequality and Threatens Democracy*. New York: Broadway Books, 2016.

Pane, John F., Elizabeth D. Steiner, Matthew D. Baird, Laura S. Hamilton, and Joseph D. Pane. "How Does Personalized Learning Affect Student Achievement?" Santa Monica. RAND Corporation, 2017. https://www.rand.org/pubs/research_briefs/RB9994.html. Accessed January 29, 2023.

Papadopoulos, Irena, et al. "A Systematic Review of the Literature Regarding Socially Assistive Robots in Pre-tertiary Education." *Computers & Education* 155 (2020).

Pardo-Guerra, Juan Pablo. *The Quantified Scholar: How Research Evaluations Transformed the British Social Sciences*. New York: Columbia University Press, 2022.

Pasquale, Frank. *The Black Box Society: The Secret Algorithms That Control Money and Information*. Cambridge, MA: Harvard University Press, 2015.

Pasquale, Frank. *New Laws of Robotics: Defending Human Expertise in the Age of AI*. Cambridge, MA: Belknap Press, 2020.

Paxton, Pamela, and Robyn Rap. "Does the Standard Voluntary Association Question Capture Informal Associations?" *Social Science Research* 60 (2016): 212–21.

Peräkylä, Anssi, Pentti Henttonen, Liisa Voutilainen, Mikko Kahri, Melisa Stevanovic, Mikko Sams, and Niklas Ravaja. "Sharing the Emotional Load: Recipient Affiliation Calms Down the Storyteller." *Social Psychology Quarterly* 78, no. 4 (2015): 301–23.

Perry, Ryan, Anders Drachen, Allison Kearney, Simone Kriglstein, Lennart E. Nacke, Rafet Sifa, Guenter Wallner, and Daniel Johnson. "Online-Only Friends, Real-Life Friends or Strangers? Differential Associations with Passion and Social Capital in Video Game Play." *Computers in Human Behavior* 79 (2018): 202–10.

Phillippo, Kate. "'You're Trying to Know Me': Students from Nondominant Groups Respond to Teacher Personalism." *The Urban Review* 44, no. 4 (2012): 441–67.

Pocock, David. "Feeling Understood in Family Therapy." *Journal of Family Therapy* 19, no. 3 (1997): 283–302.

Poland, Warren S. "The Analyst's Witnessing and Otherness." *Journal of the American Psychoanalytic Association* 48, no. 1 (2000): 17–34.

Pollmann, Monique M. H., and Catrin Finkenauer. "Investigating the Role of Two Types of Understanding in Relationship Well-Being: Understanding Is More Important Than Knowledge." *Personality and Social Psychology Bulletin* 35, no. 11 (2009): 1512–27.

Poster, Winifred R. "Who's on the Line? Indian Call Center Agents Pose as Americans for US-Outsourced Firms." *Industrial Relations: A Journal of Economy and Society* 46, no. 2 (2007): 271–304.

Power, Michael. "Counting, Control and Calculation: Reflections on Measuring and Management." *Human Relations* 57, no. 6 (2004): 765–83.

Prescod-Weinstein, Chanda. *The Disordered Cosmos: A Journey into Dark Matter, Spacetime, and Dreams Deferred.* New York: Bold Type Books, 2021.

Pressman, Sarah D., Sheldon Cohen, Gregory E. Miller, Anita Barkin, Bruce S. Rabin, and John J. Treanor. "Loneliness, Social Network Size, and Immune Response to Influenza Vaccination in College Freshmen." *Health Psychology* 24, no. 3 (2005): 297.

Pugh, Allison J. "Automated Health Care Offers Freedom from Shame, But Is It What Patients Need?" *New Yorker*. May 22, 2018.

Pugh, Allison J. "Connective Labor as Emotional Vocabulary: Inequality, Mutuality and the Politics of Feelings in Care-Work." *Signs: Journal of Women in Culture and Society* 49, no. 1 (2023): 141–64.

Pugh, Allison J. "Constructing What Counts as Human at Work: Enigma, Emotion and Error in Connective Labor." *American Behavioral Scientist* 67, no. 14 (2023): 1771–92.

Pugh, Allison J. "Emotions and the Systematization of Connective Labor." *Theory, Culture and Society* 39, no. 5 (2022): 23–42.

Pugh, Allison J. *Longing and Belonging: Parents, Children and Consumer Culture.* Berkeley: University of California Press, 2009.

Pugh, Allison J. *The Tumbleweed Society: Working and Caring in an Age of Insecurity.* New York: Oxford University Press, 2015.

Pugh, Allison J. "What Good Are Interviews for Thinking about Culture? Demystifying Interpretive Analysis." *American Journal of Cultural Sociology* 1, no. 1 (February 2013): 42–68.

Pugh, Allison J., and Sarah Mosseri. "Trust-Building versus 'Just Trust Me': Reflexivity and Resonance in Ethnography." *Frontiers in Sociology* 8 (2023): 72.

Rameson, Lian T., Sylvia A. Morelli, and Matthew D. Lieberman. "The Neural Correlates of Empathy: Experience, Automaticity, and Prosocial Behavior." *Journal of Cognitive Neuroscience* 24, no. 1 (2012): 235–45.

Ray, Raka, and Seemin Qayum. *Cultures of Servitude.* Stanford, CA: Stanford University Press, 2009.

Reeves, Byron, and Clifford Ivar Nass. *The Media Equation: How People Treat Computers, Television, and New Media Like Real People and Places.* Stanford, CA, and New York: Cambridge University Press, 1996.

Reich, Adam. "Disciplined Doctors: The Electronic Medical Record and Physicians' Changing Relationship to Medical Knowledge." *Social Science & Medicine* 74, no. 7 (2012): 1021–28.

Reich, Adam Dalton. *Selling Our Souls: The Commodification of Hospital Care in the United States.* Princeton, NJ: Princeton University Press, 2014.

Reis, Harry T., Edward P. Lemay Jr., and Catrin Finkenauer. "Toward Understanding Understanding: The Importance of Feeling Understood in Relationships." *Social and Personality Psychology Compass* 11, no. 3 (2017): e12308.

Reis, Harry T., and Patrick Shaver. "Intimacy as an Interpersonal Process." *Handbook of Personal Relationships* (1988).

Reyes, Maria R., Marc A. Brackett, Susan E. Rivers, Mark White, and Peter Salovey. "Classroom Emotional Climate, Student Engagement, and Academic Achievement." *Journal of Educational Psychology* 104, no. 3 (2012): 700–12.

Rice, Kerry Lynn. "A Comprehensive Look at Distance Education in the K–12 Context." *Journal of Research on Technology in Education* 38, no. 4 (2006): 425–48.

Rios, Victor M. *Punished*. New York: NYU Press, 2011.

Roberts, Dorothy. *Fatal Invention: How Science, Politics, and Big Business Re-create Race in the Twenty-First Century*. New York: The New Press, 2011.

Roberts, Dorothy. *Torn Apart: How the Child Welfare System Destroys Black Families—and How Abolition Can Build a Safer World*. New York: Basic Books, 2022.

Roberts, Dorothy E. "Child Protection as Surveillance of African American Families." *Journal of Social Welfare and Family Law* 36, no. 4 (2014): 426–37.

Rodriquez, Jason. *Labors of Love: Nursing Homes and the Structures of Care Work*. New York: NYU Press, 2014.

Rogers, Laura E. "'Helping the Helpless Help Themselves': How Volunteers and Employees Create a Moral Identity While Sustaining Symbolic Boundaries within a Homeless Shelter." *Journal of Contemporary Ethnography* 46, no. 2 (2017): 230–60.

Romero, Mary, and Nancy Pérez. "Conceptualizing the Foundation of Inequalities in Care Work." *American Behavioral Scientist* 60, no. 2 (2016): 172–88.

Roorda, Debora L., Helma M. Y. Koomen, Jantine L. Spilt, and Frans J. Oort. "The Influence of Affective Teacher–Student Relationships on Students' School Engagement and Achievement: A Meta-analytic Approach." *Review of Educational Research* 81, no. 4 (2011): 493–529.

Ross, Ashley. "The Surprising Way a Supermarket Changed the World." *Time*. September 9, 2016. https://time.com/4480303/supermarkets-history/. Accessed January 16, 2023.

Rousmaniere, Tony, Rodney K. Goodyear, Scott D. Miller, and Bruce E. Wampold, eds. *The Cycle of Excellence: Using Deliberate Practice to Improve Supervision and Training*. 1st ed. Hoboken, NJ: John Wiley and Sons, 2017.

Rupert, Patricia A., and David J. Morgan. "Work Setting and Burnout among Professional Psychologists." *Professional Psychology: Research and Practice* 36, no. 5 (2005): 544–50.

Safran, J. D., Peter Crocker, Shelly McMain, and Paul Murray. "Therapeutic Alliance Rupture as a Therapy Event for Empirical Investigation." *Psychotherapy* 27 (1990): 154–65.

Salem, Maha, Friederike Eyssel, Katharina Rohlfing, Stefan Kopp, and Frank Joublin. "To Err Is Human(-Like): Effects of Robot Gesture on Perceived Anthropomorphism and Likability." *International Journal of Social Robotics* 5, no. 3 (2013): 313–23.

Sandstrom, Gillian M., and Elizabeth W. Dunn. "Is Efficiency Overrated? Minimal Social Interactions Lead to Belonging and Positive Affect." *Social Psychological and Personality Science* 5, no. 4 (2014): 437–42.

Sandstrom, Gillian M., and Elizabeth W. Dunn. "Social Interactions and Well-Being: The Surprising Power of Weak Ties." *Personality and Social Psychology Bulletin* 40, no. 7 (2014): 910–22.

Saraceno, Chiara. "Social Inequalities in Facing Old-Age Dependency: A Bi-Generational Perspective." *Journal of European Social Policy* 20, no. 1 (2010): 32–44.

Sauer, Eric M., Mary Z. Anderson, Barbara Gormley, Christopher J. Richmond, and Lara Preacco. "Client Attachment Orientations, Working Alliances, and Responses to Therapy: A Psychology Training Clinic Study." *Psychotherapy Research* 20, no. 6 (2010): 702–11.

Schilbach, Leonhard, Bert Timmermans, Vasudevi Reddy, Alan Costall, Gary Bente, Tobias Schlicht, and Kai Vogeley. "Toward a Second-Person Neuroscience." *Behavioral and Brain Sciences* 36, no. 4 (2013): 393–414.

Schlechty, Phillip C. *Schools for the Twenty-First Century: Leadership Imperatives for Educational Reform*. San Francisco: Jossey-Bass, 1990.

Schulte, Fred, and Erika Fry. "Death by 1,000 Clicks: Where Electronic Health Records Went Wrong." *Kaiser Health News* and *Fortune* magazine. March 18, 2019. https://kffhealthnews.org/news/death-by-a-thousand-clicks/. Accessed May 31, 2023.

Seidman, Dov. "From the Knowledge Economy to the Human Economy." *Harvard Business Review*. November 12, 2014. https://hbr.org/2014/11/from-the-knowledge-economy-to-the-human-economy. Accessed August 7, 2023.

Seim, Josh. "Stretched Thin: Welfare Work in the Suffering City." Unpublished manuscript, 2022.

Sergeant, Jenny, and Colette Laws-Chapman. "Creating a Positive Workplace Culture." *Nursing Management* 18, no. 9 (2012): 14–19.

Serwint, Janet R., and Miriam T. Stewart. "Cultivating the Joy of Medicine: A Focus on Intrinsic Factors and the Meaning of Our Work." *Current Problems in Pediatric and Adolescent Health Care* 49, no. 12 (2019): 100665.

Shackelton, Rebecca, Carol Link, Lisa Marceau, and John McKinlay. "Does the Culture of a Medical Practice Affect the Clinical Management of Diabetes by Primary Care Providers?" *Journal of Health Services Research & Policy* 14, no. 2 (2009): 96–103.

Shah, Megha K., Nikhila Gandrakota, Jeannie P. Cimiotti, Neena Ghose, Miranda Moore, and Mohammed K. Ali. "Prevalence of and Factors Associated with Nurse Burnout in the US." *JAMA Network Open* 4, no. 2 (2021): e2036469–e2036469.

Shanafelt, Tait D., Lotte N. Dyrbye, Christine Sinsky, Omar Hasan, Daniel Satele, Jeff Sloan, and Colin P. West. "Relationship between Clerical Burden and Characteristics of the Electronic Environment with Physician Burnout and Professional Satisfaction." *Mayo Clinic Proceedings* 91, no. 7 (2016): 836–48.

Sharkey, Noel, and Amanda Sharkey. "The Crying Shame of Robot Nannies: An Ethical Appraisal." *Interaction Studies* 11, no. 2 (2010): 161–90.

Shaver, Philip, R. Mikulincer, Sahdra Mario, K. Baljinder, and Jacquelyn T. Gross. "Attachment Security as a Foundation for Kindness Toward Self and Others." In *The Oxford Handbook of Hypo-egoic Phenomena*, edited by K. W. Brown and M. R. Leary, 223–42. New York: Oxford University Press, 2017.

Sherman, Rachel. "Caring or Catering?: Emotions, Autonomy, and Subordination in Lifestyle Work." In *Caring on the Clock: The Complexities and Contradictions of Paid Care Work*, edited by Mignon Duffy, Amy Armenia, and Clare L. Stacey, 165–76. New Brunswick, NJ: Rutgers University Press, 2015.

Shi, Leiyu. "The Impact of Primary Care: A Focused Review." *Scientifica* (2012).

Shi, Leiyu, Christopher B. Forrest, Sarah Von Schrader, and Judy Ng. "Vulnerability and the Patient–Practitioner Relationship: The Roles of Gatekeeping and Primary Care Performance." *American Journal of Public Health* 93, no. 1 (2003): 138–44.

Shi, Leiyu, Lisa H. Green, and Sophia Kazakova. "Primary Care Experience and Racial Disparities in Self-Reported Health Status." *Journal of the American Board of Family Practice* 17, no. 6 (2004): 443–52.

Shim, Janet K. "Cultural Health Capital: A Theoretical Approach to Understanding Health Care Interactions and the Dynamics of Unequal Treatment." *Journal of Health and Social Behavior* 51, no. 1 (2010): 1–15.

Shiomi, Masahiro, Kayako Nakagawa, and Norihiro Hagita. "Design of a Gaze Behavior at a Small Mistake Moment for a Robot." *Interaction Studies* 14, no. 3 (2013): 317–28.

Simon, Melora, Niteesh K. Choudhry, Jim Frankfort, David Margolius, Julia Murphy, Luis Paita, Thomas Wang, and Arnold Milstein. "Exploring Attributes of High-Value Primary Care." *The Annals of Family Medicine* 15, no. 6 (2017): 529–34.

Sinsky, Christine, Lacey Colligan, Ling Li, Mirela Prgomet, Sam Reynolds, Lindsey Goeders, Johanna Westbrook, Michael Tutty, and George Blike. "Allocation of Physician Time in Ambulatory Practice: A Time and Motion Study in 4 Specialties." *Annals of Internal Medicine* 165, no. 11 (2016): 753–60.

Small, Mario L. "Weak Ties and the Core Discussion Network: Why People Discuss Important Matters with Unimportant Alters." *Social Networks* 35 (2013): 470–83.

Small, Mario Luis. *Someone to Talk To*. New York: Oxford University Press, 2017.

Smith, Anna Deavere. *Talk to Me: Travels in Media and Politics*. New York: Anchor, 2001.

Snyder, Benjamin H. *The Disrupted Workplace: Time and the Moral Order of Flexible Capitalism*. New York: Oxford University Press, 2016.

Stacey, Clare L. "Finding Dignity in Dirty Work: The Constraints and Rewards of Low-Wage Home Care Labour." *Sociology of Health & Illness* 27, no. 6 (2005): 831–54.

Stel, Mariëlle, and Roos Vonk. "Mimicry in Social Interaction: Benefits for Mimickers, Mimickees, and Their Interaction." *British Journal of Psychology* 101, no. 2 (2010): 311–23.

Stepanikova, Irena. "Racial-Ethnic Biases, Time Pressure, and Medical Decisions." *Journal of Health and Social Behavior* 53, no. 3 (2012): 329–43.

Stern, Daniel. *The Interpersonal World of the Infant: A View from Psychoanalysis and Developmental Psychology*. New York: Basic Books, 1985.

Stinson, Linda, and William Ickes. "Empathic Accuracy in the Interactions of Male Friends versus Male Strangers." *Journal of Personality and Social Psychology* 62 (1992): 787–97.

Stone, Deborah. "Why We Need a Care Movement." The Nation. March 13, 2000.

Street Jr., Richard L., Gregory Makoul, Neeraj K. Arora, and Ronald M. Epstein. "How Does Communication Heal? Pathways Linking Clinician–Patient Communication to Health Outcomes." *Patient Education and Counseling* 74, no. 3 (2009): 295–301.

Szymczak, Julia E., and Charles L. Bosk. "Training for Efficiency: Work, Time, and Systems-Based Practice in Medical Residency." *Journal of Health and Social Behavior* 53, no. 3 (2012): 344–58.

Szymczak, Julia E., Joanna Veazey Brooks, Kevin G. Volpp, and Charles L. Bosk. "To Leave or to Lie? Are Concerns about a Shift-Work Mentality and Eroding Professionalism as a Result of Duty-Hour Rules Justified?" *Milbank Quarterly* 88, no. 3 (2010): 350–81.

Taylor, Charles. "The Politics of Recognition." In *Multiculturalism: Examining the Politics of Recognition*, edited by Amy Gutmann, 25–73. Princeton, NJ: Princeton University Press, 1994.

Taylor, Megan Westwood. "Replacing the 'Teacher-Proof' Curriculum with the 'Curriculum-Proof' Teacher: Toward More Effective Interactions with Mathematics Textbooks." *Journal of Curriculum Studies* 45, no. 3 (2013): 295–321.

Theriault, Kayla M., Robert A. Rosenheck, and Taeho Greg Rhee. "Increasing Emergency Department Visits for Mental Health Conditions in the United States." *Journal of Clinical Psychiatry* 81, no. 5 (2020): 5456.

Thompson, E. P. "Time, Work-Discipline, and Industrial Capitalism." *Past & Present* 38 (1967): 56–97.

Throop, C. Jason. "On the Problem of Empathy: The Case of Yap, Federated States of Micronesia." *Ethos* 36, no. 4 (2008): 402–26.

Ticona, Julia. "Red Flags, Sob Stories, and Scams: The Contested Meaning of Governance on Carework Labor Platforms." *New Media & Society* 24, no. 7 (2022): 1548–66.

Ticona, Julia, A. Mateescu, and A. Rosenblat. "Beyond Disruption: How Tech Shapes Labor Across Domestic Work & Ridehailing." *Data & Society* (2018): 58. https://datasociety.net/wp-content/uploads/2018/06/Data_Society_Beyond_Disruption_FINAL.pdf.

Ticona, Julia, and Alexandra Mateescu. "Trusted Strangers: Carework Platforms' Cultural Entrepreneurship in the On-Demand Economy." *New Media & Society* 20, no. 11 (2018): 4384–4404.

Timmermans, Stefan, and Steven Epstein. "A World of Standards but Not a Standard World: Toward a Sociology of Standards and Standardization." *Annual Review of Sociology* 36 (2010): 69–89.

Tolentino, Julia C., Weidun Alan Guo, Robert L. Ricca, Daniel Vazquez, Noel Martins, Joan Sweeney, Jacob Moalem, et al. "What's New in Academic Medicine: Can We Effectively Address the Burnout Epidemic in Healthcare?" *International Journal of Academic Medicine* 3, no. 3 (2017): 1.

Trotter, LaTonya J. *More than Medicine: Nurse Practitioners and the Problems They Solve for Patients, Health Care Organizations, and the State*. Ithaca, NY: Cornell University Press, 2020.

Twenge, Jean M., Roy F. Baumeister, C. Nathan DeWall, Natalie J. Ciarocco, and J. Michael Bartels. "Social Exclusion Decreases Prosocial Behavior." *Journal of Personality and Social Psychology* 92, no. 1 (2007): 56.

Turkle, Sherry. *Alone Together: Why We Expect More from Technology and Less from Each Other*. New York: Basic Books, 2011.

Turkle, Sherry. "Authenticity in the Age of Digital Companions." *Interaction Studies* 8, no. 3 (December 2007): 501–17.

Vaidyam, Aditya Nrusimha, Hannah Wisniewski, John David Halamka, Matcheri S. Kashavan, and John Blake Torous. "Chatbots and Conversational Agents in Mental Health: A Review of the Psychiatric Landscape." *Canadian Journal of Psychiatry* 64, no. 7 (2019): 456–64.

Vale, Mira D., and Denise White Perkins. "Discuss and Remember: Clinician Strategies for Integrating Social Determinants of Health in Patient Records and Care." *Social Science & Medicine* 315 (2022): 115548.

Vallas, Steven, and Juliet B. Schor. "What Do Platforms Do? Understanding the Gig Economy." *Annual Review of Sociology* 46 (2020): 273–94.

Valli, Linda, and Daria Buese. "The Changing Roles of Teachers in an Era of High-Stakes Accountability." *American Educational Research Journal* 44, no. 3 (2007): 519–58.

Vellekoop, Heleen, Matthijs Versteegh, Simone Huygens, Isaac Corro Ramos, László Szilberhorn, Tamás Zelei, Balázs Nagy, et al. "The Net Benefit of Personalized Medicine: A Systematic Literature Review and Regression Analysis." *Value in Health* (2022).

Verghese, Abraham. "The Importance of Being." *Health Aff* 35, no. 10 (2016): 1924–27.

Vinson, Alexandra H. "'Constrained Collaboration': Patient Empowerment Discourse as Resource for Countervailing Power." *Sociology of Health & Illness* 38, no. 8 (2016): 1364–78.

Vinson, Alexandra H., and Kelly Underman. "Clinical Empathy as Emotional Labor in Medical Work." *Social Science & Medicine* 251 (2020): 112904.

Wang, Fan, Yu Gao, Zhen Han, Yue Yu, Zhiping Long, Xianchen Jiang, Yi Wu, et al. "A Systematic Review and Meta-analysis of 90 Cohort Studies of Social Isolation, Loneliness and Mortality." *Nature Human Behaviour* (2023): 1–13.

Wajcman, Judy. "How Silicon Valley Sets Time." *New Media & Society* 21, no. 6 (2019): 1272–89.

Wajcman, Judy. *Pressed for Time: The Acceleration of Life in Digital Capitalism*. Chicago: University of Chicago Press, 2020.

Waytz, Adam, John Cacioppo, and Nicholas Epley. "Who Sees Human? The Stability and Importance of Individual Differences in Anthropomorphism." *Perspectives on Psychological Science* 5, no. 3 (2010): 219–32.

Weisband, Suzanne, and Sara Kiesler. "Self-Disclosure on Computer Forms: Meta-analysis and Implications." In *Proceedings of the SIGCHI Conference on Human Factors in Computing Systems*, edited by Michael J. Tauber, 3–10. New York: Association for Computing Machinery, 1996.

Wells, Kate. "An Eating Disorders Chatbot Offered Dieting Advice, Raising Fears About AI in Health." NPR. June 9, 2023.

White, Ronald D. "Kicks, Pranks, Dog Pee: The Hard Life of Food Delivery Robots." *Los Angeles Times*. March 17, 2022.

White, W. D. "Reason, Rationalization, and Professionalism in the Era of Managed Care." *Journal of Health Politics, Policy and Law* 29, no. 4 (2004): 853–68.

Wikan, Unni. *Resonance: Beyond the Words*. Chicago: University of Chicago Press, 2013.

Wilce, James M., and Janina Fenigsen. "Emotion Pedagogies: What Are They, and Why Do They Matter?" *Ethos* 44, no. 2 (2016): 81–95.

Williams, Christine L. *Gaslighted: How the Oil and Gas Industry Shortchanges Women Scientists*. Berkeley: University of California Press, 2021.

Williamson, Ben. "Psychodata: Disassembling the Psychological, Economic, and Statistical Infrastructure of 'Social-Emotional Learning.'" *Journal of Education Policy* 36, no. 1 (2021): 129–54.

Wingfield, Adia Harvey. "Are Some Emotions Marked 'Whites Only'? Racialized Feeling Rules in Professional Workplaces." *Social Problems* 57, no. 2 (2010): 251–68.

Winnicott, D. W. *Playing and Reality*. New York: Routledge, 1989.

Wolkomir, Michelle, and Jennifer Powers. "Helping Women and Protecting the Self: The Challenge of Emotional Labor in an Abortion Clinic." *Qualitative Sociology* 30, no. 2 (2007): 153–69.

World Economic Forum. "The Future of Jobs Report 2020." Geneva: World Economic Forum, 2020. https://www.weforum.org/reports/the-future-of-jobs-report-2020/.

Wright, Patrice. "Affective Burdens: Racialized Emotions in Reproductive Healthcare." University of Virginia. Unpublished manuscript, 2022.

Yuval-Davis, Nira. "Belonging and the Politics of Belonging." *Patterns of Prejudice* 40, no. 3 (2006): 197–214.

Zaki, Jamil, Niall Bolger, and Kevin Ochsner. "It Takes Two: The Interpersonal Nature of Empathic Accuracy." *Psychological Science* 19, no. 4 (2008): 399–404.

Zapf, Dieter, Claudia Seifert, Barbara Schmutte, Heidrun Mertini, and Melanie Holz. "Emotion Work and Job Stressors and Their Effects on Burnout." *Psychology & Health* 16, no. 5 (2001): 527–45.

Zeavin, Hannah. *The Distance Cure: A History of Teletherapy*. Cambridge, MA: MIT Press, 2021.

Zelizer, Viviana. *The Purchase of Intimacy*. Princeton, NJ: Princeton University Press, 2005.

Zelizer, Viviana A. "How I Became a Relational Economic Sociologist and What Does That Mean?" *Politics & Society* 40, no. 2 (2012): 145–74.

Zuboff, Shoshana. *The Age of Surveillance Capitalism: The Fight for a Human Future at the New Frontier of Power*. London: Profile Books, 2019.

Zulman, Donna M., Marie C. Haverfield, Jonathan G. Shaw, Cati G. Brown-Johnson, Rachel Schwartz, Aaron A. Tierney, Dani L. Zionts, et al. "Practices to Foster Physician Presence and Connection with Patients in the Clinical Encounter." *JAMA* 323, no. 1 (2020): 70–81.

INDEX

A NOTE ON THE TYPE

This book has been composed in Arno, an Old-style serif typeface in the classic Venetian tradition, designed by Robert Slimbach at Adobe.